Communications in Computer and Information Science 1614

More information about this series at https://link.springer.com/bookseries/7899

Mayank Singh · Vipin Tyagi · P. K. Gupta ·
Jan Flusser · Tuncer Ören (Eds.)

Advances in Computing and Data Sciences

6th International Conference, ICACDS 2022
Kurnool, India, April 22–23, 2022
Revised Selected Papers, Part II

Springer

Editors
Mayank Singh
University of KwaZulu-Natal
Durban, South Africa

P. K. Gupta
Jaypee University of Information Technology
Waknaghat, India

Tuncer Ören
University of Ottawa
Ottawa, ON, Canada

Vipin Tyagi
Jaypee University of Engineering
and Technology
Guna, India

Jan Flusser
Institute of Information Theory
and Automation
Prague, Czech Republic

ISSN 1865-0929   ISSN 1865-0937 (electronic)
Communications in Computer and Information Science
ISBN 978-3-031-12640-6   ISBN 978-3-031-12641-3 (eBook)
https://doi.org/10.1007/978-3-031-12641-3

This Springer imprint is published by the registered company Springer Nature Switzerland AG
The registered company address is: Gewerbestrasse 11, 6330 Cham, Switzerland

Preface

Computing techniques like big data, cloud computing, machine learning, the Internet of Things (IoT), etc. are playing a key role in the processing of data and retrieval of advanced information. Several state-of-art techniques and computing paradigms have been proposed based on these techniques. This volume contains papers presented at 6th International Conference on Advances in Computing and Data Sciences (ICACDS 2022) held during April 22–23, 2022, at G. Pullaiah College of Engineering and Technology (GPCET), Kurnool, Andhra Pradesh, India. The conference was organized specifically to help bring together researchers, academicians, scientists, and industry experts and to derive benefits from the advances of next-generation computing technologies in the areas of advanced computing and data sciences.

The Program Committee of ICACDS 2022 is extremely grateful to the authors who showed an overwhelming response to the call for papers, with over 411 papers submitted. All submitted papers went through a double-blind peer-review process, and finally 69 papers were accepted for publication in the Springer CCIS series. We are thankful to the reviewers for their efforts in finalizing the high-quality papers.

The conference featured many distinguished personalities such as Buddha Chandrashekhar, Advisor and CCO, All India Council for Technical Education, India; D V L N Somayajulu, Director, Indian Institute of Information Technology Design and Manufacturing, Kurnool, India; Shailendra Mishra, Majmaah University, Saudi Arabia; Dimitrios A. Karras, National and Kapodistrian University of Athens, Greece; Athanasios (Thanos) Kakarountas, University of Thessaly, Greece; Huiyu Zhou, University of Leicester, UK; Antonino Galletta, University of Messina, Italy; Arun Sharma, Indira Gandhi Delhi Technical University for Women, India; and Sarika Sharma, Symbiosis International (Deemed University), India; among many others. We are very grateful for the participation of all speakers in making this conference a memorable event.

The Organizing Committee of ICACDS 2022 is indebted to Sri. G. V. M. Mohan Kumar, Chairman, GPCET, India, for the confidence that he gave to us during organization of this international conference, and all faculty members and staff of GPCET, India, for their support in organizing the conference and for making it a grand success.

We would also like to thank Vamshi Krishna, CDAC, India; Sandip Swarnakar, GPCET, India; Sameer Kumar Jasra, University of Malta, Malta; Hemant Gupta, Carleton University, Canada; Nishant Gupta, Sharda University, India; Archana Sar, MGM CoET, India; Arun Agarwal, University of Delhi, India; Mahesh Kumar, Divya Jain, Kunj Bihari Meena, Neelesh Jain, Nilesh Patel, and Kriti Tyagi, JUET Guna, India; Vibhash Yadav, REC Banda, India; Sandhya Tarar, Gautam Buddha University, India; Vimal Dwivedi, Abhishek Dixit, and Vipin Deval, Tallinn University of Technology, Estonia; Sumit Chaudhary, Indrashil University, India; Supraja P, SRM Institute of Science and Technology, India; Lavanya Sharma, Amity University, Noida, India; Poonam Tanwar and Rashmi Agarwal, MRIIRS, India; Rohit Kapoor, SK Info Techies,

Noida, India; and Akshay Chaudhary and Tarun Pathak, Consilio Intelligence Research Lab, India, for their support.

Our sincere thanks to Consilio Intelligence Research Lab, India; the GISR Foundation, India; SK Info Techies, India; Adimaginz Marketing Services, India; and Print Canvas, India, for sponsoring the event.

May 2022   Mayank Singh
                                                                  Vipin Tyagi
                                                                  P. K. Gupta
                                                                  Jan Flusser
                                                                  Tuncer Ören

Organization

Steering Committee

Alexandre Carlos Brandão Ramos	UNIFEI, Brazil
Mohit Singh	Georgia Institute of Technology, USA
H. M. Pandey	Edge Hill University, UK
M. N. Hooda	BVICAM, India
S. K. Singh	IIT BHU, India
Jyotsna Kumar Mandal	University of Kalyani, India
Ram Bilas Pachori	IIT Indore, India
Alex Norta	Tallinn University of Technology, Estonia

Chief Patrons

G. V. M. Mohan Kumar	GPCET, India
G. Pullaiah	GPCET, India

Patrons

C. Srinivasa Rao	GPCET, India
K. E. Sreenivasa Murthy	RCEW, India

Honorary Chairs

Shailendra Mishra	Majmaah University, Saudi Arabia
M. Giridhar Kumar	GPCET, India
S. Prem Kumar	GPCET, India

General Chairs

Jan Flusser	Institute of Information Theory and Automation, Czech Republic
Mayank Singh	Consilio Research Lab, Estonia

Advisory Board Chairs

P. K. Gupta	JUIT, India
Vipin Tyagi	JUET, India

Technical Program Committee Chairs

Tuncer Ören	University of Ottawa, Canada
Viranjay M. Srivastava	University of KwaZulu-Natal, South Africa
Ling Tok Wang	National University of Singapore, Singapore
Ulrich Klauck	Aalen University, Germany
Anup Girdhar	Sedulity Group, India
Arun Sharma	IGDTUW, India
Mahesh Kumar	JUET, India

Conference Chair

N. Ramamurthy	GPCET, India

Conference Co-chairs

M. Rama Prasad Reddy	GPCET, India
G. Ramachandra Reddy	Ravindra College of Engineering for Women, India

Conveners

Sandip Swarnakar	GPCET, India
Sameer Kumar Jasra	University of Malta, Malta
Hemant Gupta	Carleton University, Canada

Co-conveners

Arun Agarwal	Delhi University, India
Suprativ Saha	Brainware University, India
B. Madhusudan Reddy	Ravindra College of Engineering for Women, India
Gaurav Agarwal	IPEC, India
Ghanshyam Raghuwanshi	Manipal University, India
Prathamesh Churi	NMIMS, India
Lavanya Sharma	Amity University, India

Organizing Chairs

Shashi Kant Dargar	KARE, India
K. C. T. Swamy	GPCET, India

Organizing Co-chairs

Abhishek Dixit	Tallinn University of Technology, Estonia
Vibhash Yadav	REC Banda, India
Nishant Gupta	MGM CoET, India
Nileshkumar Patel	JUET, India
Neelesh Kumar Jain	JUET, India

Organizing Secretaries

Akshay Kumar	CIRL, India
Rohit Kapoor	SKIT, India
Syed Afzal Basha	GPCET, India

Creative Head

Tarun Pathak	Consilio Intelligence Research Lab, India

Program Committee

A. K. Nayak	Computer Society of India, India
A. J. Nor'aini	Universiti Teknologi MARA, Malaysia
Aaradhana Deshmukh	Aalborg University, Denmark
Abdel Badeeh Salem	Ain Shams University, Egypt
Abdelhalim Zekry	Ain Shams University, Egypt
Abdul Jalil Manshad Khalaf	University of Kufa, Iraq
Abhhishek Verma	IIITM Gwalior, India
Abhinav Vishnu	Pacific Northwest National Laboratory, USA
Abhishek Gangwar	Center for Development of Advanced Computing, India
Aditi Gangopadhyay	IIT Roorkee, India
Adrian Munguia	AI MEXICO, Mexico
Amit K. Awasthi	Gautam Buddha University, India
Antonina Dattolo	University of Udine, Italy
Arshin Rezazadeh	University of Western Ontario, Canada
Arun Chandrasekaran	National Institute of Technology Karnataka, India
Arun Kumar Yadav	National Institute of Technology Hamirpur, India
Asma H. Sbeih	Palestine Ahliya University, Palestine
Brahim Lejdel	University of El-Oued, Algeria
Chandrabhan Sharma	University of the West Indies, West Indies
Ching-Min Lee	I-Shou University, Taiwan
Deepanwita Das	National Institute of Technology Durgapur, India
Devpriya Soni	Jaypee Institute of Information Technology, India

Sotiris Kotsiantis	University of Patras, Greece
Subhasish Mazumdar	New Mexico Tech, USA
Sudhanshu Gonge	Symbiosis International University, India
Tomasz Rak	Rzeszow University of Technology, Poland
Vigneshwar Manoharan	Bharath Corporation, India
Xiangguo Li	Henan University of Technology, China
Youssef Ouassit	Hassan II University of Casablanca, Morocco

Sponsor

Consilio Intelligence Research Lab, India

Co-sponsors

GISR Foundation, India
Print Canvas, India
SK Info Techies, India
Adimaginz Marketing Services, India

Contents – Part II

Contents – Part I

An Ensemble Feature Selection Framework for the Early Non-invasive Prediction of Parkinson's Disease from Imbalanced Microarray Data

Jisha Augustine[✉] and A. S. Jereesh

Bioinformatics Lab, Department of Computer Science,
Cochin University of Science and Technology, Kochi 682022, Kerala, India
{jishaaugustine,jereesh}@cusat.ac.in

Abstract. Early identification of Parkinson's disease (PD) using a non-invasive method is essential to slow down the disease progression with appropriate therapy. This can be accomplished by analysing microarray gene expression data from blood samples. This study proposes a computational framework for predicting PD from blood-based microarray gene expression data. Pre-processing, data balancing and feature reduction, and prediction are the three stages of the proposed system. In the pre-processing stage, annotation, cross-platform normalisation, and integration were performed. Balanced subsets were created using k-means clustering on majority samples and random undersampling. The ANOVA filter extracted critical features from balanced subsets in the feature reduction stage, and various cost-sensitive classification models and an ensemble model were built in the prediction stage. The method could achieve an AUC of 82.6% using the cost-sensitive Logistic regression classifier and 83.2% using the ensemble model on independent test data. The experimental results indicate that the suggested framework could effectively diagnose PD at the early stages.

Keywords: Parkinson's disease · Microarray gene expression · Imbalanced data · Majority sample clustering · Machine learning

1 Introduction

Parkinson's disease (PD) is a prevalent neurodegenerative disease that primarily affects the elderly. As the disease progresses, the motor and non-motor symptoms of PD deteriorate, lowering the quality of life for the patient [1]. PD is currently diagnosed using clinical symptoms and neuroimaging approaches. These approaches detect PD in its later phases when about 70% to 80% of dopaminergic neurons have degenerated [2]. At the moment, all existing treatments for Parkinson's disease do not halt disease progression but do enhance patients' quality of life. Early detection of PD may aid in providing appropriate treatment and serve as a helpful tool for selecting patients while researching novel medications [2]. Several studies analysed gene expression data from PD patients to identify uncovered valuable patterns [3, 4]. Most of these data were obtained from biopsy or autopsy-based samples, which are difficult to obtain [3, 4]. Gene expression data from blood samples, the proteins produced by these genes, and the pathways implicated may provide crucial clues to our insight into the molecular pathogenesis of PD in a cost-effective, non-invasive, and less complex manner.

Numerous investigations on the gene expression profiles of people with PD in blood have found significant differences in gene expression between PD and healthy controls (HC), suggesting the possibility of developing a blood test for PD prediction. [5–7]. Machine learning has achieved a significant role in analysing gene expression data. Several machine learning methods have been developed to identify molecular markers and predict PD from blood-based gene expression data [7–10]. A recent study identified 29 genes which effectively classify PD from integrated blood-based gene expression datasets using two-layer embedded-wrapper feature selection and classification [8]. Another study performed a meta-analysis of four distinct blood-based microarray data sets and identified 59 signature genes using collinearity recognition followed by recursive feature elimination [9]. The meta-analysis of three independent blood gene expression datasets conducted by Jiang et al. [10] selected nine genes using Least absolute shrinkage and selection operator (LASSO) and Random forest (RF) feature selection. Shamir et al. [7] identified 87 gene-signature for idiopathic PD using computational analysis of whole blood gene expression data. Moreover, studies involving the detection of PD at the early stages are limited.

This study proposed a new framework for the early prediction of PD from integrated blood-based microarray gene expression data. Here, imbalanced data handling and reduction of feature space are challenging. The proposed system consists of three stages. Data annotation, cross-platform normalisation, and data balancing were performed in the preprocessing stage. K-means clustering on majority samples with random undersampling (RUS) was applied to generate a balanced data set. In the feature reduction stage, we combined relevant features selected by Analysis of variance (ANOVA) from each balanced subset. In the prediction stage, four cost-sensitive base classification models, including Logistic regression (LR), Random forest (RF), Support vector machine-Radial (SVMR), Support vector machine-Linear (SVML), and a soft ensemble of LR and SVML, were used.

In the rest of this paper, Sect. 2 illustrates the materials and proposed methodology, including dataset collection and a description of three phases. Section 3 describes the evaluation criteria. In Sect. 4, findings and comparative analysis are given. Lastly, Sect. 5 concludes our study.

2   Materials and Methods

This study aims to detect Parkinson's disease early using microarray gene expression data from blood samples. The initial stage was to gather blood-based data sets for Parkinson's disease. Annotation, cross-platform normalisation, and integration were performed on the data. The combined data was divided into two sets: training and test. Clustering on majority samples and resampling were used to construct balanced subsets. From the balanced subsets, an ensemble feature selection was made. The training set with decreased features was used to create classification models. The suggested system's detailed workflow is depicted in Fig. 1.

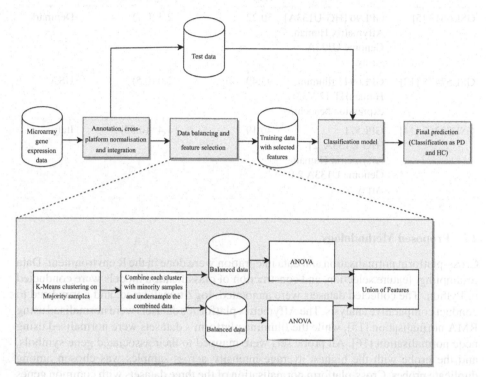

Fig. 1. The workflow of the proposed system

2.1 Dataset Collection

Our work used the following selection conditions to obtain gene expression data from blood samples from the GEO database [11]. (a) "Homo sapiens" is selected as the organism; (b) The samples are blood-based; (c) Selected cohorts have at least 50 samples, and those samples are from PD patients and healthy control; (d) The average Hoehn and Yahr score is less than 2.5, which indicates the early stage of PD [12]. The datasets GSE6613, GSE57475, and GSE72267 [5, 13, 14] were chosen based on the criteria. The description of selected datasets is given in Table 1.

Table 1. Description of selected datasets

Dataset	Platform	Sample size (PD: HC)	Hoehn and Yahr stage, mean (SD)	Country
GSE6613 [5]	GPL96 [HG-U133A] Affymetrix Human Genome U133A Array	50:22	2.3 (0.7)	Denmark
GSE57475 [13]	GPL6947 Illumina HumanHT-12 V3.0 expression beadchip	93:49	2.0 (0.5)	USA
GSE72267 [14]	GPL571 [HG-U133A_2] Affymetrix Human Genome U133A 2.0 Array	40:19	1.45 (0.55)	Italy

2.2 Proposed Methodology

Cross-platform normalisation and data integration were done in the R environment. Data resampling, feature selection, and construction of classification models were conducted in Python. The collected datasets were annotated, log2-transformed, and normalised to conduct comparative analysis. The Affymetrix platform's datasets were normalised using RMA normalisation [15], while the Illumina platform's datasets were normalised using neqc normalisation [16]. All probe sets were mapped to their associated gene symbols, and the probe with the highest average intensity across samples was chosen among duplicate probes. Cross-platform normalisation of the three datasets with common genes was conducted using Z-score transformation [17]. The training set (80% of the total data) was used for feature selection and model creation, while the remaining 20% was used for independent validation.

To handle data imbalance, we used a combination of clustering and resampling techniques. The majority samples (M_j) were clustered using k-means clustering. The optimum number of clusters was calculated using the average silhouette method, which identifies the optimum k based on the average silhouette coefficient of observations for different values of k [18]. After obtaining the optimum k-value, the majority samples were clustered into k sub-clusters, $M_j = \{C_1, C_2, C_3 \ldots, C_k\}$ where C_i is the sub-cluster. As illustrated in Eq. 1, these sub-clusters were combined with minority samples (M_n) to generate k subsets.

$$S_i = C_i \cup M_n, i = 1, 2, 3 \ldots k \tag{1}$$

Random undersampling (RUS) was employed to balance these subsets, which randomly selected a set of majority class samples and removed these samples from the imbalanced data [19].

The high dimensionality of gene expression data is one of the primary issues. We employed an ensemble of ANOVA, a univariate feature selection method, to minimise the features (here genes) [20]. ANOVA prioritises the best features based on the f-ratio derived by univariate statistical tests. The f-ratio denotes the ratio of between-class variance to within-class variance. The magnitude of the f-ratio provides an indicator of class separation. The formula for the f-ratio is specified in Eq. 2.

$$f - ratio = \frac{BCV}{WCV} \tag{2}$$

where the f-ratio is, the score of feature and a high score indicates the ability to identify the sample.

BCV is the sample variance between the groups, and the expression for calculating BCV is specified in Eq. 3.

$$BCV = \frac{\sum_{k=1}^{C} n_k (\overline{x_k} - \overline{x})^2}{C - 1} \tag{3}$$

where n_k is the number of samples in the kth class and C is the total number of classes. $\overline{x}_k$ denotes the mean of the kth class and $\overline{x}$ denotes the overall mean.

WCV is the sample variance within the groups and the formula is specified in Eq. 4.

$$WCV = \frac{(\sum_{k=1}^{C} \sum_{l=1}^{n_k} (x_{kl} - \overline{x})^2) - (\sum_{k=1}^{C} n_k (\overline{x_k} - \overline{x})^2)}{N - C} \tag{4}$$

where x_{kl} is the lth measurement of the kth class and N is the total number of samples.

Our analysis selected all significant features with a p-value $<= 0.05$ based on the False Positive Rate (FPR) test as the most informative features. A reduced feature set was created by combining selected features from each balanced dataset.

Following the selection of features, PD was predicted by learning a model in the smaller feature space. One of the appropriate strategies for classifying imbalanced data is cost-sensitive learning [21, 22]. A cost-sensitive learning strategy is to penalise the minority class for incorrect classification by increasing the weight of the class while decreasing the weight of the majority class. Different class weighted base classifiers were evaluated, including LR, SVML, SVMR, and RF. Furthermore, a soft ensemble of the LR and SVML model was developed for evaluation purposes. The training data was subjected to three times 10-fold cross-validation, and the test data was used for independent evaluation.

3 Evaluation Criteria

It is well recognised that overall accuracy (Accuracy) in skewed recognition tasks typically gives biased evaluation; consequently, in addition to accuracy, several other specialised evaluation measures, including F-measure, AUC, Precision, and Recall, are required to quantify a learner's classification performance [22]. The metrics are determined using Eqs. 5, 6 and 7.

$$Precision = \frac{TruePositive}{TruePositive + FalsePositive} \tag{5}$$

$$Recall = \frac{TruePositive}{TruePositive + FalseNegative} \tag{6}$$

$$F - measure = \frac{2 * Recall * Precision}{Recall + Precision} \tag{7}$$

The AUC is the area under the ROC curve, which depicts a classifier's performance based on the False Positive Rate and True Positive Rate pairs [23]. It has been shown to be a reliable performance metric for the problem of class imbalance [22].

4   Results and Discussion

With adequate therapy, an early diagnosis of Parkinson's disease may help delay the condition's progression [2]. The use of gene expression from blood samples allows for a more precise and straightforward diagnosis of Parkinson's disease. We employed feature selection and classification approaches to attain this goal. The results of our method for distinguishing PD from HC are presented in this section.

The systematic analysis of public databases identified three different datasets (GSE6613, GSE57475, and GSE72267). After annotation and summarisation, 11958 unique Gene IDs common to all datasets were included for the analysis. Figure 2 demonstrates the distribution of genes among the three datasets. We combined the Z-transformed datasets to create a single dataset with 273 samples (patients) and 11958 features (genes). Out of these samples, 183 were PD and 90 HC. The combined dataset was split into training (80%) and test (20%) sets. The training set was employed for data balancing, feature selection, and model construction. The test set was utilised for validating the prediction capability of developed models.

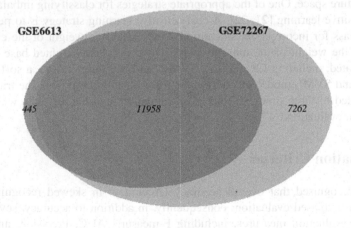

Fig. 2. The distribution of genes among the datasets

The data imbalance and high dimensionality of these integrated datasets provide a significant difficulty. To correct the data imbalance, we employed majority clustering along with resampling. K-means clustering was employed to cluster the majority samples of the training data. Figure 3 depicts the average silhouette score for various k values of k-means clustering. The optimum k was determined to be two based on the average silhouette score. Two datasets were formed by integrating the two subclusters with the minority samples. The RUS approach was utilised to balance these datasets.

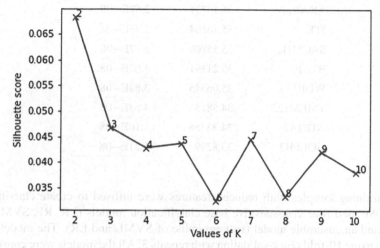

Fig. 3. Silhouette analysis for optimal k

The ANOVA feature selection was performed on the balanced datasets, and the union of the selected features was taken. The ANOVA selected 5039 features based on p-value $<= 0.05$. The top twenty genes selected based on the p-value is given in Table 2.

Table 2. Top 20 genes with f-score and p-value

Gene	f-score	p-value
GIMAP6	45.7827	7.08E−10
ARHGEF3	42.95641	1.97E−09
SUN1	40.72622	4.48E−09
FNTA	39.59819	6.83E−09
SRSF3	38.97864	8.62E−09
TMEM243	38.27689	1.12E−08
CNIH1	37.90305	1.29E−08
OTUD4	37.76967	1.36E−08

(continued)

Table 2. (*continued*)

Gene	f-score	p-value
RYK	37.49862	1.51E−08
PRPS2	36.66642	2.07E−08
MTO1	36.18827	2.49E−08
UPF3A	36.17276	2.50E−08
AKAP11	36.10731	2.57E−08
ITK	35.76204	2.93E−08
SACM1L	35.55808	3.17E−08
BTAF1	35.21391	3.62E−08
WDR11	35.06345	3.84E−08
TMEM123	34.5815	4.63E−08
ATP2A2	34.33356	5.10E−08
GOLPH3	33.8299	6.21E−08

The training samples with reduced features were utilised to create classification models. We used four cost-sensitive base classification models (LR, RF, SVML, and SVMR) and an ensemble model (soft ensemble of SVML and LR). The models were evaluated using 10-fold cross-validation with repeats 3. All the models were constructed using sklearn-1.0.1 with default parameters. The comparative performance of the models on training data in AUC is presented in Table 3.

Table 3. Comparison of the classifiers of the proposed model

Classifier	AUC (mean (SD))
LR	0.833 (0.090)
RF	0.667 (0.112)
SVMR	0.738 (0.108)
SVML	0.829 (0.090)
Ensemble	0.832 (0.085)

Figure 4 shows the AUC of different classifiers on the test set. The cost-sensitive LR classifier and the ensemble model outperform with the AUC of 82.6% and 83.2%, respectively.

Table 4 summarises the other performance measures including F-measure, Recall, Precision, and Accuracy metrics prediction on the test set. The cost-sensitive LR classifier and ensemble model had the best results, with an F-measure of 85% indicating that these models can accurately discriminate between PD and HC.

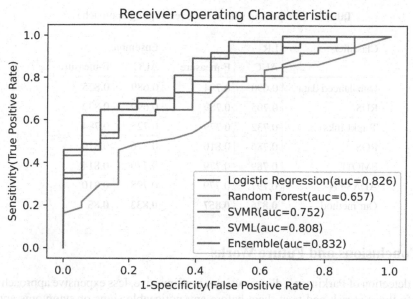

Fig. 4. Receiver operating characteristic (ROC) curve of classification models on the test set

Table 4. Performance of the classifiers on test data

Classifier	F-measure	Recall	Precision	Accuracy
LR	0.857	0.891	0.825	0.8
RF	0.813	1.0	0.685	0.690
SVMR	0.763	0.783	0.743	0.672
SVML	0.847	0.972	0.75	0.763
Ensemble	0.85	0.918	0.790	0.781

To verify the benefits of ensemble feature selection based on majority sample clustering, the proposed system is compared to (1) ANOVA feature selection on unbalanced data, (2) ANOVA feature selection on balanced data utilising resampling techniques such as Random Undersampling (RUS), TomeKLinks [24], Random Oversampling (ROS) [19], Synthetic Minority Oversampling Technique (SMOTE) [25] and Adaptive Synthetic Sampling (ADSYAN) [26]. Table 5 shows the comparative outcome. Our strategy outperformed several resampling techniques, according to the results.

Table 5. Comparison with various resampling approaches

Classifier	LR		Ensemble	
	AUC	F-measure	AUC	F-measure
Unbalanced data	0.690	0.794	0.689	0.825
RUS	0.705	0.789	0.699	0.839
TomkLinks	0.732	0.779	0.725	0.784
ROS	0.786	0.810	0.776	0.810
SMOTE	0.789	0.779	0.779	0.814
ADASYN	0.797	0.779	0.798	0.810
Our method	**0.826**	**0.857**	**0.832**	**0.85**

5   Conclusions and Future Works

Early detection of Parkinson's disease using a non-invasive, less expensive approach can identify those at risk and treat them before any noticeable signs or symptoms appear [27]. With the advent of massively parallel profiling of genes from blood samples and improved machine learning technologies to handle large and heterogeneous datasets, the early diagnosis of PD is possible.

This work provided a computational approach for predicting PD early using blood-based microarray gene expression data from several platforms. Cross-platform normalisation and integrating datasets, majority sample clustering with random undersampling to balance the data, ensemble feature selection to minimise feature space, and model generation are part of the computational architecture. The experiment analysis has proven that the classifier can work quite efficiently after data balancing with ensemble feature selection. We predicted PD with a high AUC and F-measure on the test set as part of the validation.

Our approach can effectively identify PD in its early stages and can be used for disease diagnosis. In the future, as the volume of data grows, improved machine learning and deep learning algorithms may be employed to provide a more accurate diagnosis of Parkinson's disease.

References

1. Blauwendraat, C., Nalls, M.A., Singleton, A.B.: The genetic architecture of Parkinson's disease. Lancet Neurol. **19**(2), 170–178 (2020)
2. Karlsson, M.K., et al.: Found in transcription: accurate Parkinson's disease classification in peripheral blood. J. Parkinson's Dis. **3**(1), 19–29 (2013)
3. Keo, A., et al.: Transcriptomic signatures of brain regional vulnerability to Parkinson's disease. Commun. Biol. **3**(1), 1–12 (2020)
4. Benoit, S.M., et al.: Expanding the search for genetic biomarkers of Parkinson's disease into the living brain. Neurobiol. Dis. **140**, 104872 (2020)
5. Scherzer, C.R., et al.: Molecular markers of early Parkinson's disease based on gene expression in blood. Proc. Natl. Acad. Sci. **104**(3), 955–960 (2007)

6. Pinho, R., et al.: Gene expression differences in peripheral blood of Parkinson's disease patients with distinct progression profiles. PLoS ONE **11**(6), e0157852 (2016)
7. Shamir, R., et al.: Analysis of blood-based gene expression in idiopathic Parkinson disease. Neurology **89**(16), 1676–1683 (2017)
8. Augustine, J., Jereesh, A.S.: Blood-based gene-expression biomarkers identification for the non-invasive diagnosis of Parkinson's disease using two-layer hybrid feature selection. Gene **823**, 146366 (2022)
9. Falchetti, M., Prediger, R.D., Zanotto-Filho, A.: Classification algorithms applied to blood-based transcriptome meta-analysis to predict idiopathic Parkinson's disease. Comput. Biol. Med. **124**, 103925 (2020)
10. Jiang, F., Qianqian, W., Sun, S., Bi, G., Guo, L.: Identification of potential diagnostic biomarkers for Parkinson's disease. FEBS Open Bio **9**(8), 1460–1468 (2019)
11. Barrett, T., et al.: NCBI GEO: archive for functional genomics data sets—update. Nucleic Acids Res. **41**(D1), D991–D995 (2012)
12. Hoehn, M.M., Yahr, M.D.: Parkinsonism: onset, progression and mortality. Neurology, **50**, 318–318 (2001)
13. Locascio, J.J., et al.: Association between α-synuclein blood transcripts and early, neuroimaging-supported Parkinson's disease. Brain **138**(9), 2659–2671 (2015)
14. Calligaris, R., et al.: Blood transcriptomics of drug-naive sporadic Parkinson's disease patients. BMC Genomics **16**(1), 1–14 (2015)
15. Irizarry, R.A., Bolstad, B.M., Collin, F., Cope, L.M., Hobbs, B., Speed, T.P.: Summaries of Affymetrix GeneChip probe level data. Nucleic Acids Res. **31**(4), e15–e15 (2003)
16. Shi, W., Oshlack, A., Smyth, G.K.: Optimizing the noise versus bias trade-off for Illumina whole genome expression BeadChips. Nucleic Acids Res. **38**(22), e204–e204 (2010)
17. Cheadle, C., Vawter, M.P., Freed, W.J., Becker, K.G.: Analysis of microarray data using Z score transformation. J. Mol. Diagn. **5**(2), 73–81 (2003)
18. Rousseeuw, P.J.: Silhouettes: a graphical aid to the interpretation and validation of cluster analysis. J. Comput. Appl. Math. **20**, 53–65 (1987)
19. Batista, G.E.A.P.A., Prati, R.C., Monard, M.C.: A study of the behavior of several methods for balancing machine learning training data. ACM SIGKDD Explor. Newsl. **6**(1), 20–29 (2004)
20. Johnson, K.J., Synovec, R.E.: Pattern recognition of jet fuels: comprehensive GC× GC with ANOVA-based feature selection and principal component analysis. Chemom. Intell. Lab. Syst. **60**(1–2), 225–237 (2002)
21. Elkan, C.: The foundations of cost-sensitive learning. In: International Joint Conference on Artificial Intelligence, vol. 17, no. 1, pp. 973–978 (2001)
22. He, H., Garcia, E.A.: Learning from imbalanced data IEEE transactions on knowledge and data engineering, vol. 21, pp. 1263–1284 (2009)
23. Fawcett, T.: An introduction to ROC analysis. Pattern Recogn. Lett. **27**(8), 861–874 (2006)
24. Tomek, I.: Two modifications of CNN. IEEE Trans. Syst. Man Cybern. **6**, 769–772 (1976)
25. Chawla, N.V., Bowyer, K.W., Hall, L.O., Philip Kegelmeyer, W.: SMOTE: synthetic minority over-sampling technique. J. Artif. Intell. Res. **16**, 321–357 (2002)
26. He, H., Bai, Y., Garcia, E.A., Li, S.: ADASYN: adaptive synthetic sampling approach for imbalanced learning. In: 2008 IEEE International Joint Conference on Neural Networks (IEEE World Congress on Computational Intelligence), pp. 1322–1328. IEEE (2008)
27. Gopar-Cuevas, Y., et al.: Pursuing multiple biomarkers for early idiopathic Parkinson's disease diagnosis. Mol. Neurobiol. **58**(11), 5517–5532 (2021). https://doi.org/10.1007/s12035-021-02500-z

Optimization Enabled Neural Network for the Rainfall Prediction in India

Ananda R. Kumar Mukkala[1]([⊠]), S. Sai Satyanarayana Reddy[1], P. Praveen Raju[2], Mounica[2], Chiranjeevi Oguri[2], and Srinivasu Bhukya[2]

[1] Department of CSE, Sreyas Institute of Engineering and Technology, Hyderabad 500068, India
anandranjit@gmail.com
[2] Department of ECE, Sreyas Institute of Engineering and Technology, Hyderabad 500068, India

Abstract. Rainfall prediction plays a major role in ensuring the livelihood of many people especially, for the farmers. Heavy and irregular flow of rainfall can cause flood, landslide and much other destruction. To prevent this, rainfall should be predicted in a periodic manner. As a contribution, the proposed Spotted Hyena based nonlinear autoregressive model (SH-NARX) prediction model effectively predicts the rainfall in a yearly, monthly and quarterly manner using the Indian rainfall dataset. The data is collected and trained using the NARX neural network, which is a non linear autoregressive network that is optimized using the spotted hyena optimization for rainfall prediction. The performance of the prediction model is analyzed based on RMSE and PRD that are minimal, highlighting the higher accuracy rates.

Keywords: NARX neural network · Rainfall prediction · Spotted hyena · Indian rainfall data · RMSE

1 Introduction

Water plays a major role in the human life and no life can be pertained in the earth without water. One of the important sources of water is rainfall. Rainfall can be defined as a quasi-periodic signal with recurrent periodic fluctuations that can occurs at different levels of varying noises [8, 9, 3]. Rainfall is necessary to recharge the depleting ground sources as well as for agriculture. The agricultural productivity of the country entirely depends on the rainfall which is directly proportional to the growth of economy. The developing country such as India and Vietnam completely relies upon the monsoon rainfall for the agricultural productivity [10, 11, 12, 2]. The accurate prediction of the rainfall greatly helps in the futuristic planning of the agricultural activities as well as the severity of the natural disasters such as flood, landslide, and drought could be identified by predicting the rainfall [13, 14, 2]. It is quite difficult to forecast the rainfall due to the rapid change and complexity of the weather conditions.

Various researches are employed for the accurate detection of rainfall but in a broader perspective it is quite difficult to accurately predict the occurrence of rainfall at most of the times. Basically, the prediction of rainfall is performed using the empirical and

M. Singh et al. (Eds.): ICACDS 2022, CCIS 1614, pp. 12–23, 2022.
https://doi.org/10.1007/978-3-031-12641-3_2

dynamic methods. The empirical approach utilizes the previous information or historical prediction and this approach is most commonly used in the regression and artificial neural networks. The dynamic approach is utilized by the physical and statistical methods. The advancement of technology in recent years promotes the prediction of rainfall using the techniques regression, support vector machine (SVM), and K-nearest neighbor (KNN). Deep learning models are useful in examining large dataset and provide factual information and it is useful in computational applications [1]. Numerous methods are used in the prediction of rainfall and are categorized into three groups such as: statistical, dynamic and satellite based methods [15, 16] and the statistical methods are frequently used due to its inexpensive and time consuming nature [3].

The main objective of the research is to develop a rainfall prediction model to effectively predict the rainfall so that the destruction caused by the rainfall can be greatly avoided. As a contribution the data are collected from the rainfall predicted from the India dataset and then the SH-NARX model is used for the prediction which is optimized using the spotted hyena optimization. The NARX is a nonlinear autoregressive model that predicts the output based on the past values as well as current values. Then the final output are predicted and the main contribution of the proposed SH-NARX model are as follows.

- *Spotted Hyena optimization:* The spotted hyena optimization consists of faster convergence, and the search of the solution will be presented in a globalized search manner. Due to its simplicity and ease of use this optimization is widely used.
- *SH-NARX model:* The SH-NARX model has the capability to train the samples with less number of training cycles. The prediction error is also greatly reduced using this model.
- The PRD and RMSE values are improved to prove the significance of the proposed SH-NARX model

The alignment of the manuscript is as follows: The existing experimentations along with its advantages and disadvantages are enumerated in Sect. 2, the proposed optimization enabled SH-NARX model along with its architecture are depicted in Sect. 3. The brief discussions of the results achieved are enumerated in Sect. 4 band finally the rainfall prediction model is concluded in Sect. 5.

2 Motivation

In this context, the review of the existing prediction models with the challenges of the research is presented, which motivated the researcher in designing a prediction model based on ANN and optimization. R. Venkatesh *et al.* [1] introduced a rainfall prediction system using generative adversarial networks which performs well in both monthly and annual averages but the drawback is that it consumes more time and there is a need for the GPU-based computational resources. Duong Tran Anh *et al.* [3] developed a novel hybrid models for monthly rainfall prediction which predicts the rainfall in more accurate manner compared with conventional methods but it consists of a complex structure. Joao Trevizoli Esteves *et al.* [4] established an automatic ANN modeling that has the capability to find the global optimum solution but the results obtained has an average accuracy. Hatem Abdul-Kader et al. [6] predicted the rainfall using long short term memory (LSTM) which is strong enough to predict the rainfall in the Sidoarjo but the prediction mainly concentrated on single place. S. Dhamodharavadhani and R. Rathipriya [7] presented a Map Reduce-based exponential smoothing for the region wise rainfall prediction that showed efficient runtime improvement but the smoothing methods represented can be furthermore optimized. The major challenges of the research is enumerated as follows:

- The meteorological parameters considered for the prediction of rainfall is stochastic in nature which is a challenging task to be resolved [2].
- When there is a necessity for local-scale projections, the modeling of the variabilites present in the rainfall events are difficult to be modeled [3].
- It is a challenging task to obtain reliable and accurate prediction models because of the uncertainity and variability present in the prediction models that spatially forecast the rainfall for a short period of time [4].

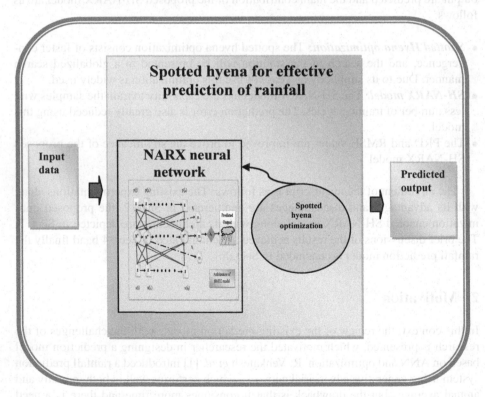

Fig. 1. Systematic representation of rainfall prediction model

3 Proposed Prediction Model Using the Optimization Enabled Neural Network Classifier

The main aim of the research is to develop a rainfall prediction model that effectively predicts the rain in a particular region in a periodic manner either monthly or yearly. The input data obtained from a particular region on a specific time is trained using the NARX neural network and the optimization of the neural network is performed using the spotted hyena optimization. The NARX neural network has the capability to predict the error between the predicted data and the ground output. Finally the probability of occurrence of the amount of rainfall is predicted using this network. The representation of the proposed model is shown in Fig. 1

3.1 Hybrid Optimization for Classifier Training

Spotted hyenas are skillfull animals that perform hunting in a well-organized preplanned manner. The spotted hyena characteristics are observed as the feature, which renders the optimal solution. The optimization features of the spotted hyena are used for declaring the best solution for the classifier, which is used for tuning of the internal model parameters. The social relation and the adaptability of the spotted hyenas are taken into account for the optimization problems, which is mathematically expressed in different phases.

Surrounding Phase: Initially the hyena start searching for the prey and once the prey is detected then it starts updating the information to different hyenas. The hyena which locates near to the location of the prey is considered as the current best solution and the information updated by the hyena can be mathematically expressed as follows

$$\vec{G_a} = \mid \vec{P} . \vec{M_b}(v) - \vec{M}(v) \mid \tag{1}$$

$$\vec{M}(v+1) = \vec{M_b}(v) - \vec{L} . \vec{G_a} \tag{2}$$

Here $\vec{G_a}$ represents the distance between the target prey and the spotted hyena, v denotes the current iteration. $\vec{P}$ and $\vec{L}$ are the coefficient vectors, $\vec{M_b}$ indicates the position of the prey and $\vec{P}$ indicates rthe position of the spotted hyena. $\mid \mid$ and $\cdot$ absolute value and multiplication with respect to the vectors. The coefficient vectors $\vec{P}$ and $\vec{L}$ are given as follows

$$\vec{P} = 2.h\vec{w_1} \tag{3}$$

$$\vec{L} = 2\vec{q} . h\vec{w_2} - \vec{q} \tag{4}$$

$$\vec{q} = 5 - \left(iteration * {}^{5} \middle/ \max_{iteration} \right) \tag{3}$$

Here $iteration = 1, 2, 3 ...... \max_{iteration}$. While performing the maximum number of iterations the $\vec{q}$ linearly decreases from the value 5 to 0. $h\vec{w_1}$ and $h\vec{w_2}$ are the random vectors has the value [0, 1] in the 2-dimensional environment.

Chasing Phase: The spotted hyenas always hunt in group with trusted members. The hyenas know the location of the prey where the prey is present. The hyena which knows the location of the prey is termed as optimal solution and all the other solutions are also updated. The hyenas started updating their positions can be represented by

$$\vec{G_a} = |\vec{P} . \vec{M_a} - \vec{M_g}| \tag{4}$$

$$\vec{M_g} = \vec{M_a} - \vec{L} . \vec{G_a} \tag{5}$$

$$\vec{E_a} = \vec{M_g} + \vec{M_{g+1}} + ... + \vec{M_{g+k}} \tag{6}$$

$\vec{M_a}$ defines the position of the first hyena who spotted the prey, $\vec{E_a}$ indicates the other spotted hyenas. Here K indicates the number of spotted hyenas are computed as follows

$$k = count_{numbers}\left(\vec{M_a}, \vec{M_{a+1}}, \vec{M_{a+2}} + ... \left(\vec{M_a} + \vec{J}\right)\right) \tag{7}$$

Here $\vec{J}$ refers to the random vector that has the values ranges from [0.5, 1], $count_{numbers}$ denotes the number of solutions and the count of all candidate solution after summing with $\vec{J}$, which is similar to the far best solution in the provided search space. $\vec{E_a}$ consists of number of optimal solutions in the group of clusters.

Bombarding Phase: When the value of $\vec{q}$ decreases the attacking of the prey is initiated. The vector $\vec{L}$ also gets decreased from 5 to 0. If the condition $L < 1$ is satisfied the hyena starts attacking the prey and is mathematically expressed as

$$\vec{M_{v+1}} = \frac{\vec{E_a}}{K} \tag{8}$$

$\vec{M_{v+1}}$ saves the best solution and updates the position of other search agents according to their positions.

Scouting Phase: The search of the prey for the hyenas always depends on the $\vec{E_a}$. They diverge from each other to search the prey and unite to attack the prey depending upon the vector $\vec{L}$ ranges between -1 and 1. $\vec{P}$ is a random vector that assigns weight to the prey which makes the searching more efficient.

3.2 Architecture of the Proposed Hybrid Optimization Enabled NN Classifier

The significant network used in the prediction for non-linear time series is NARX neural network. It is a recurrent feed forward network that has effective learning rate, and converges faster to the solution. It also enchants multi-layer network, recurrent loop and time delay in the non-linear prediction. The prediction is based on the past values and the present input values. The architecture of the NARX network consists of the input layer, hidden layer and the output layers. The output obtained from the NARX rainfall prediction is mathematically expressed as,

$$p(j+1) = F\big[(p(j), \ldots \ldots, p(j_{z1}); \ u(i), \ldots \ldots u(i_{z2}))\big] \tag{9}$$

where, $p(j)$ is the raindrop data in the jth time series and $p(j_z)$ represents the rainfall data at j_z time series. z_1 and z_2 are the delay factors employed for the prediction (Fig. 2).

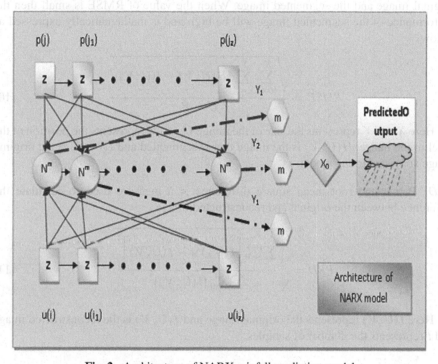

Fig. 2. Architecture of NARX rainfall prediction model

4 Results and Discussion

The results obtained using the proposed spotted hyena based optimization in the NARX model is summarized and evaluated in the following section as follows:

4.1 Experimental Setup

The research is accomplished using the software MATLAB in windows 10 operating system with 8 gb RAM. The dataset used for the execution is Rainfall in India to predict the rainfall.

4.2 Dataset Description

The dataset utilized for the experiment is collected from the National Data Sharing and Accessibility Policy (NDSAP) and the rainfall predicted from the India are taken into account from the year 1901 to 2015. For the detailed enumeration the dataset is classified into yearly, monthly and quarterly basis.

4.2.1 Performance Metrics

RMSE: Root mean square error is used to measure the variation occurred between the original image and the segmented image. When the value of RMSE is small then the performance of the segmented image will be high and is mathematically expressed as follows

$$RMSE = \sqrt{\frac{\sum_{U=1}^{X} \sum_{V=1}^{Y} [H[U,V] - I[U,V]]^2}{U \times V}} \tag{10}$$

Here X and Y represents the size of the image U and V represents the position of the pixel in the image. $H(U,V)$ is the image that is segmented and $I(U,V)$ is the original image subjected to the model.

PRD: Percentage root mean square difference is a measure used to calculate the difference between the original and reconstructed image.

$$PRD = \sqrt{\frac{\sum_{z=0}^{t-1} \left[I(U,V) - \hat{I}(U,V) \right]^2}{\sum_{Z=0}^{t-1} [I(U,V)]^2}} \tag{11}$$

Here $I(U,V)$ represents the original image and $\hat{I}(U,V)$ is the reconstructed image and t represents the number of samples.

4.3 Comparative Methods

The methods used to compare the proposed SH-NARX network model are LM-NARX, Linear regression modeling, and RBF-HPSOGA. The values are measured for varying sizes but for the simplified version several values are enumerated in the below sections.

4.3.1 Comparative Analysis

Analysis Based on RMSE Values: Initially the RMSE values are measured based on the size of the training data in an yearly basis and are shown in Fig. 3a.The RMSE value for the methods LM-NARX, Linear regression modeling, RBF-HPSOGA and the proposed SH-NARX prediction model for the training size of 0.65 are 0.054645613, 0.131040967, 0.053178225, 0.006395029 respectively. Similarly for the training size of 0.7 for the methods LM-NARX, Linear regression modeling, RBF-HPSOGA and the proposed SH-NARX prediction model are 0.054386483, 0.128716118, 0.03784572, and 0.005845776 respectively. Furthermore the RMSE values of the methods LM-NARX, Linear regression modeling, RBF-HPSOGA and the proposed SH-NARX prediction model for the training percentage 0.8 are 0.05399651, 0.119272397, 0.02033044, and 0.003127474 respectively.

The RMSE values for the methods in a monthly basis is measured and shown in Fig. 3b. The RMSE value for the methods LM-NARX, Linear regression modeling, RBF-HPSOGA and the proposed SH-NARX prediction model for the training size of 0.55 are 1.567359981, 18.07018917, 17.05821744, and 0.108860232 respectively. Similarly the RMSE value of the data under the training size 0.45 for the methods LM-NARX, Linear regression modeling, RBF-HPSOGA and the proposed SH-NARX prediction model is 2.520388226, 31.2543726, 30.70843581, 0.138727057 respectively. Correspondingly the values of the methods LM-NARX, Linear regression modeling, RBF-HPSOGA and the proposed SH-NARX prediction model for the training size 0.45 are 2.071838111, 26.768377, 23.2582918, and 0.126290269 respectively.

The RMSE values in a quarterly manner are measured and exhibited in the Fig. 3c. The RMSE value of the methods LM-NARX, Linear regression modeling, RBF-HPSOGA and the proposed SH-NARX prediction model for the training size 0.5 are 0.110449739, 0.242172466, 0.556730604, and 0.013896454 respectively. Similarly for the training size of 0.65, the RMSE values are measured for the methods LM-NARX, Linear regression modeling, RBF-HPSOGA and the proposed SH-NARX prediction model and are enumerate as 0.074979355, 0.155438121, 0.220683831, and 0.011657799 respectively. Likewise the RMSE values for the methods LM-NARX, Linear regression modeling, RBF-HPSOGA and the proposed SH-NARX prediction model for the training size 0.8 are listed as 0.057285533, 0.127414519, 0.107858025, and 0.009971462 respectively.

Analysis Based on PRD Values: The PRD values are also measured for the methods LM-NARX, Linear regression modeling, RBF-HPSOGA and the proposed SH-NARX prediction model and are revealed in Fig. 4a. Initially the values are measured for the yearly basis for varying training sizes of data and the values are enumerated. The PRD values for the methods LM-NARX, Linear regression modeling, RBF-HPSOGA and the proposed SH-NARX prediction model for the training size 0.7 are 5.272694952, 9.619634919, 3.917220226, and 1.535235903 respectively. In a similar manner the values of the training size 0.6 for the methods LM-NARX, Linear regression modeling, RBF-HPSOGA and the proposed SH-NARX prediction model are 5.296588236, 9.765313608, 5.145555559, and 1.65364854 respectively. Similarly the PRD values for the training size 0.45 are 5.331479133, 10.01870928, 9.247074244, and 1.869263315 respectively.

Secondly the PRD values are measured in a monthly basis and are shown in Fig. 4b. The PRD values for the methods LM-NARX, Linear regression modeling, RBF-HPSOGA and the proposed SH-NARX prediction model for the training size of 0.8 are 3.197236148, 17.74855856, 17.20523995, and 0.20707524 respectively. For the training size of 0.4 of the methods LM-NARX, Linear regression modeling, RBF-HPSOGA and the proposed SH-NARX prediction model is given by 3.464095582, 185.7834588,

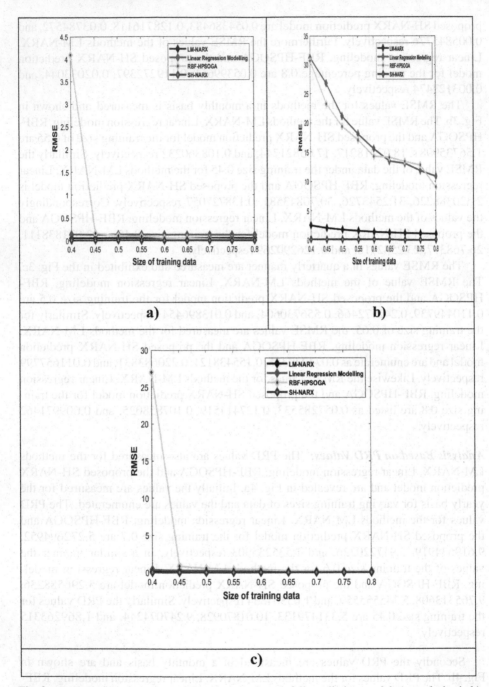

Fig. 3. Analysis based on the RMSE values in the rainfall prediction model a) yearly basis b) monthly basis c) quareterly basis

138.234133, and 0.954183783 respectively. For the training percentage 0.6 for the methods LM-NARX, Linear regression modeling, RBF-HPSOGA and the proposed SH-NARX prediction model the values are enumerated as 3.312602169, 41.12819639, 33.25213038, and 0.868705142 respectively.

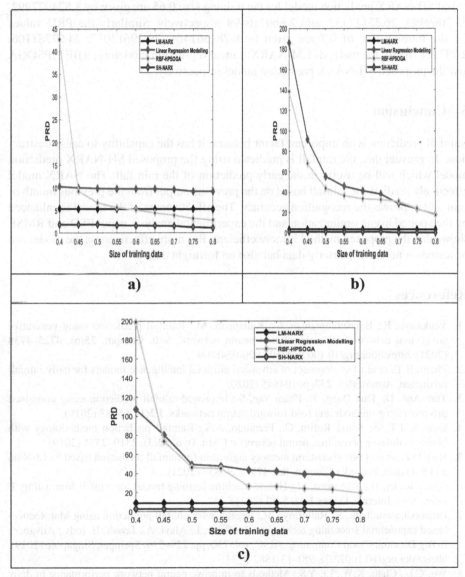

Fig. 4. Analysis based on the PRD values in the rainfall prediction model a) yearly basis b) monthly basis c) quarterly basis

Finally the PRD values are measured in a quarterly basis and are depicted in Fig. 4c. The PRD values for the training size of 0.8 for the methods LM-NARX, Linear regression modeling, RBF-HPSOGA and the proposed SH-NARX methods are 8.694244079, 35.76247282, 19.7700456, and 2.201340922 respectively. Similarly the PRD values for the methods LM-NARX, Linear regression modeling, RBF-HPSOGA and the proposed SH-NARX prediction model for the training size 0.65 are given by 8.824927992, 42.2069493, 26.35117132, and 2.660116499 respectively. Similarly the PRD values for the training size of 0.7 are given by 8.781661461, 39.29613037, 24.07781106, 2.387528763 for the methods LM-NARX, Linear regression modeling, RBF-HPSOGA and the proposed SH-NARX prediction model respectively.

5 Conclusion

Rainfall prediction is an important factor because it has the capability to cause destruction. To prevent this, the rainfall is predicted using the proposed SH-NARX prediction model which will be useful in the early prediction of the rain fall. The NARX model effectively predicts the rainfall based on the previous output from the previous month or year and improves the recognition accuracy. The effectiveness of the model is enhanced by the spotted hyena optimization and the experimental analysis using PSD and RMSE shows that the proposed method is more efficient. For more effectiveness the model can be tested on not only quarterly data but also on fortnight data also.

References

1. Venkatesh, R., Balasubramanian, C., Kaliappan, M.: Rainfall prediction using generative adversarial networks with convolution neural network. Soft. Comput. **25**(6), 4725–4738 (2021). https://doi.org/10.1007/s00500-020-05480-9
2. Pham, B.T., et al.: Development of advanced artificial intelligence models for daily rainfall prediction. Atmos. Res. **237**, p. 104845 (2020)
3. Tran Anh, D., Duc Dang, T., Pham Van, S.: Improved rainfall prediction using combined pre-processing methods and feed-forward neural networks. J **2**(1), 65–83 (2019)
4. Esteves, J.T., de Souza Rolim, G., Ferraudo, A.S.: Rainfall prediction methodology with binary multilayer perceptron neural networks. Clim. Dyn. **52**(3), 2319–2331 (2019)
5. Haq, D.Z., et al. Long short-term memory algorithm for rainfall prediction based on El-Nino and IOD data. Procedia Comput. Sci. **179**, 829–837 (2021)
6. Abdul-Kader, H., Mohamed, M.: Hybrid machine learning model for rainfall forecasting. J. Intell. Syst. Internet Things **1**(1), 5–12 (2021)
7. Dhamodharavadhani, S., Rathipriya, R.: Region-wise rainfall prediction using MapReduce-based exponential smoothing techniques. In: Peter, J., Alavi, A., Javadi, B. (eds.) Advances in Big Data and Cloud Computing. AISC, vol. 750, pp. 229–239. Springer, Singapore (2019). https://doi.org/10.1007/978-981-13-1882-5_21
8. Wu, C.L., Chau, K.W., Li, Y.S.: Methods to improve neural network performance in daily flows prediction. J. Hydrol. **372**, 80–93 (2009)
9. Sang, Y.F.: A review on the applications of wavelet transform in hydrology time series analysis. Atmos. Res. **122**, 8–15 (2013)
10. Sahai, A.K., Soman, M.K., Satyan, V.: All India summer monsoon rainfall prediction using an artificial neural network. Clim. Dyn. **16**(4), 291–302 (2000)

11. Trinh, T.A.: The impact of climate change on agriculture: findings from households in Vietnam. Environ. Resour. Econ. **71**(4), 897–921 (2018)
12. Le, L.M., et al.: Development and identification of working parameters for a lychee peeling machine combining rollers and a pressing belt. AgriEngineering **1**(4), 550–566 (2019)
13. Abbot, J., Marohasy, J.: Skilful rainfall forecasts from artificial neural networks with long duration series and single-month optimization. Atmos. Res. **197**, 289–299 (2017)
14. Navone, H.D., Ceccatto, H.A.: Predicting Indian monsoon rainfall: a neural network approach. Clim. Dyn. **10**(6–7), 305–312 (1994)
15. Davolio, S., Miglietta, M.M., Diomede, T., Marsigli, C., Morgillo, A., Moscatello, A.: A meteo-hydrologicalprediction system based on a multi-model approach for precipitation forecasting. Nat. Hazards Earth Syst. Sci. **8**, 143–159 (2008)
16. Diomede, T., et al.: Discharge prediction based on multi-model precipitation forecasts. Meteorol. Atmos. Phys. **101**, 245–265 (2008)

Application Initiated Data Session as a Service for High Reliable and Safe Railway Communications

Evelina Pencheva[1]([✉]) and Ivaylo Atanasov[2]

[1] Todor Kableshkov University of Transport, Sofia, Bulgaria
evelina.nik.pencheva@gmail.com
[2] Technical University of Sofia, Sofia, Bulgaria
iia@tu-sofia.bg

Abstract. Future Railway Mobile Communication System which is critical for European Train Control Systems is expected to provide highly reliable, safe, and secure services and applications. The paper studies how mobile edge cloud computing may support the requirements of railway applications, discussing virtualization of rail functions and co-existence with mobile edge services. A new mobile edge service is designed that provides core functionality for many railways use cases enabling application-initiated and manipulated data sessions. Using the service interfaces, on-board and trackside applications can establish data sessions e.g. in case of emergency. The service is described by typical use cases, data types, resources and supported methods, and some implementation issues are discussed.

Keywords: Future Railway Mobile Communication System · Network function virtualization · Multi-Access Edge Computing · Service oriented architecture

1 Introduction

High speed railways are expected to provide comfortable, flexible, safe, and environmentally sustainable transport service. They will feature improved capacity, connectivity, and sustainability. The ever increasing passenger and freight traffic call for enhancement of line capacity including train top speed, acceleration and braking rates and train control allowance time. Connectivity brings together Internet of Things (IoT), Artificial Intelligence (AI) and Machine Learning (ML). IoT objects equipped with sensors on the train onboard systems, rolling stock and trackside equipment can monitor different parameters, AI can install the required intelligence for data mining and ML can interfere about tasks to be executed. Sustainability means that the transport service must be provided with minimizing the energy consumption and in impact of greenhouse gas emissions [1].

Harmonizing the telecommunications in railways aims full interoperability and the Future Railway Mobile Communications System (FRMCS) standard is the next step in this direction [2, 3]. FRMCS will provide high reliable and low latency mobile communication between train and track for voice and signaling data and control commands.

© The Author(s), under exclusive license to Springer Nature Switzerland AG 2022
M. Singh et al. (Eds.): ICACDS 2022, CCIS 1614, pp. 24–37, 2022.
https://doi.org/10.1007/978-3-031-12641-3_3

By adopting future proof technologies like fifth generation (5G) and beyond, and AI, FRMCS will enable advance railway operations that improve passenger experience and efficiency and will unlock intelligence by strengthening security measures and using data to increase situational awareness [4, 5].

While the most research in the area focuses on techniques for increasing the capacity of wireless communication technologies, less attention is paid on railway-specific communication services [6–10]. Some of the applications, that can differentiate rail operators, include communications-based train control, video surveillance systems, and passenger information systems. The strong requirements of such applications for ultra-high reliability and low latency can be satisfied by deployment of Multi-access Edge Computing (MEC) technology at the edge of the railway mobile communication network. MEC converges the network function virtualization and software-defined paradigms in the vicinity of end users and can enable AI applications to optimize railway operations by advanced traffic management, remote driving, railway service continuity, and follow-me infotainment in railway scenarios [11–14].

In this paper, it is studied how MEC may be integrated with FRMCS. The capabilities of standardized MEC services to support the requirements of typical FRMCS use cases, described in [15] and [16], are evaluated, and a new MEC service, which enables initiation and management of data sessions by an application, is proposed. The service may be used as a building block by different applications and thus it facilitates the interoperability.

The paper is structured as follows. Next section presents the system architecture of FRMCS and the proposed MEC deployment in railway networks. Section 3 describes the main FRMCS use cases and the support of standardized MEC services which illustrate the research motivation. Section 4 provides a definition of a new MEC service for application initiated data sessions with typical use cases and Sect. 5 discusses some implementation issues. Finally, the paper is concluded.

2 Mobile Cloud at the Edge of the Railway Network

European Telecommunications Standard Institute (ETSI) has defined a logical architecture of FRMCS system, which considers the railway requirements for future flexible radio communications, and the key reference points [3]. The implications of specific challenges such as positioning, security, and smooth migration, on the FRMCS architecture are analyzed. In this section, the ETSI FRMCS logical architecture is presented and a possible MEC deployment scenario is described.

The FRMCS logical architecture is shown in Fig. 1.

The FRMCS logical architecture make differentiation between so-called Railway Application Stratum, Service Stratum and Transport Stratum. The Railway Application Stratum uses the service offered the Service Stratum and provides railway-specific applications. The Service Stratum provides communication services for information exchange between service users and complementary services which support communication services and railway applications, such as user positioning. The Transport Stratum provides access and core functionality for the FRMCS system. This separation enables independent stratum evolving.

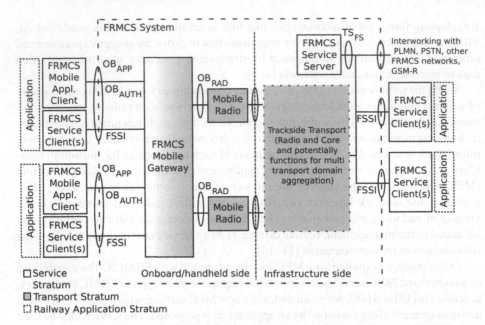

Fig. 1. A high level view of FRMCS logical architecture and system boundaries [3]

As shown in Fig. 1, the railway applications make the use of a FRMCS Mobile Application Client, which enables application authorization, and one or more FRMCS Service Clients, which make possible usage of services provided by the Service Stratum. The FRMCS Mobile Gateway provides access to the FRMCS Transport Stratum for humans or machines that use communication and complementary services. The FRMCS Clients access the FRMCS Mobile gateway through the OB$_{APP}$ reference point, which is composed of OB$_{APP}$ reference point used for application authorization and FSSI reference point used for provisioning of mission critical communication functions and for common functions such as location and policy management, configuration management, registration and service authorization. In addition, FRMCS Mobile Gateway monitors the Mobile radio unit(s) operation through the OB$_{RAD}$ reference point, which reflects the standard user plane between User Equipment (UE) and an application. The Mobile Radio units expose UE functionality. The Trackside Transport provides radio access and core functionality. The FRMCS Service Server as an end point of service level sessions with FRMCS Service Clients, provides authorization, location management, user profiles and control, as well as interworking with legacy communication systems. The TS$_{FS}$ reference point corresponds to the 5G N5 and N6 interfaces [17].

The FRMCS standards define just the logical architecture and user requirements based on survey of different use cases. Defining of building blocks and interfaces, provisioning of communication services to application layer and ensuring interoperability are considered as future activities. In this paper, a new communication service is designed and the supported application programming interfaces are described. The service may be used as a brick in different railway applications sharing the service code and thus facilitating interoperability.

FRMCS will apply the principles of Software Defined Networking (SDN), Network Function Virtualization (NFV) and Cloudification. SDN, NFV and MEC virtualize and centralize networking into data centers at the edge of the mobile network. So, the implementation of FRMCS may deploy distributed core network functionality the edge of the railway network with virtualized functions. To meet the specific FRMCS requirements, the 5G core functions decomposed into a number of Service-base Architecture elements may be virtualized and installed within MEC cloud infrastructure. The MEC host may provide the virtualization infrastructure for both core network functions and for the MEC platform which provides MEC services, and for railway applications. Thus, the FRMCS Service Server may be implemented as a railway-specific MEC cloud providing MEC applications and core applications, as shown in Fig. 2. For example, the railway-specific MEC cloud may be deployed in the area on Radio Block Centre (RBC), which is responsible for the management and control of railway circulation through the System of Command and Control [18]. As a future step, the railway-specific MEC cloud may provide virtualized environment for RBC functions running as containers.

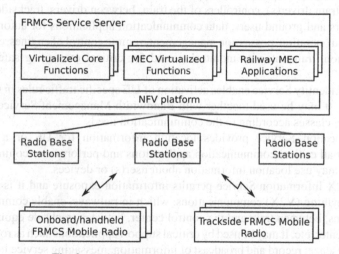

Fig. 2. Mobile edge cloud in the railway network

The distributed core network functionality may also include virtualized IMS (Internet Protocol Multimedia Subsystem) functions which provide the required functionality for mission critical voice, data, and messaging services. IMS is a complex subsystem, and it is unlikely that IMS virtualized functions will be deployed at the end of the network, e.g. in the area of an RBC. So, many of the FRMCS communication services may be implemented as MEC services to improve efficiency and to minimize latency.

The next section provides a brief description of FRMCS use cases and their support by standardized MEC services.

3 MEC Services Support for FRMCS Use Cases

The technology independent FRMCS user requirements are defined in a form of individual railway-specific applications, each of which is justified by an identified use case [16]. The applications are classified as critical, performance and business ones. The critical communications applications are essential for train safety and movements, emergency communications, trackside maintenance, shunting, presence, automatic train control, etc. The performance related communication applications help to improve the railway operation maintenance, such as telemetry and train departure. The business communication applications support the railway business operation, e.g. wireless internet/intranet, etc. Standardized MEC services and their support to FRMCS applications are as follow.

MEC Bandwidth Management Service aggregates the application specific bandwidth requirements for bandwidth size, bandwidth priority, or both, and optimizes bandwidth usage. It may be used to provide reliable communication bearer by critical voice communication applications and by performance communication applications which require service continuity. Examples of such applications include on-train voice communication between the train driver(s), controllers of the train, between drivers, trackside workers, shunting users and ground users, data communication applications for automatic train protection and operation, possession management, remote control of engines, emergency communication, train integrity monitoring, and for critical messaging safety related services.

MEC UE identity Service enables activation of UE specific traffic rules in the mobile edge system. It may be used together with Bandwidth Management Service to assign different QoS classes according to the communication needs.

MEC Location Service provides location information for UE or a group of UEs. Almost all critical communication applications and performance communication applications may use location information about user(s) or devices.

MEC V2X Information Service permits information exposure and it is to support vehicle to anything (V2X) communications, which in railways, enable communication between trains, between train and the control center, between trackside equipment and the control center, etc. It may be used by critical support applications such as roll management and presence, record and broadcast of information, messaging service to exchange information among railway users, train departure train communications, etc.

MEC Radio Network Information Service gathers radio network information on current radio conditions and exposes it to applications. It may be used by on-train telemetry applications to increase performance or to support operation management, by infrastructure telemetry applications to support equipment supervision and demand forecasting and response, and information help point for public.

MEC WLAN Information Service provides WLAN access related information. It may be used by business communication applications including wireless internet for passengers on platforms, wireless on-train data communication for train staff where an access to intranet/internet is required for operational purposes or for customer satisfaction.

MEC Fixed Access Information Service provides contextual information about fixed access network and may be used by all critical communication applications and performance communication applications which require reliable fixed bearers for voice, data, and messaging.

The brief review on standardized MEC services shows that these services do not support functionality for application initiated sessions which is essential for many railway applications. The next section describes the proposed new MEC service for creating and managing data sessions initiated by critical communication and performance communication applications (application initiated data session).

The service may be used by FRMCS applications for:

- data session initiation by trackside maintenance warning system with trackside maintenance workers,
- data session between a ground based or train based system and the infrastructure system to monitor and control signal and indicators, train detection, level crossing elements, lighting controls and alarms and others,
- data session initiated by driver safety device to a ground user,
- data session initiated by train integrity monitoring system between component monitoring train integrity.
- data session establishment between on-train systems for on-train telemetry communications,
- data sessions between infrastructure systems and/or ground for infrastructure telemetry communications.

4   Service Description

The proposed new Application Initiated Data Session (AppIDS) service enables an application to set up a data session usually with at least two participants. The application may subsequently add, remove, transfer data session participants. The application may retrieve the data session or session participant status. The application may also force data session to be terminated for all participants.

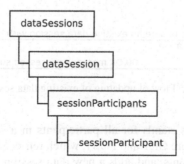

Fig. 3. Resources supported by the AppIDS service

The AppIDS service design follows the REpresentational State Transfer (REST) architectural style, where logical and physical entities are represented as resources. Figure 3 shows the structure of AppIDS service resources. The resources are organized in a tree structure and their Uniform Resource Identifiers (URIs) follow the root URI where the AppIDS service is registered. The resources can be manipulated using HTTP methods.

The dataSessions resource represents all data sessions created by an application. It supports HTTP method GET, used to retrieve information about a list of dataSession resource, and HTTP method POST, which is used to create a dataSession resource. The dataSession resource represents an individual data session, and the applicable HTTP methods are GET, used to retrieve information about a specific data session, PUT, used to update information about a specific data session, DELETE, used to terminate a specific data session by removing all its participants.

Figure 4 illustrates the flow of creating a new data session resource. The application sends a request to create a new data session with the data session information as described below. The AppIDS service responds with an approval sending the ID of the created data session.

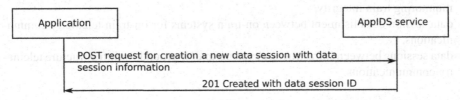

Fig. 4. Flow of creating of a new resource representing a data session

Figure 5 illustrates the flow of update information about existing data session. The application sends a request to update a specific data session and the AppIDS service responds with an update approval.

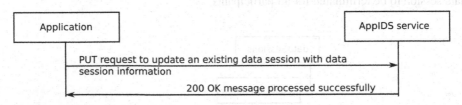

Fig. 5. Flow of updating of existing data session

The sessionParticipants stands for all participants in a specific data session. The supported HTTP methods are GET and POST, which retrieve information about the list of all data session participants, and adds a new data session participant, respecttively. The sessionParticipant resource represents a particular session participant. The HTTP method GET applied to this resource retrieves information about the participant, the

PUT method updates information about the resource, the PATCH method to modify the resource representation, and the DELETE method removes the specific data participant.

Figure 6 illustrates the flow of adding a new participant to existing data session. The application requests to create a new resource representing data session participant providing the participant information as described below. The AppIDS service responds with request approval and the participant ID.

Figure 7 illustrates the flow of retrieving information about a specific participant in an existing data session. The application request information about specific participant in an existing data session providing the participant ID. The AppIDS service responds with request approval delivering the requested participant info.

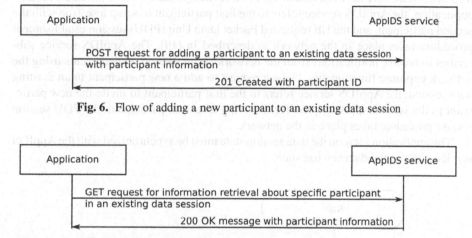

Fig. 6. Flow of adding a new participant to an existing data session

Fig. 7. Flow of retrieving information about specific participant in an existing data session

Figure 8 illustrates the flow of removing a participant from an existing data session. The application sends a request to remove a participant from existing data session to the AppIDS service. The AppIDS service responds with removal approval.

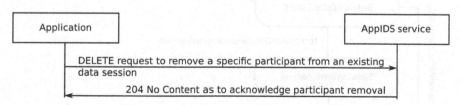

Fig. 8. Flow of removing a participant from existing data session

The dataSessionInformation data type contains the address (URI) of the first and the second data session participants, optionally the name of the session initiator, e.g. the name on whose behalf the data session is being established, charging information, media information which identifies one or more media types for the session to be applied to the session participants, and information about whether it is allowed for session participant to change the media type during the session.

The participantInformation data type contains the existing data session identifier, session participant address of the user to be added to the data session, and media information identifying one or more media tyles for the session participant. For each media type the media direction (uplink, downlink or bidirectional) has to be specified.

5   Implementation Issues

The AppIDS service implementation requires co-location of MEC platform with virtualized distributed core network functions. Upon request of data session initiation form an application, the AppIDS service refers to the first participant to setup a session with the second participant, and the UE requested Packet Data Unit (PDU) session establishment procedure takes place in the network as described in [19]. The AppIDS service subscribes to receive notifications from the network about session related events using the network exposure functionality. Upon a request to add a new participant to an existing data session, the AppIDS service refers to the first participant to invite the new participant to the session. Upon request for removing a session participant, the PDU session release procedure takes place in the network.

The application view on the data session state must be synchronized with the AppIDS service view on the data session state.

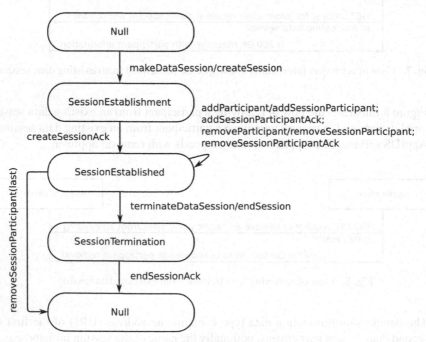

Fig. 9. The application's model of the data session state

Figure 9 shows the simplified data session state model supported by an application. Due to external events such as an alarm in the trackside maintenance warning system, the application initiates a session between the trackside maintenance warning system and the main trackside maintenance worker. The application may decide to add another trackside maintenance worker in the appropriate area to the established data session, to remove a trackside maintenance worker from the session, or to terminate the data session. In each state, the application may query about the data session and participant state (not shown in Fig. 9).

Figure 10 shows the simplified data session state model supported by the AppIDS service. Upon requests from application, the AppIDS service refers to the first session participant to request a network procedure corresponding to the request and subscribes to receive notification about data session related events. Upon event notification the AppIDS service responds to the application requests.

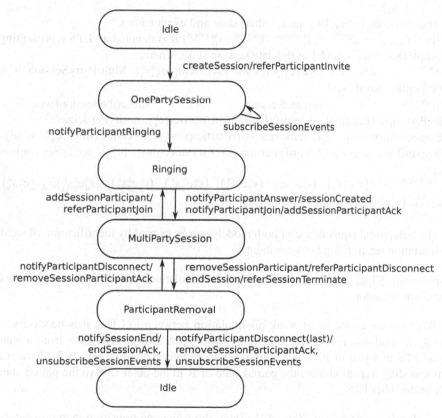

Fig. 10. The AppIDS service's model of the data session state

Figure 9 and Fig. 10 are simplified as they do not reflect the behavior when the application requests are not approved, or the request timeout inspires.

To prove the state models' synchronization, i.e. exposure of equivalent behavior, both models are formally described as Labelled Transition Systems (LTS), where an LTS is a quadruple of a set of states, a set of events, a set of transitions and an initial state.

By $ST_{app} = (S^{app}, E^{app}, T^{app}, s_0^{app})$ it is denoted an LTS representing the application's model of the data session state, where

$S^{app} = \{$Null $[s_1^a]$, SessionEstablishment $[s_2^a]$, SessionEstablished $[s_3^a]$, SessionTermination $[s_4^a]\}$;

$E^{app} = \{$makeDataSession $[e_1^a]$, createSessionAck $[e_2^a]$, addParticipant $[e_3^a]$, addSessionParticipantAck $[e_4^a]$, removeParticipant $[e_5^a]$, removeSessionParticipantAck $[e_6^a]$, removeSessionParticipantAck(last) $[e_7^a]$, terminateDataSession $[e_8^a]$, endSessionAck $[e_9^a]\}$;

$T^{app} = \{(s_1^a e_1^a s_2^a), (s_2^a e_2^a s_3^a), (s_3^a e_3^a s_3^a), (s_3^a e_4^a s_3^a), (s_3^a e_5^a s_3^a), (s_3^a e_6^a s_3^a), (s_3^a e_7^a s_1^a), (s_3^a e_8^a s_4^a), (s_4^a e_9^a s_1^a)\}$;

$s_0^{app} = s_1^a$.

The notations in brackets are for short state and event names.

By $ST_{service} = (S^{service}, E^{service}, T^{service}, s_0^{service})$ it is denoted an LTS representing the AppIDS service's model of the data session state, where.

$S^{service} = \{$Idle $[s_1^s]$, OnePartySession $[s_2^s]$, Ringing$[s_3^s]$, MultiPartySession $[s_4^s]$, ParticipantRemoval $[s_5^s]\}$;

$E^{service} = \{$createSession $[e_1^s]$, subscribeSessionEvents$[e_2^s]$, notifyParticipantRinging$[e_3^s]$, notifyParticipantAnswer$[e_4^s]$, notifyParticipantJoin $[e_5^s]$, addSessionParticipant $[e_6^s]$, removeSessionParticipant $[e_7^s]$, endSession $[e_8^s]$, notifyParticipantDisconnect $[e_9^s]$, notifyParticipantDisconnect(last) $[e_{10}^s]$, notifySessionEnd $[e_{11}^s]\}$;

$T^{service} = \{(s_1^s e_1^s s_2^s), (s_2^s e_2^s s_2^s), (s_2^s e_3^s s_3^s), (s_3^s e_4^s s_4^s), (s_3^s e_5^s s_4^s), (s_4^s e_6^s s_3^s), (s_4^s e_7^s s_5^s), (s_5^s e_9^s s_4^s), (s_4^s e_8^s s_5^s), (s_5^s e_{10}^s s_1^s), (s_5^s e_{11}^s s_1^s)\}$;

$s_0^{service} = s_1^s$.

The behavioral equivalence of both models can be proved by identification of weak bi-simulation relationship between them.

Proposition: ST_{app} and $ST_{service}$ have a weak bi-simulation relationship, i.e. they expose equivalent behavior.

To prove the existence of weak bi-simulation between to LTSs, it is necessary to identify a relationship R between pairs of states such as each transition from a state in one LTS in a pair of R, which terminates in a state in another pair of R, there is a corresponding transition from the paired state of R in the other LTS to the paired state of R in the other LTS.

Proof: Let R $= \{(s_1^a, s_1^s), (s_3^a, s_4^s)\}$. Then, the following transition mapping can be identified:

1. The application creates a data session with two parties: for $\forall (s_1^a e_1^a s_2^a) \sqcap (s_2^a e_2^a s_3^a) \exists \{(s_1^s e_1^s s_2^s) \sqcap (s_2^s e_2^s s_2^s), (s_2^s e_3^s s_3^s) \sqcap (s_3^s e_4^s s_4^s)$.
2. The application adds a data session participant: for $\forall (s_3^a e_3^a s_3^a) \exists (s_4^s e_6^s s_3^s) \sqcap (s_4^s e_6^s s_3^s)$.

3. The application removes a data session participant but not the last one: for $\forall(s_3^a e_5^a s_3^a) \sqcap (s_3^a e_6^a s_3^a) \exists (s_4^s e_7^s s_5^s) \sqcap (s_5^s e_9^s s_4^s)$.
4. The application removes the last participant from the data session: for $\forall(s_3^a e_6^a s_1^a) \exists (s_4^s e_7^s s_5^s) \sqcap (s_5^s e_{10}^s s_1^s)$.
5. The application terminates the data session: for $\forall(s_3^a e_8^a s_4^a) \sqcap (s_4^a e_9^a s_1^a) \exists (s_4^s e_8^s s_5^s) \sqcap (s_5^s e_{11}^s s_1^s)$.

Therefore, R is a weak bi-simulation relationship, and ST_{app} and $ST_{service}$ expose equivalent behavior. ∎

The formal model description and the proof for their equivalent behavior may be used in conformance tests of service implementation against the service specification.

The emulation of AppIDS application programming interface is based on a RESTful solution and Cassandra as a NoSQL database. The experiment setup includes multi-threaded REST clients, implemented in Java, at the application side, a docker instance 1, which consists of a REST service part and a Cassandra client part, and a docker instance 2 for the Cassandra server part. Eclipse Vert.x is used for integration of Java implemented service part within the Cassandra client of the service.

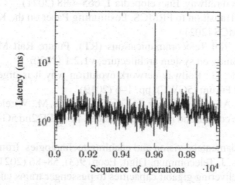

Fig. 11. Sequence of latency values

The HTTP traffic is generated by multi-threated clients as consecutive POST requests. In order to evaluate the latency each request is ornamented by an experimental header that bears the moment of submission in nanoseconds. At the moment of response receipt another nano-timestamp is taken and the difference, forming the latency sample, is collected into a sequence of such values. The total traffic volume is 20 thousand requests/responses and the recorded latency values for the 10-th group of one thousand values is shown in Fig. 11. The results point out that the injected latency is about 2 ms, with some spikes that might be regarded as temporary micro-outage. Moreover, the sequence of raw latency values might be treated as a stochastic process and making the probability mass function for each time-window of thousand operations gives hint that the process is not stationary, thus, it is hard to be predicted.

6 Conclusion

The paper studies how MEC technology, which is regarded as a 5G key component can provide railway applications which require high transmission rate, low latency, and high availability. After discussion on how standardized MEC services may support the requirements of railway application, a new MEC service is presented. The proposed service enables applications to initiate and manipulate data sessions, which is a key requirement for many critical communication and performance communication railway applications. The service is described by typical use cases and some implementation issues are considered. The service may be used as a building block in different applications enabling cost reduction and interoperability.

Acknowledgments. This research was supported by the Bulgarian National Science Fund under Grant No. KP-06-H37/33.

References

1. Watson, I.: High speed railway. Encyclopedia **1**, 665–688 (2021)
2. UNIFE: Successful Transition to FRMCS, Positioning Paper on the Key Success Factors for the transition to FRMCS (2021)
3. ETSI: TR 103 459 Rail Telecommunications (RT); Future Rail Mobile Communication System (FRMCS); Study on system architecture, v1.2.1 (2020)
4. Pierre, T., Christophe, G.: Railway network evolution why it is urgent to wait for 5G? In: IEEE 2nd 5G World Forum (5GWF), pp. 1–6 (2019)
5. Härri, J., Arriola, A., Aljama, P., Lopez, I., Fuhr, U., Straub, M.: Wireless technologies for the next-generation train control and monitoring system. In: IEEE 2nd 5G World Forum (5GWF), 179–184 (2019)
6. Soldani, D.: 6G fundamentals: vision and enabling technologies, from 5G to 6G trustworthy and resilient systems. J. Telecommun. Digit. Econ. **9**(3), 58–86 (2021)
7. Jamaly N., et al.: Delivering gigabit capacities to passenger trains tales from an operator on the road to 5G, pp. 1–7 (2021). arXiv:2105.06898 [cs.NI]
8. Berbineau M., et al.: Zero on site testing of railway wireless systems: the Emulradio4Rail platforms. In: IEEE 93rd Vehicular Technology Conference (VTC2021-Spring), pp. 1–5 (2021)
9. Allen, B.H., Brown, T.W.C., Drysdale, T.D.: A new paradigm for train to ground connectivity using angular momentum. In: IEEE 2nd 5G World Forum (5GWF), pp. 185–188 (2019)
10. Saideh, M., Alsaba, Y., Dayoub, I., Berbineau, M.: Performance evaluation of multi-carrier modulation techniques in high speed railway environment with impulsive noise. In: 2019 IEEE 2nd 5G World Forum (5GWF), pp. 243–248 (2019)
11. Atanasov, I., Pencheva, E., Nametkov, A.: Handling mission critical calls at the network edge. In: International Conference on Mathematics and Computers in Science and Engineering (MACISE), pp. 6–9 (2020)
12. Santi, N., Mitton. N.: A resource management survey for mission critical and time critical applications in multi access edge computing. ITU J. Geneva Int. Telecommun. Union **2**(2) (2021)
13. Feng L., et al.: Energy-efficient offloading for mission-critical IoT services using EVT-embedded intelligent learning. IEEE Trans. Green Commun. Netw. **5**(3), 1179–1190 (2021)

14. Pencheva, E., Atanasov, I., Vladislavov, V.: Mission critical messaging using multi-access edge computing. Cybern. Inf. Technol. **19**(4), 73–89 (2019)
15. International Union of Railways: Future Railway Mobile Communication System, Use cases, v2.0.0 (2020)
16. International Union of Railways: Future Railway Mobile Communication System, User Requirements Specification, v5.0.0 (2020)
17. 3GPP TS 23.501 3rd Generation Partnership Project; Technical Specification Group Services and System Aspects: System architecture for the 5G System (5GS); Stage 2; Release 17, v17.03.0 (2021)
18. ERA UNISIG EEIG ERTMS USERS GROUP: System Requirements Specification, Chapter 2, Basic System Description, SUBSET-026-2 3.6.0 (2016)
19. 3GPP TS 23.502 3rd Generation Partnership Project; Technical Specification Group Services and System Aspects: Procedures for the 5G System (5GS); Stage 2; Release 17, v17.03.0 (2021)

Acute Lymphoblastic Leukemia Disease Detection Using Image Processing and Machine Learning

Abhishek D. Chavan, Anuradha Thakre, Tulsi Vijay Chopade[✉], Jessica Fernandes, Omkar S. Gawari, and Sonal Gore

Department of Computer Engineering, Pimpri Chinchwad College of Engineering, Pune, India
{abhishek.chavan19,anuradha.thakare,tulsi.chopade19, jessica.fernandes19,omkar.gawari19}@pccoepune.org

Abstract. Acute Lymphoblastic Leukemia (ALL) is a cancer type in which there is an increase of white blood cells (WBCs) in our body. This article presents a method that detects the presence of these abnormal cells in the bloodstream using machine learning and image processing algorithms. A methodology to identify ALL using machine learning classification techniques like Convolutional Neural Network (CNN), Artificial Neural Network (ANN), Logistic Regression, and Support Vector Machine (SVM) using the existing dataset (ALL-IDB2) is discussed. The outcome of the paper is to analyze the ALL IDB2 dataset and predict the output as ALL infected or not. According to the experimental results, it is observed that the performance of CNN supersites other machine learning classifiers for the proposed classification in terms of accuracy.

Keywords: Image acquisition · Image pre-processing · Image segmentation · Feature extraction · Machine learning · Classification · Artificial neural networks · Support vector machine · Convolutional neural network · Logistic regression · Naïve Bayes · K-nearest neighbors · Decision tree · Random Forest algorithms

1 Introduction

Leukemia is a cancer-type disease in which abnormal cells are formed. It is caused due to the rise in WBC in the body. Leukemia begins its infection from the bone marrow. ALL affects the lymphocytes, a white blood cell type that is essential for our immune system. As a result of this, the immune system gets weak and it decreases the capacity of bone marrow for the supply of red blood cells (RBCs) and platelets.

Doctors mostly face challenges in examining colored microscopic blood images. The process is time-consuming, less effective, and not appropriate for analyzing a huge number of cells.

M. Singh et al. (Eds.): ICACDS 2022, CCIS 1614, pp. 38–51, 2022.
https://doi.org/10.1007/978-3-031-12641-3_4

Machine learning (ML) methods have become a popular tool for medical researchers. From this, it can be determined whether a person has Leukemia or not. In many cases, the leukemia cells remain in the bone marrow. So you need a Bone marrow test. But everything is done manually so sometimes there can be flaws in the result.

A solution for this is automated detection of Leukemia by providing simple blood sample images which in turn process the image and provide flawless reports.

2 Literature Review

In [1] (Rajpurohit, Patil, Choudhary, Gavasane and Kosamkar 2018) (Kumar, Mishra and Asthana 2018) algorithm in which they used image processing and classification techniques for identification. K-means clustering algorithm is used for segmentation. Feature extraction is done by considering the color, shape, and texture features. For classification. They have used KNN and Naive Bayes classifier. The proposed algorithm gives effective accuracy for ALL detection.

In [2] (Rajpurohit, Patil, Choudhary, Gavasane and Kosamkar 2018), they developed a system in which they used image processing and classification techniques for early detection. OpenCV and skimage are used for image processing and different classifiers like feed Forward Network (FNN), Convolutional Neural Network (CNN), K Nearest Neighbors (KNN), and Support Vector Machine (SVM) are used for classification. CNN shows the maximum accuracy with 98.33%. Also, a web-based application is developed with an effective user interface.

In [3] (Karthikeyan and N. 2017), They have proposed a method in which first they performed image acquisition then image pre-processing which consists of noise removing and Contrast enhancement using adaptive histogram equalization. Gabor Texture Extraction method is used to get color features and for classification, SVM is used. The Fuzzy c-means method shows 90% accuracy.

In [4] (Patil and Raskar 2015), they have proposed a method in which they convert the image from RGB color space to HSV domain and perform the median filtering. Then, thresholding is done by Otsu's method. Grey Level Co-occurrence Matrix (GLCM) is used for feature extraction. SVM is used for classification. It's fast, simple, and got good results.

In [5] (Moradiamin, Samadzadehaghdam, Talebi and Kermani 2015), they have proposed a system in which an RGB image is converted to HSV and then the histogram technique is used for feature demonstration. Segmentation is performed by fuzzy c-means clustering technique and then by using a watershed algorithm to separate overlapping objects. Geometric feature for size and shape of a nucleus, statistical features for grayscale image histogram. SVM is used for classification. This technique has achieved good accuracy by using fuzzy c-means with SVM.

In [6] (Singhal 2015), they have applied the HSI color-based segmentation method for segmentation. Then, shape features are obtained from segmented images. They have used correlation-based feature selection. Classification is done by SVM. 92.30% accuracy is obtained.

In [7] (Moradiamin, Kermani, Talebi and Oghli 2015), a method for the detection of cancerous cells, using only features extracted from the image of their nucleus is proposed.

First, they performed image acquisition, then image pre-processing operations in which they convert the image from RGB to HSV Color space, and then Histogram equalization on the V band. Nucleus segmentation is performed by the k-means clustering algorithm. Then, feature extraction is done. SVM classifier is used for the classification.

In [8] (Pathirage, Marapana, Chandrananda and Amarathunga 2016), They have proposed a system that consists of pre-processing which involves resizing and contrasting the image, image segmentation using k-means clustering, feature extraction in which 4 types of features are extracted namely shape, texture, color, fractal dimension, and classification using BP – Scaled Conjugate Gradient (SCG) neural network.

In [9] (Chand and Vishwakarma 2020), They have proposed different steps consisting of image enhancement, segmentation of image, texture analysis, classification, and calculation of accuracy. Image classifications rely on the texture analysis results, while texture analysis relies on the segmentation results. This method has obtained 91.54% of the highest accuracy.

In [10] (Bodzas, Kodytek and Zidek 2020), they have proposed a method that includes taking images as input and its pre-processing. After that leukocyte segmentation consists of 4 stages: thresholding, three-phase filtration, detection of adjacent cells, and cell separation. External and graphical methods are used for feature extraction. Classification is done by using SVM and ANN. The overall accuracy is 97.52% designations.

3   Existing Techniques

1. Data collection/acquisition: Snapshots are accumulated by a digital camera. Then store it in relevant format (.jpeg, .png, etc.).
2. Pre-processing: In this, image formatting is done. This is used to remove the historical noise and to reduce the unwanted conflicts present in the image. In this, colorful images are converted to grayscale. The sample image is represented in Fig. 1

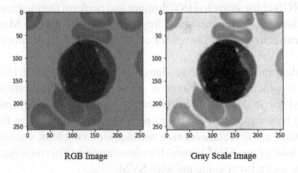

RGB Image   Gray Scale Image

Fig. 1. RGB to grayscale

3. Image segmentation: This step includes differentiating images into different parts according to their properties and features. In this, labels are assigned to the characteristics accordingly. Different techniques like Edge-Based Segmentation, Neural Networks for Segmentation, cluster-based segmentation, region-based segmentation, Otsu Segmentation, and Thresholding segmentation are used for image segmentation. The sample segmented image is represented in Fig. 2.

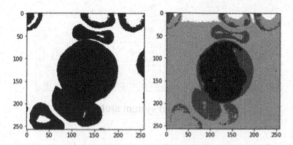

Fig. 2. Segmented image

4. Feature extraction: The procedure of extracting the matching records from the entire photo and transferring the information into a set of elements with their labels for the feature extraction. This step, primarily based upon the elements like color, shape, and texture features are extracted.
5. Classification: Procedure of labeling vectors or image pixels for differentiating them so that it will be easy to classify them immediately. Different classifiers like artificial neural networks, decision trees, convolutional neural networks, Logistic Regression, K- nearest neighbor, Support Vector Machine, etc. are used for classification.

4 Dataset

ALL_IDB2 is the dataset used for detecting acute lymphoblastic leukemia. This dataset contains a total of 260 images in the.tif format. Images labeled with 0 are of healthy people and images labeled with 1 are of ALL infected patients. The size of all the images is 257 × 257.

5 Proposed Architecture

This section discusses the proposed architecture for Acute Lymphoblastic Leukemia Disease Detection Using Image Processing and Machine Learning algorithms which are represented in Fig. 3.

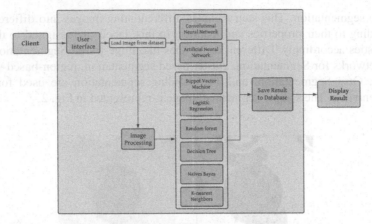

Fig. 3. Proposed system architecture

The proposed architecture consists of two main modules, i.e. image processing and classification. First, the client visits the interface and enters the image in the proper format (.jpeg, .png, etc.). Now, the user needs to upload the image through his local system, and then he can see the image with the label on it as ALL infected or healthy accordingly.

The interior design of the system consists of CNN, ANN SVM, Logistic Regression, Random Forest, Decision Tree, Naive Bayes, and KNN classifiers from which the classification is done and the appropriate results are obtained. These results are stored in a particular database and then visible to the user. The results obtained are discussed in the next section.

6   Results and Discussion

This section discusses the stepwise results for Acute Lymphoblastic Leukemia Disease Detection Using Image Processing and Machine Learning algorithms to decrease image Processing.

This module takes the image from the IDB2 dataset and stores it in the CSV file. The specific features like length, area, perimeter, aspect ratio, etc. are taken and segmentation is done. Feature extraction is done using inbuilt functions or specific formulae with the help of the OpenCV library in python. Features extracted from this get stored in the CSV format. This CSV file is given to the classifier which is the input for that classifier.

Classification Algorithms
This module takes the image as input in the CSV file format consisting of extracted features. The classifier function is to assign a label to the images according to which class they belong. It is done using the predefined extracted features. Finally, the result is obtained in the Boolean digits, either 0 or 1. 1 indicates the person is infected with ALL, and 0 is of a healthy person.

Component heads identify the different components of your paper and are not topically subordinate to each other. Examples include Acknowledgments and References and, for these, the correct style to use is "Heading 5". Use "figure caption" for your Figure captions, and "table head" for your table title. Run-in heads, such as "Abstract", will require you to apply a style (in this case, italic) in addition to the style provided by the drop-down menu to differentiate the head from the text.

Convolutional Neural Network (CNN): A Convolutional Neural Network is a deep learning method. It has 3 layers which are input, hidden, and output layers. Input to the input layer of an array structure which is obtained from the pixels of the image. The hidden layer performs feature extraction with calculation using the sigmoid function. It includes convolution, pooling, and flattening operations. The last output layer identifies the object in the image. Tensor Flow, the Keras framework can be used for CNN-based classification. The network contains 3 convolutions, 3 Max pooling, 1 Flattening, and 2 dense layers.

Artificial Neural Network (ANN): Artificial Neural Network is also a deep learning method. It differs from CNN in its image processing techniques. It uses perceptrons which is a simple network with a single neuron. It uses feed-forward and backpropagation algorithms for calculation. It is implemented using the MLP classifier imported from sci-kit learn.

Support Vector Machine (SVM): SVM is a supervised machine learning method. It divides the multiple classes using hyperplanes. It uses the kernel concept to shift the data into its required form in space. It is implemented using the Sklearn model. This model has in-built features that need to be extracted.

Logistic Regression: Logistic regression is a supervised machine learning method. It is used for binary classification. It is a statistical method. It measures the probability of an event going to occur and is used to predict the result for only two classes. It uses the logit function for calculating the probability of a binary event.

Decision Tree: A decision tree is a supervised machine learning method. It is a flow chart-like structure. In this, the inner nodes consider the feature, branches consider the decision rule, and the outer nodes consider the outcome. It uses the entropy concept for removing impurities present in the input dataset. Entropy is nothing but the impurities present in the input examples. It uses the gain ratio and Gini index method for selecting the partitioning attribute. Here we have used the decision tree model imported from the Sklearn library.

Random Forest: Random Forest is a supervised machine learning method. In this, the random samples are collected from the given dataset and then the construction of a decision tree on each sample is done. Then the result of each sample is taken and voted for each result. The final evaluation is done using the maximum number of votes it gets. It is implemented using a model imported from sci-kit learn which gives one extra variable.

Naive Bayes: Naive Bayes is a supervised machine learning method. It is a statistical method that uses the Bayes theorem for calculation. In this method, the probability is calculated for the prior class and then the probability is calculated using the Bayes theorem for the posterior class. Then the input is given to the maximum occurrence class. It is implemented using a model imported from sci-kit learn.

K nearest neighbor (KNN): K nearest neighbor is an unsupervised machine learning method. It is based on a lazy algorithm. That means model training does not require training data, it does work in testing data. It is implemented using a model imported from sci-kit learn.

Figure 5, 6, 7, 8, 9, 10, 11, 12 and Fig. 13 represents the results of various machine learning algorithms' model accuracy.

Results of CNN
See Fig. 4.

```
Model: "sequential"

Layer (type)                    Output Shape              Param #
=================================================================
conv2d (Conv2D)                 (None, 198, 198, 16)      448

max_pooling2d (MaxPooling2D)    (None, 99, 99, 16)        0

conv2d_1 (Conv2D)               (None, 97, 97, 32)        4640

max_pooling2d_1 (MaxPooling2    (None, 48, 48, 32)        0

conv2d_2 (Conv2D)               (None, 46, 46, 64)        18496

max_pooling2d_2 (MaxPooling2    (None, 23, 23, 64)        0

flatten (Flatten)               (None, 33856)             0

dense (Dense)                   (None, 512)               17334784

dense_1 (Dense)                 (None, 1)                 513
=================================================================
Total params: 17,358,881
Trainable params: 17,358,881
Non-trainable params: 0
```

Fig. 4. CNN model architecture

Here a sequential model is developed which contains three convolutional layers, 3 pooling layers, one flatten, and 2 dense layers with a total of 17,358,881 parameters from which all are the trainable parameters.

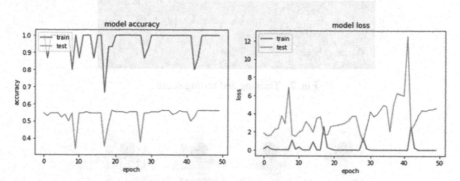

Fig. 5. Model accuracy and loss

Results of CNN

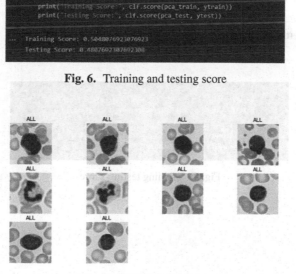

Fig. 6. Training and testing score

Fig. 6a. Identified images

Results of SVM

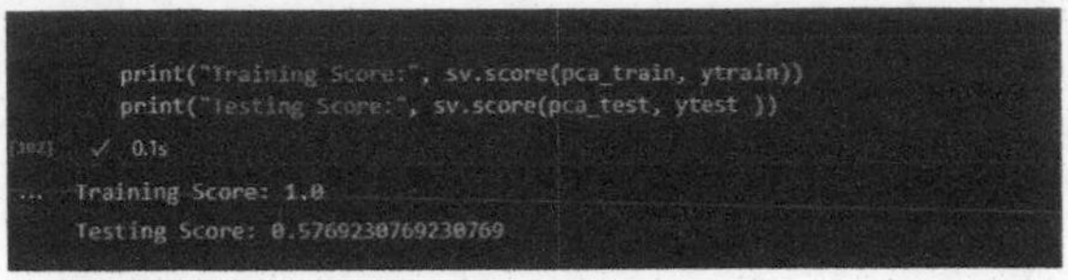

```
print("Training Score:", sv.score(pca_train, ytrain))
print("Testing Score:", sv.score(pca_test, ytest ))
[102]  ✓ 0.1s
...  Training Score: 1.0
     Testing Score: 0.5769230769230769
```

Fig. 7. Training and testing score

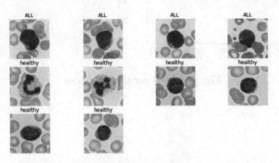

Fig. 7a. Identified images

Results of Logistic Regression

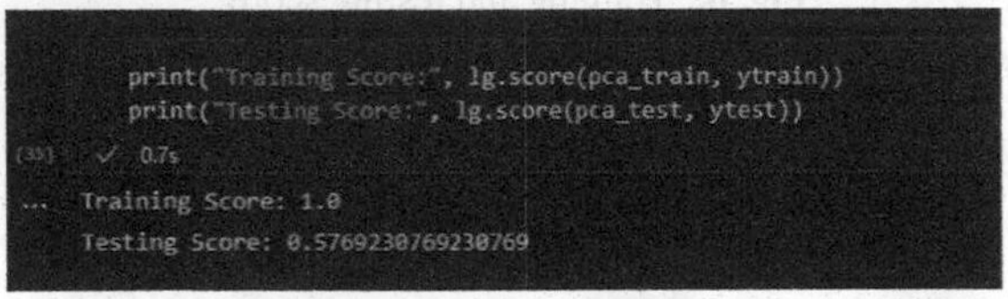

```
print("Training Score:", lg.score(pca_train, ytrain))
print("Testing Score:", lg.score(pca_test, ytest))
[39]  ✓ 0.7s
...  Training Score: 1.0
     Testing Score: 0.5769230769230769
```

Fig. 8. Training testing score

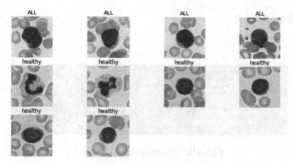

Fig. 8a. Identified images

Results of Decision Tree

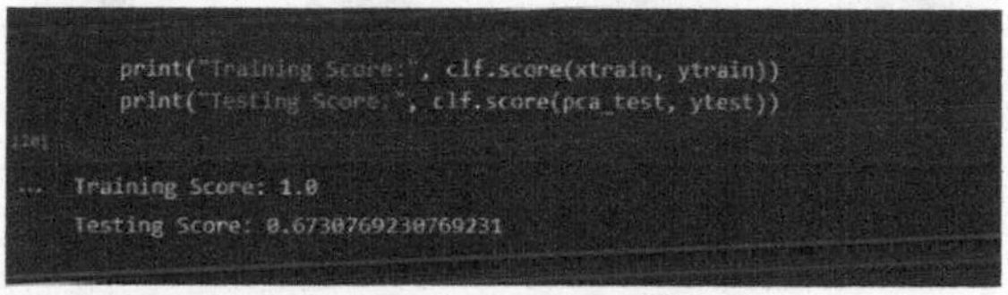

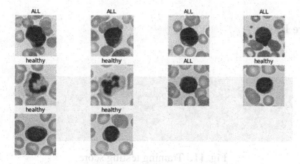

Fig. 9. Training testing score

Fig. 9a. Identified images

Results of Random Forest

Fig. 10. Training testing score

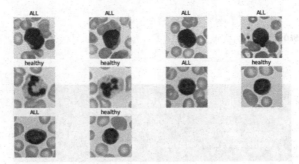

Fig. 10a. Identified images

Results of Naïve Bayes

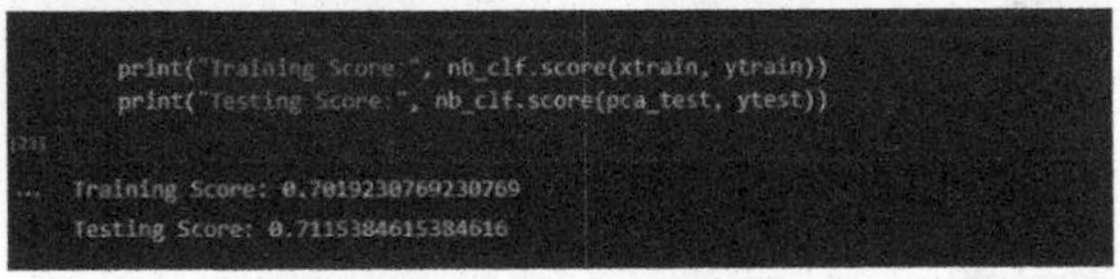

Fig. 11. Training testing score

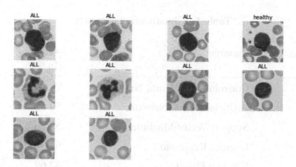

Fig. 11a. Identified images

Results of K Nearest Neighbor

```
print("Training Score:", knn_clf.score(xtrain, ytrain))
print("Testing Score:", knn_clf.score(pca_test, ytest))

Training Score: 0.8125
Testing Score: 0.8269230769230769
```

Fig. 12. Training testing score

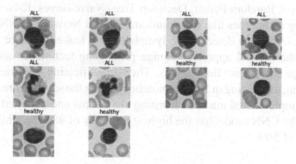

Fig. 12a. Identified images

Table 1. Classification accuracy

Sr. no.	Classifier	Accuracy (In %)
1	Convolutional Neural Network	98.95
2	Artificial Neural Network	50.52
3	Support Vector Machine	55.76
4	Logistic Regression	65.38
5	Random Forest	82.69
6	Decision Tree	69.23
7	Naïve Bayes	71.15
8	K-Nearest Neighbors	80.76

Table 1 depicts the results of all the classifiers in terms of accuracy. It is observed that classification with CNN resulted in maximum accuracy, while ANN has the minimum accuracy compared to other classifiers.

7 Conclusion

We have implemented different machine learning classification algorithms like SVM, Logistic regression, Random Forest, Decision Tree, Naive Bayes, K-Nearest Neighbor, and deep learning techniques like Convolutional Neural Network (CNN) and Artificial Neural Network (ANN) to detect acute lymphoblastic leukemia more effectively and efficiently than the manual approach. Image processing techniques are performed to process the image and extract the features. Then, classification is carried out. We have performed training and testing in Jupyter Notebook using the sci-kit learn library through which we have implemented machine learning algorithms and obtained the testing and training score. The CNN model has the highest accuracy of 98.95% while ANN has the lowest accuracy of 50%.

References

1. Kumar, S., Mishra, S., Asthana, P., Pragya: Automated detection of acute leukemia using k-mean clustering algorithm. In: Bhatia, S., Mishra, K., Tiwari, S., Singh, V. (eds.) Advances in Computer and Computational Sciences. AISC, vol. 554, pp. 655–670. Springer, Singapore (2018). https://doi.org/10.1007/978-981-10-3773-3_64
2. Rajpurohit, S., Patil, S., Choudhary, N., Gavasane, S., Kosamkar, P.: Identification of acute lymphoblastic leukemia in microscopic blood image using image processing and machine learning algorithms, September 2018
3. Karthikeyan, T., Poornima, N.: Microscopic image segmentation using fuzzy C means for leukemia diagnosis. Int. J. Adv. Res. Sci. Eng. Technol. 4(1), 3136–3142 (2017)
4. Patil, T.G., Raskar, V.B.: Automated leukemia detection by using contour signature method. Int. J. Adv. Found. Res. Comput. 2 (2015)

5. Moradiamin, M., Samadzadehaghdam, N., Talebi, A., Kermani, S.: Enhanced recognition of acute lymphoblastic leukemia cells in microscopic images based on feature reduction using principle component analysis, December 2015
6. Singhal, V.: Correlation-based feature selection for diagnosis of acute lymphoblastic leukemia. In: Third International Symposium on Women in Computing and Informatics, August 2015
7. Moradiamin, M., Kermani, S., Talebi, A., Oghli, M.G.: Recognition of acute lymphoblastic leukemia cells in microscopic images using k-means clustering and support vector machine classifier. J. Med. Signals Sens. **5**, 49 (2015)
8. Pathirage, S., Marapana, S., Chandrananda, S., Amarathunga, N.: Detection of leukemia using image processing and machine learning, October 2016
9. Chand, S., Vishwakarma, V.P.: Comparison of segmentation algorithms for leukemia classification. In: Proceedings of the First International Conference on Advanced Scientific Innovation in Science, Engineering and Technology, ICASISET 2020, pp. 16–17, May 2020
10. Alexander Bodzas, A., Kodytek, P., Zidek, J.: Automated detection of acute lymphoblastic leukemia from microscopic images based on human visual perception. Front. Bioeng. Biotechnol. **8,** 1005 (2020)
11. Ratley, A., Minj, J., Patre, P.: Leukemia disease detection and classification using machine learning approaches: a review. In: ICPC2T (20202)
12. Vaghela, H.P., Modi, H., Pandya, M., Potdar, M.B.: Leukemia detection using digital image processing techniques. Int. J. Appl. Inf. Syst. (IJAIS) (2015)
13. Sigit, R., Bachtiar, M.M., Fikri, M.I.: Identification of leukemia diseases based on microscopic human blood cells using image processing. In: International Conference on Applied Engineering (ICAE), 20 December 2018
14. Khobragade, S., Mor, D.D., Patil, C.Y.: Detection of leukemia in microscopic white blood cell images. In: International Conference on Information Processing, 13 June 2016
15. Bagasjvara, R.G., Candradewi, I., Hartali, S., Harjoko, A.: Automated detection and classification techniques of acute leukemia using image processing: a review. In: 2nd International Conference on Science and Technology Computer (ICST), 16 March 2017

Analysis of COVID-19 Detection Algorithms Based on Convolutional Neural Network Models Using Chest X-ray Images

Archana R. Nair[✉] and A. S. Remya Ajai[✉]

Department of Electronics and Communication Engineering, Amrita Vishwa Vidyapeetham,
Amritapuri, India
archanarn5974@gmail.com, remya@am.amrita.edu

Abstract. COVID-19, a disease caused by corona virus is a worldwide pandemic which put millions into death. Not only on lives of people but also it has affected all countries in terms of development and wealth. The main challenge is to detect COVID-19 effectively with high accuracy. A fast classification algorithm can help the health professionals in many ways. This work focuses on the implementation analysis of various Convolutional Neural Network (CNN) which are pre-trained in detecting the disease from chest X-ray images with highest accuracy. Transfer learning is utilized, and fine tuning is performed to get a reliable classification of the image data. For the pre-trained CNN model Mobilenet-V2, highest accuracy of 94.298% and precision of 89.76% obtained.

Keywords: COVID-19 · CNN · Transfer learning · Classification · Fine tuning · Performance metrics

1 Introduction

The COVID-19 disease is treated as a global pandemic. On 31st December 2019, World Health Organization (WHO) gave an alert on a special case of respiratory illness with symptoms like fever and cough. By January 30, 2020 WHO promulgated it as a serious global health emergency and gave the name to the disease as COVID-19 [1]. It snatched millions of lives and still countries are suffering from this pandemic. Early detection of disease is the only way to reduce the spread [2, 3].

RTPCR (Reverse Transcription-Polymerase Chain Reaction) and antigen test are proved to be efficient in detecting COVID-19 [4]. But insufficient test kits with exponentially increasing patients compelled health professionals to look for an alternative. Today, detection of disease with radiology images is getting more attraction [5]. Although CT scans are more effective in detection, the increasing number of patients and extra burden on radiologists caused to move from CT scans to Chest X-rays [6].

After COVID-19 outbreak, researchers began to explore the possibilities of deep learning to detect and classify the disease. Machine learning is considered as a subset of Artificial Intelligence (AI) containing approaches to improve the capability of machines

M. Singh et al. (Eds.): ICACDS 2022, CCIS 1614, pp. 52–63, 2022.
https://doi.org/10.1007/978-3-031-12641-3_5

in doing various tasks with experience. It gives the ability for computers to learn and perform. Using Machine Learning approach, patterns from large, noisy and complex datasets can be extracted. This feature of machine learning is what makes it highly suitable for biomedical detection and classification applications [7–9]. Artificial Neural Network (ANN) constitutes a major area under machine learning. It is a brain inspired technique which uses artificial neurons which mimics the actual human brain neurons. Deep learning is a part of ANN which gains major attraction these days. Classical approaches for biomedical image classification consist of hand-crafted feature extraction and then recognition [10]. But in popular deep learning models like Convolutional Neural Network (CNN), manually doing feature extraction is not needed.

Another neural network gaining more focus these days is Spiking Neural Network (SNN) [11]. The important feature of SNN is that it closely mimics a biological neuron. The problem in using SNN is that it lacks a learning method developed specifically for training. Also, it is computationally intensive. For image classification purposes, there is a need to transform the images into spikes. It is found that it will be a lossy process. Also, a good performance cannot be expected. Because of all these reasons, CNN is preferred over SNN for this project.

A pretrained model approach is employed in this paper. For training a neural network from beginning, large dataset is required. In the case of COVID-19 classification, getting a huge dataset is practically impossible. Transfer learning is the method to use a pretrained neural network. Training time is largely reduced by using this approach. It serves as a better initial model and offers faster training rate. With a good initial point and elevated learning rate, the machine learning model converge at a higher performance level. It results in highly accurate output after training.

Here seven pretrained networks are selected for analysis which exhibited good performance as given by various literatures. They are Googlenet, Alexnet, Squeezenet, Mobilenetv2, Resnet-50, Resnet-101 and Vgg19. The mentioned neural networks are analysed based on different performance parameters like accuracy, specificity, precision, sensitivity, and f-score. All implementation and analysis were based on Kaggle dataset. There were two categories of images: normal and COVID-19 affected chest X-ray images. The task is to accurately predict whether a chest X-ray image fed to the model is of a COVID-19 affected person or that of a normal man.

In the pretrained network, fine tuning is performed in the last layers to suit the current classification task. Fine tuning upgrades the accuracy of the model by a large margin. A technique called data augmentation is also employed to combat non availability of huge dataset. Data augmentation includes flipping, rotation and scaling of images.

Detailed literature review is given in Sect. 2. Section 3 describes the methodology adopted and about performance parameters used in this work. Results obtained and its discussion is presented in Sect. 4 and the conclusion is given in Sect. 5.

2 Literature Review

Sharma et al. put forwards the performance assessment of various CNN architectures for object recognition from real time video such as Alexnet, Googlenet and ResNet50. The analysis shows that Googlenet and ResNet50 showed advancement in performance while comparing with Alexnet. CIFAR 10 and CIFAR 100 datasets were used for this project and only prediction accuracy is considered as performance metric. With CIFAR 100, the model obtained accuracy as follows. For AlexNet: 44.10%, GoogleNet: 64.40%, and ResNet 50: 59.82%. By using CIFAR 10, the accuracy values obtained were, AlexNet: 36.12%, GoogleNet: 71.67%, ResNet 50: 78.10% [12].

Maeda-Gutiérrez et al. did the work to classify diseases caused to tomato plant into ten classes. This work focuses on fine tuning of CNN model. Here, analysis is done using AlexNet, GoogleNet, Inception V3, ResNet 18, and ResNet 50. Nine distinct classes of tomato plant diseases and another healthy class from PlantVillage are used as dataset. This dataset is created by the owners of the paper itself by directly taking images from farm. This work used so many performance metrics for evaluation, like accuracy, precision, sensitivity, specificity, F-score, Area Under the Curve (AUC), and time. Work done is mainly concentrated on fine tuning, that is performed in the last 3 layers of pre trained models to improve performance. The inferences obtained from this paper are as follows. Performance of GoogleNet is better when compared with other similar networks. Inception V3 is one of the deepest CNN architectures. But it shows poor performance [13].

Analysis on six classifiers like: Alexnet, Googlenet, VGG-16, Resnet, InceptionResNet-V2 and Darknet 19 were carried out by Benali et al. In the analysis, it is found that Inception-ResNet-V2 network outperformed all other nets while considering Top-1 and Top-5 accuracy which are 80.3% and 95.1% respectively. Inferred that, architectures rooted in residual concept attain elevated accuracy because of negligible number of parameters while comparing with other architectures. Alexnet underperformed in terms of accuracy and number of MACs are very less compared to Inception-Resnet-V2. Best performers in object detection application are found to be VGG and ResNets [14].

Xiao et al. did a work based on detecting the masks wore by workers. VGG-19 is used in this project. It is proposed that three fully connected layers of this network should be restored with one Flatten layer and two fully connected layers. The advanced version of training the model mainly used the fine-tuning method in transfer learning. Using this technique, parameters of pre-trained VGG-19 CNN model is transferred to the convolution layer, pooling layer and fully connected layer of the proposed model to detect masks [15].

Demir et al. seek to find application of their work in biomedical field. Here, for detecting cancer at initial stages, CNN models like Inception-V3 and ResNet101 are implemented with the help of datasets containing skin cancer images. Dataset is formed with 2437 images having size $224 \times 224 \times 3$. It is composed of malignant and benign images which is taken from ISIC-Archive database. For ResNet101 and Inception-V3, the results from this work stated an accuracy of 84.09% and 87.42% respectively [16].

Lungs disease classification can be an application of CNN and is explained by Shin et al. In this paper, for identifying the interstitial lung disease, 5 distinct methods of neural networks are presented which are CNN based. Dataset consists of CT scan slices (2D). It carries images with a count of 905 of 120 people having 6 different kind of lung tissue conditions. Model is trained with the help of architectures like CifarNet, GoogleNet, AlexNet, and ImageNet. Also implementation of various architectures is done to detect the Thoracoabdominal Lymph node and made keen evaluation [17].

3 Methodology

The workflow is presented as a block diagram and is given in Fig. 1.

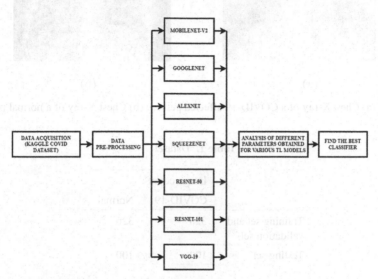

Fig. 1. Basic block diagram showing workflow

COVID-19 dataset is obtained from Kaggle database [18, 19]. Only images satisfying a minimum quality is selected for training the model. Then the images are subjected to the training phase. For that, seven renowned classifiers are selected which include Mobilenet-v2, Googlenet, Alexnet, Squeezenet, Resnet-50, Resnet-101 and VGG-19. The task is to find which classifier is best to classify COVID-19. To find this, several performance metrics are used. As it's a biomedical application, more focus will be given to test accuracy and sensitivity.

3.1 Data Preprocessing

This work focuses on COVID-19 disease classification. The dataset count is 3616. It consists of chest X-ray images of patients affected with COVID-19 and 10,000 images of normal chest X-rays. Sample images for each category is given as Fig. 2. The image count chosen for training and testing purpose is shown in Table 1.

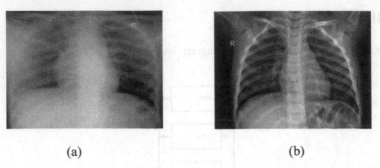

(a)   (b)

Fig. 2. (a) Chest X-ray of a COVID-19 affected person. (b) Chest X-ray of a normal person

Table 1. Dataset

	Image count	
	COVID-19	Normal
Training set and validation set	326	326
Testing set	100	100

It is clear from the above table, the image count in each category is balanced. Entire images contained in dataset are in Portable Network Graphics(png) file format. 1024 × 1024 pixels or 256 × 256 pixels are the resolution of images. All the pretrained models selected need images of size 224 × 224 × 3 pixels. So, the images are resized prior to feeding into the model. Training and validation images are selected randomly with a 7:3 ratio respectively. Data augmentation is the method used over the images to raise the dataset count. Data augmentation includes flipping, rotation and scaling of images.

3.2 Transfer Learning

Transfer learning can be viewed as a technique of reutilizing a pre-trained model for a new piece of work [20]. It saves training time and improves accuracy. In this paper, seven pretrained models are analysed which includes Googlenet, Alexnet, Squeezenet, Mobilenetv2, Resnet-50, Resnet-101 and Vgg19.

Alexnet
Alexnet holds 60 million parameters and 650000 neurons [21]. It consists of eight layers in which five are convolutional layers and next three are fully connected layers. There is max pooling layer and softmax classification layer. Inorder to prevent overfitting, a dropout ratio of 0.5 is also provided. Instead of using saturating activation functions like Tanh or Sigmoid, the authors used a special activation function called Rectified Linear

Unit (ReLU) non-linearity inorder to have a faster training. ReLU is present in the first seven layers of Alexnet.

Googlenet
Unlike Alexnet, Googlenet possess only seven million parameters [22]. It has 4 convolutional layers, 9 inception modules, 4 max pooling and 3 average pooling layers, 5 fully connected layers and 3 softmax classifier layers. It is a wider CNN with 22 layers. It uses a dropout ratio of 0.7 and ReLU activation function in convolutional layers. It can be said that Googlenet is a powerhouse having increased computational efficiency.

Residual Network (ResNet)
ResNet is based on deep architectures, and it uses a technique called skip connections [23]. That is, training from a few layers is skipped and it connects the intermediate layers directly to the output. It upgrades the performance of the architecture to a large extent. The residual unit contains convolutional, max pooling and fully connected layers. ResNet can have 18, 34, 50, 101, 152 and 1202 deep layers. In this project, ResNet-50 and ResNet-101 are selected for analysis. Computing resources and training time rises with rise in the number of deep layers.

Squeezenet
Squeezenet architecture is a smaller CNN architecture which doesn't show compromise in accuracy [24]. It has 50 times fewer parameters than Alexnet while achieving the same accuracy as that of Alexnet. It can be easily implemented on FPGAs having limited memory. The authors were able to compress the model to less than 0.5 MB. The base of Squeezenet architecture is the fire module which is made up of a squeeze layer and an expand layer. The main features include it uses 1×1 filters instead of 3×3 filters and pooling is done at late stage only resulting in large activation maps for convolutional layers.

Mobilenetv2
For mobile and resource constrained environments, a new architecture called Mobilenetv2 was introduced [25]. Here, number of parameters and memory requirement is greatly reduced while maintaining the accuracy. A regular convolution operation will do both filtering and combining in a single go, but in Mobilenetv2 these operations are done as separate steps. It is called depthwise separable convolution. ReLU6 is used as the activation function. This architecture uses very less computation power to run or for transfer learning purposes.

VGG-19
VGG-19 can be considered as the successor of Alexnet. It has 19 layers comprising of sixteen convolutional layers, 3 fully connected layers and a final softmax layer [26]. The model achieved good classification accuracy. This architecture can be used for a variety of complex tasks like face recognition.

In a CNN, the convolutional layers present will extract image features which are low level and high level. Final learnable layer and classification layer will classify the applied feed in image based on the output from convolutional layers. When pretrained networks are used, these final layers are replaced with new layers which are adapted to new classification task and dataset. Fully connected layer acts as final layer with learnable weights in most of the networks. In the new classification task, this layer is superseded with a novel fully connected layer having outputs equal to the number of classes in new task. Generally learning rate is increased in the new layers while performing transfer learning. Freezing of layers is performed to speed up the network training. Freezing means setting of learning rate in the earlier layers of a network to zero. Freezing of layers also has an advantage that it prevents overfitting if the new dataset is small. Data augmentation is performed for resizing the training images to the desired size as required by the network. Also, data augmentation is used to increase dataset by performing flipping, translating and scaling of images. No learning is there in initial layers as it is frozen. Middle layers having slower learning and final layers equipped with fast learning is preferable to speed up training process. Training options make use of stochastic gradient descent with momentum. The advantage of using this technique is that it converges better with longer training time.

3.3 Performance Parameters

The performance of the network is evaluated by generating the confusion matrix. Confusion matrix gives the information about predicted class and actual class [27].

The efficiency indicators such as accuracy, specificity, precision, sensitivity, and F score are evaluated [13]:

Accuracy
It gives the fraction of predictions that the trained model gives as correct. For e.g., an accuracy of 80% means 2 of every 10 labels is incorrect and remaining 8 are correctly predicted.

$$Accuracy = \frac{TP + TN}{TP + TN + FP + FN}. \tag{1}$$

where TP-True Positive, FN-False Negative, FP-False Positive, TN-True Negative.

Specificity
It gives the number of correct predictions from all X-rays which are actually normal. For e.g., a specificity of 70% means 3 of every 10 normal people are mislabeled as COVID-19 affected and 7 are correctly labelled as normal.

$$Specificity = \frac{TN}{TN + FP}. \tag{2}$$

Precision
It gives an estimation about the proportion of positive identifications which are actually right. For e.g., if precision is found to be 50%, it means when an X-ray is predicted as COVID-19 affected, it is correct 50% of time.

$$Precision = \frac{TP}{TP + FP}. \tag{3}$$

Sensitivity (Recall)
It tells what proportion of actual positives were identified as such. For eg, a sensitivity of 60% means the model correctly identifies 60% of all COVID-19 affected X-rays.

$$Sensitivity = \frac{TP}{TP + FN}. \tag{4}$$

F score
It considers both precision and sensitivity. Here, harmonic mean of precision and recall is calculated.

$$F\ score = 2 * \frac{Precision * Recall}{Precision + Recall}. \tag{5}$$

4 Results and Discussion

The required chest X-ray images of COVID-19 affected and healthy were collected from Kaggle dataset. This project is done with the help of MATLAB 2021a. The images were stored in an image datastore. Used 70% of images for training and 30% for validation purpose. The test images were stored as a separate set. As mentioned earlier, seven pre-trained networks were utilized to find the best classifier. All models were trained first and then tested to find the performance parameters. Table 2 showing summarized results for each pre trained network is given below.

Table 2. Result summary

Net name	Test accuracy	Validation accuracy	Specificity	Precision	Sensitivity	F score
Mobilenet-v2	94.298	98.19	88.6	89.76	100	94.61
Googlenet	93.859	98.19	87.72	89.06	100	94.21
Alexnet	92.982	97.59	85.96	87.69	100	93.44
Squeezenet	93.421	98.19	86.84	88.37	100	93.83
Resnet-50	93.859	98.19	87.72	89.06	100	94.21
Resnet-101	93.716	98.19	88.6	89.66	98.84	94.02
VGG-19	93.859	97.59	87.72	89.06	100	94.21

From analyzing above results, it can be found that Mobilenet-v2 is the best classifier for COVID-19 disease classification. For biomedical applications, the important performance metrics are accuracy and sensitivity. Mobilenet-v2 has a test accuracy of 94.298 and sensitivity of 100%. It means all COVID-19 affected X-rays are identified correctly as COVID-19 affected itself. No COVID-19 cases are missed here even though some normal X-rays were found as COVID-19 affected. Except for Resnet-101, all classifiers gave a sensitivity of 100%. All classifiers showed a training accuracy of 100% as expected.

The comparison of test accuracy, specificity, precision, sensitivity, and f-score are presented in Fig. 3, 4, 5, 6 and Fig. 7.

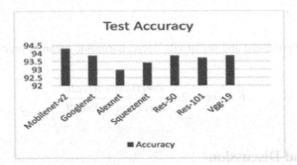

Fig. 3. Comparison of test accuracy values obtained for considered CNN models.

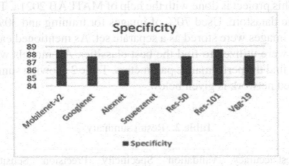

Fig. 4. Comparison of specificity values obtained for considered CNN models.

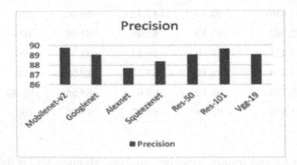

Fig. 5. Comparison of precision values obtained for considered CNN models.

Mobilenet-v2 has fewer parameters (3.5 million) when compared with other pre trained networks and it leads to faster training and reduced complexity. It also has a

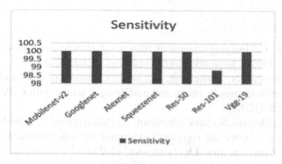

Fig. 6. Comparison of sensitivity values obtained for considered CNN models.

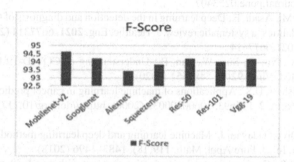

Fig. 7. Comparison of F-score values obtained for considered CNN models.

lower memory requirement of 13 MB. It makes it very suitable for implementation of CNN in a FPGA board.

5   Conclusion

In this paper, an algorithm is implemented to identify the best classifier for classifying COVID-19 disease. Seven classifiers namely Mobilenet-v2, Googlenet, Alexnet, Squeezenet, Resnet-50, Resnet-101 and VGG-19 were studied and implemented in MATLAB 2021a based on their relevance. For analysis purpose, efficiency evaluation metrics like test accuracy, validation accuracy, specificity, precision, sensitivity and f-score for each pre trained network are calculated. All classifiers performed well and thus proves that deep learning networks are promising candidates for complex classification problems. Among seven classifiers, Mobilenet-v2 is found to be the best classifier to classify random chest X-ray images into COVID-19 affected and normal. Utilization of transfer learning and fine tuning of layers are the principal players in this project.

However, the features of images used for training and testing purpose greatly affects the performance of classifiers. In classifying biomedical images, low quality images and the number of images available cause hindrance for building efficient deep learning models. This work can be further broadened to find different lung disorders along with COVID-19. The analysis can also be done with various datasets available where new COVID variants are also considered.

References

1. Chung, M., et al.: CT imaging features of 2019 novel coronavirus (2019-nCoV). Radiology **295**(1), 202–207 (2020). https://doi.org/10.1148/radiol.2020200230
2. Toussie, D., Voutsinas, N., Finkelstein, M., et al.: Clinical and chest radiography features determine patient outcomes in young and middle-aged adults with COVID-19. Radiology **297**(1), E197–E206 (2020)
3. Manikandan, A., Shriram, S., Sarathchandran, C., Palaniappan, S., Rohith, N.D.: Prediction of COVID-19 with supervised regression algorithm through minimum variance unbiased estimator. Int. J. Curr. Res. Rev. **13**(11), s55–s62 (2021)
4. Love, J., et al.: Comparison of antigen- and RT-PCR-based testing strategies for detection of SARS-CoV-2 in two high-exposure settings. PLoS ONE **16**(9), e0253407 (2021). https://doi.org/10.1371/journal.pone.0253407
5. Ghaderzadeh, M., Asadi, F.: Deep learning in the detection and diagnosis of COVID-19 using radiology modalities: a systematic review. J. Healthc. Eng. **2021**, 6677314 (2021). https://doi.org/10.1155/2021/6677314
6. Sun, P., Lu, X., Xu, C., Sun, W., Pan, B.: Understanding of COVID-19 based on current evidence. J. Med. Virol. **92**(6), 548–551 (2020)
7. Cruz, J.A., Wishart, D.S.: Applications of machine learning in cancer prediction and prognosis. Cancer Inform. **2**, 117693510600200030 (2006). https://doi.org/10.1177/117693510600200030
8. Kusuma, S., Divya Udayan, J.: Machine learning and deep learning methods in heart disease (HD) research. Int. J. Pure Appl. Math. **119**(18), 1483–1496 (2018)
9. Remya Ajai, A.S., Gopalan, S.: Analysis of active contours without edge-based segmentation technique for brain tumor classification using SVM and KNN classifiers. In: International Conference on Communication Systems and Networks, ComNet 2019, Thiruvananthapuram, pp. 1–10 (2019)
10. Indumathi, T.V., Sannihith. K., Krishna, S., Remya Ajai, A.S.: Effect of co-occurrence filtering for recognizing abnormality from breast thermograms. In: Proceedings of the Second International Conference on Electronics and Sustainable Communication Systems (ICESC-2021), Coimbatore, pp. 1170–1175 (2021)
11. Vaila, R., Chiasson, J., Saxena, V.: Feature extraction using spiking convolutional neural networks, pp. 1–8, July 2019. https://doi.org/10.1145/3354265.3354279
12. Sharma, N., Jain, V., Mishra, A.: An analysis of convolutional neural networks for image classification. Procedia Comput. Sci. **132**, 377–384 (2018)
13. Maeda-Gutiérrez, V.,et al.: Comparison of convolutional neural network architectures for classification of tomato plant diseases. Appl. Sci. **10**(4), 1245 (2020)
14. Benali Amjoud, A., Amrouch, M.: Convolutional neural networks backbones for object detection. In: El Moataz, A., Mammass, D., Mansouri, A., Nouboud, F. (eds.) ICISP 2020. LNCS, vol. 12119, pp. 282–289. Springer, Cham (2020). https://doi.org/10.1007/978-3-030-51935-3_30
15. Xiao, J., Wang, J., Cao, S., Li, B.: Application of a novel and improved VGG-19 network in the detection of workers wearing masks. J. Phys. Conf. Ser. **1518**(1), 012041 (2020)
16. Demir, A., Yilmaz, F., Kose, O.: Early detection of skin cancer using deep learning architectures: resnet-101 and inception-v3. In: 2019 Medical Technologies Congress (TIPTEKNO), Izmir, Turkey, pp. 1–4 (2019)
17. Shin, H.C., et al.: Deep: convolutional neural networks for computer-aided detection: CNN architectures, dataset characteristics, and transfer learning. IEEE Trans. Med. Imaging **35**(5), 1285–1298 (2016)

18. Chowdhury, M.E.H., et al.: Can AI help in screening viral and COVID-19 pneumonia? IEEE Access **8**, 132665–132676 (2020)
19. Rahman, T., et al.: Exploring the effect of image enhancement techniques on COVID-19 detection using chest X-ray images. arXiv preprint arXiv:2012.02238 (2020)
20. Tamil Priya, D., Divya Udayan, J.: Transfer learning techniques for emotion classification on visual features of images in the deep learning network. Int. J. Speech Technol. **23**(2), 361–372 (2020). https://doi.org/10.1007/s10772-020-09707-w
21. Krizhevsky, A., Sutskever, I., Hinton, G.E.: ImageNet classification with deep convolutional neural networks. In: Advances in Neural Information Processing Systems, New York, pp. 1097–1105. Curran Associates, Inc. (2012)
22. Szegedy, C., et al.: Going deeper with convolutions. In: Proceedings of the IEEE Conference on Computer Vision and Pattern Recognition, Boston, MA, USA, 7–12 June 2015, pp. 1–9 (2015)
23. He, K., Zhang, X., Ren, S., Sun, J.: Deep residual learning for image recognition. In: Proceedings of the IEEE Conference on Computer Vision and Pattern Recognition, Las Vegas, NV, USA, 26 June–1 July 2016, pp. 770–778 (2016)
24. Iandola, F.N., Han, S., Moskewicz, M.W., Ashraf, K., Dally, W.J., Keutzer, K.: SqueezeNet: AlexNet-level accuracy with 50x fewer parameters and <0.5MB model size. arXiv:1602.07360 [cs], November 2016
25. Sandler, M., Howard, A., Zhu, M., Zhmoginov, A., Chen, L.-C.: MobileNetV2: inverted residuals and linear bottlenecks. arXiv:1801.04381 [cs], March 2019
26. Simonyan, K., Zisserman, A.: Very deep convolutional networks for large-scale image recognition. arXiv:1409.1556 [cs], April 2015
27. Bhagya, T., Anand, K., Kanchana, D.S., Remya Ajai, A.S.: Analysis of image segmentation algorithms for the effective detection of leukemic cells. In: Proceedings of the Third International Conference on Trends in Electronics and Informatics (ICOEI 2019), Tirunelveli, pp. 1232–1236 (2019)

Data Analysis Using NLP to Sense Human Emotions Through Chatbot

Fardin Rizvy Rahat, Md. Shahriar Mim, Aqil Mahmud, Ishraqul Islam,
Mohammed Julfikar Ali Mahbub, Md Humaion Kabir Mehedi[✉],
and Annajiat Alim Rasel

Department of Computer Science and Engineering, Brac University, 66 Mohakhali,
Dhaka 1212, Bangladesh
{fardin.rizvy.rahat,md.shahriar.mim,mahmud,ishraqul.islam,
mohammed.julfikar.ali.mahbub,humaion.kabir.mehedi}@g.bracu.ac.bd,
annajiat@bracu.ac.bd

Abstract. Years of Technological progress have made machines capable of understanding basic human emotions from their tonality. Written text is also an important method of communication where a wide range of emotions can be expressed. Thus, it is imperative for machines to be also able to understand the emotions being portrayed in the text messages. For this study, an analysis was done on a dataset known as GoEmotions for the potential development of a chatbot. This dataset contains 58 thousand selected Reddit comments collected from various subreddits that use the English language. The results shows that, these were then classified and categorized based on the principal preserved component analysis (PPCA) method into four main emotions; positive, negative, ambiguous, and neutral successfully. And for this task, Natural Language Processing (NLP) pre-training was applied, which was a transformer based machine learning technique.

Keywords: Machine learning · NLP · Emotion recognition · Reddit · Chatbot

1 Introduction

Studying human emotions has become a really hot topic for research in the field of Artificial intelligence. As we know, humans can communicate through a wide range of complex feelings with only a few words, so that is why it has been a goal since the dawn of technology to teach machines how to interpret effects and emotions. Such works can be seen since the 1960s when computer programs were being conceptualized to study communication of natural language between man and machine [1]. Since then, the journey of human speech recognition began through machine learning. Now, through the advancement of technology and with the help of natural language processing we have found new ground to apply artificial intelligence in our daily life. Nowadays, text speech bots are being used in multiple service-based websites which have eliminated the need for human interactions online. At present, We know some mainstream voice assistants like Alexa, Siri, Google Assistant have become a part of our life [2]. However, these assistants were nothing but command-based operators in our smartphones in their early days [3]. These speech recognizing systems were previously used only to understand certain forms of speech over voice and their capabilities were very limited. Now, they are even able to determine the subtle differences in the tonality of speech and they are gradually becoming closer to understanding user emotions through the way the speech is delivered [4]. This demonstrates great signs of advancements towards the vision of proper machine interpretation of human emotions.

Communication between a man and a computer through emotion and speech recognition has immense benefits. As we know human emotions have a substantial influence on the brain such as the ability of learning, memorizing, problem-solving, perception and different emotional interactions that occur during a conversation affects a person's life and their thinking process. And this is very crucial in modern health care while handling a patient who has stress and depression-related problems through artificial intelligence [5]. Moreover, if it could understand what the user is trying to say as clearly as possible it would really help rehabilitate the patients when they are adapting to their emotional state, thus motivating them and leading the patients to a swift recovery. So, we can say depending upon the state of a patient different approaches can be taken for curing different problems.

Here, in this paper, we studied GoEmotions [6], which is the largest human-annotated dataset containing 58k carefully selected comments extracted from the popular English subreddits on Reddit, where they have been either labeled for 27 emotion categories or as Neutral. Our main goal here is to analyze the dataset in order to use it for building a chatbot that interacts with users' emotions and understands the positive and negative aspects of their delivered speech and tries to promote positivity more in the interactions and remove the negatively perceived emotions from the communications between the chatbot and the user.

The objectives of this paper are mentioned as follows:

- Analysis of all different types of emotion displayed in the dataset.
- Classification, grouping and separation of the data points into binary classes of negative and positive emotions.
- Preparation of the analyzed dataset for chatbot development.

Henceforth, a chatbot can be developed through fulfilling all the objectives, across which human emotions could be understood better in order to help promote more positivity among people.

2   Related Works

With a view to improving the health of Human beings many technological systems and services have been invented. Among them, providing responsive services in a fraction of time, Chatbots developed using Artificial Intelligence is a very significant invention. Chatbots is a computer program that automatically generates a response in a form of text or speech according to the inputs in texts or speech of the user.

In 1966, J Weizenbaum at MIT invented ELIZA, the world's first Chatbot [1]. ELIZA could run on MIT's MAC time-sharing system only and allowed for some specific types of natural language communication between people and machines. ELIZA evaluated input sentences using decomposition rules that were activated by keywords in the input text. Reassembly rules associated with selected decomposition rules were used to generate the responses. However, ELIZA had some technical problems. For ages, researchers tried to develop more dynamic, robust, and efficient chatbots. Eventually, researchers started to use the Chabot for mental health counseling.

D Lee et al. suggested a novel chatbot system for psychiatric counseling services. They argued that more rational and dynamic sentiment analysis will result in better mental health [7]. They used Natural Language Processing (NLP) to understand the contents of the conversation. They also developed a customized counseling response by taking input from Mental Health Experts. They collected their data from mass media such as scripts of plays and radio shows which contain manually described emotions.

T Kamita et al. [9] proposed the use of smartphones and chatbots to execute a self-guided mental healthcare course. They stated, it is convenient to use smartphones and chatbots for mental healthcare services and works as a motivation for regular and continuous use. They used the SAT method invented by T Munakata which is an interview-style counseling and treatment methodology [8]. SAT is a structured imaging treatment that uses a subjunctive mood to convey people back in time to their ancestors and, finally, to their particles in order to foster the rebuilding of stable attachment links. However, T Kamita et al. had found that using a chatbot may boost a user's motivation and assistance in reduction of stress while also being useful in a self-guided mental health course.

Among all the mental illnesses suicidal thoughts can be very devastating. But thanks to the NLP. In 2016, researcher B L Cook et al. combined the applications of NLP and ML to predict suicidal thoughts and psychiatrist symptoms [13]. Suicidal thoughts and acute psychiatric symptoms were predicted using natural language processing and machine learning among patients recently released from being hospitalized or in emergency hospital settings in Madrid, Spain. At numerous follow-up stages, participants replied to standardized mental and physical health tests. Suicidal thoughts and psychiatric symptoms were the outcome variables of interest (GHQ-12). They contrasted NLP-based algorithms that used an unstructured query with logistic regression estimation techniques that used structured data. Positive Predictive Value, specificity, and sensitivity for NLP-based suicidal ideation models were 0.61, 0.57 and 0.56 compared to 0.73, 0.62, and 0.76 for structured databased models. NLP-based models of elevated mental symptoms had Positive Predictive Value, specificity, and sensitivity of 0.56, 0.60, and 0.59 compared to 0.79, 0.85, and 0.79 in structured models. All in all, they have stated that based merely on replies to a basic general mood inquiry,

NLP-based models were able to yield quite good prediction values. These models have the potential to quickly identify people at risk of suicide or psychiatric distress, and they might be a low-cost screening option in situations where longer structured item surveys are not possible.

Furthermore, in 2021, researcher A Tewari et al. proposed a chatbot and a software named MentalEase and also compared existing chatbots for psychotherapy [14]. According to them, MentalEase is a smartphone application that employs NLP techniques to give not just conversational assistance but also a toolbox of useful features to help people maintain their mental health. Patients with mild anxiety and depression can benefit from integrating mental health evaluation tools into the chatbot interface, as well as a regular treatment. They used fundamental NLP techniques such as word embeddings, sentiment analysis, and models like the Sequence-to-Sequence model and attention mechanism. They compared two chatbots named Wysa and Woebot. Both of them are used for virtual mental therapy. They stated that Wysa, in order to communicate effectively, needs more information and asks a lot of open-ended inquiries. On the other hand, Woebot's communication may appear to be one-sided. It appears to be contrived and does not address many subtle topics. Additionally, they proposed a new chatbot architecture. The suggested system is an Android mobile application that combines the power of Deep Learning and NLP to give not just conversational aid in the form of a chatbot, but also a toolbox of valuable features to help people maintain their mental health. They chose OpenSubtitles Corpus and Cornell Movie Dialog Corpus for the data collection.

The covid-19 pandemic situation causes mental health breakdowns, globally [10]. Researchers felt the urge to develop a chatbot that can do conversations with a concerned user daily. For that reason, Rodrigo et al. [11] proposed a chatbot system that can chat with Students daily to improve their mental health. They used the PERMA model, developed by Martin Seligman [12], to assess a student's well-being and offer coping techniques for maintaining positive mental health and emotions. Their chatbot system communicates with users through a text-based interface.

In our work, we have used a chatbot based on NLP to provide mental counseling determining the user's inputs, since NLP gives the ability to the machines in understanding and process human language so that they may execute repetitive activities automatically and eventually this key functionality of NLP can result in a more robust, accurate, efficient and instantaneous.

3 Dataset

The dataset utilized in this paper is Google's GoEmotions which is obtained from Kaggle [6]. It for the most part centers around conversational information, where feeling is a basic part of the correspondence. Since the Reddit stage offers an enormous, freely accessible content which incorporates direct discussion among users, it is an important asset for emotion investigation. Thus, we assembled GoEmotions utilizing Reddit remarks dated from the beginning of Reddit in

2005 to January 2019, obtained from the subreddits with a minimum of 10k remarks, barring erased data and remarks of other languages.

In order to empower constructing comprehensive agent feeling models, information curation measures were applied to guarantee the dataset does not build up broad, nor feeling explicit, language inclinations. This was especially significant in light of the fact that Reddit is known to have a young demographic predisposition inclining towards the youthful male, which is not indicative of an around the world diverse populace.

Positive		Negative		Ambiguous
admiration 🖐	joy 😃	anger 😠	grief 😧	confusion 😕
amusement 😄	love 🖤	annoyance 😒	nervousness 😬	curiosity 🤔
approval 👍	optimism 🤞	disappointment	remorse 😔	realization 💡
caring 🤗	pride 😌	disapproval 👎	sadness 😞	surprise 😲
desire 😍	relief 😅	disgust 🤢		
excitement 😆		embarrassment 😳		
gratitude 🙏		fear 😨		

Fig. 1. GoEmotions taxonomy including 28 emotion categories

From Fig. 1 we can see GoEmotions taxonomy which consists of 28 emotion categories beforehand. These categories are found after the compilation of all fifty eight thousands selected Reddit comments which was labeled for these 28 categories of emotion. This taxonomy includes a large number of positive, negative and ambiguous emotion categories making it easier to understand tasks that require understanding of emotion such as analysis of customers feedback.

The stage additionally presents a slant towards toxic, hostile language. To address these worries, hurtful remarks were distinguished utilizing predefined terms for hostile grown-up and disgusting substance, and for personality and religion, which was utilized for information separating and veiling. Information was also shifted to diminish irreverence, limit message length, and equilibrium for addressed feelings and opinions. To stay away from over-portrayal of famous subreddits and to guarantee the remarks likewise reflect less dynamic subreddits, the information among subreddit networks was additionally adjusted.

Scientific categorization was made to mutually augment three goals:

- Give the best inclusion of the feelings communicated in Reddit information.
- Give the best inclusion of sorts of passionate articulations.
- Limit the general number of feelings and their cross-over.

Such a scientific classification permits fine-grained understanding that is information-driven and additionally tending to probable information scarcity for particular sentiments.

4 Methodology

In this paper, we will discuss the overall architecture of our model, where the chatbot will feel the emotions from text-based sentences. Here we carried out data pre-processing, feature engineering. The next few paragraphs will demonstrate how the dataset is divided into certain types of emotions such as positive, negative, neutral and ambiguous.

4.1 Dataset Pre-processing

In the stage of preprocessing, we mainly focused on the data cleaning from the raw dataset which excludes id and unnecessary punctuation removal which allows us to get more accurate data.

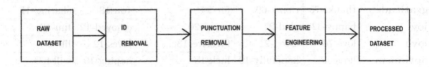

Fig. 2. Steps for data pre-processing

After cleaning the data we carried out feature engineering which in turn allowed us to extract the different features, which is also known as the emotion that will be determined from the sentences via the help of a chatbot. The sequence of steps taken for the pre-processing have been illustrated in Fig. 2.

4.2 Feature Engineering

In this segment of feature extraction from our new processed dataset after pre-processing, we will use this method to extract the different features present.

Table 1. Ambiguous emotions tabulated

confusion	confused-18	why-11	sure-10	what-10	understand 8
curiosity	curious-22	what-18	why-13	how-11	did-10
realization	realize-14	realized-12	realised-7	realization-6	thought-6
surprise	wow-23	surprised-21	wonder-15	shocked-12	omg-11

To carry out the procedure after the process of filtering at least one label amounts to 93% of the raw data. So to carry out training and testing, we split the data into 80% for training and 20% for testing. Thus the final result is based on the test dataset even though we filtered our dataset there were a certain amount of examples that fall under the emotion ambiguous text.

Table 2. Positive emotions tabulated.

admiration	great-24	awesome-32	amazing-30	good-28	beautiful-23
amusement	lol-66	haha-32	funny-27	lmao-21	hilarious-18
approval	agree-24	not-13	don't-12	yes-12	agreed-11
caring	you-12	worry-11	careful-9	stay-9	your-8
desire	wish-29	want-8	wanted-6	could-6	ambitious-4
excitement	excited-21	happy-8	cake-8	wow-8	interesting-7
gratitude	thanks-75	thank-69	for-24	you-18	sharing-17
joy	happy-32	glad-27	enjoy-20	enjoyed-12	fun-12
love	love-76	loved-21	favorite-13	loves-12	like-9
optimism	hope-45	hopefully-19	luck-18	hoping-16	will-8
pride	proud-14	pride-4	accomplishment-4		
relief	glad-5	relieved-4	relieving-4	relief-4	

Table 3. Negative emotions tabulated.

anger	fuck-24	hate-18	fucking-18	angry-11	dare-10
annoyance	annoying-14	stupid-13	fucking-12	shit-10	dumb-9
disappoint-ment	disappointing-11	disappoint-ed-10	bad-9	disappoint-ment-7	unfortunately-7
disapproval	not-16	don't-14	disagree-9	nope-8	doesn't-7
disgust	disgusting-22	awful-14	worst-13	worse-12	weird-9
embarrass-ment	embarrassing-12	shame-11	awkward-10	embarrass-ment-8	embarrassed-7
fear	scared-16	afraid-16	scary-15	terrible-12	terrifying-11
grief	died-6	rip-4			
nervous-ness	nervous-8	worried-8	anxiety-6	anxious-4	worrying-4
remorse	sorry-39	regret-9	apologies-7	apologize-6	guilt-5
sadness	sad-31	sadly-16	sorry-15	painful-10	crying-9

After the completion of training and testing, we found a certain list of words from the list and grouped them into emotions of four categories: ambiguous (Table 1), positive (Table 2), negative (Table 3), and neutral with neutral emotions not used. Hence we will further evaluate the performance of the model on this hierarchy. From the tables above, we are mainly focusing on the positive and negative emotions of our dataset in this paper, that is Table 2 and 3.

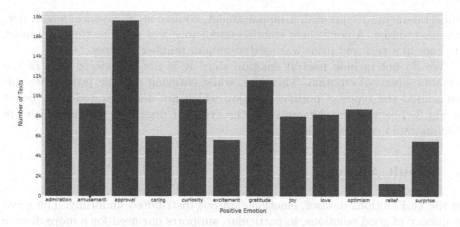

Fig. 3. Positive emotion histogram

In Fig. 3 we can see representation of positive emotions in the form of bar chart which shows a comparison of how many times these emotions where used in texts.

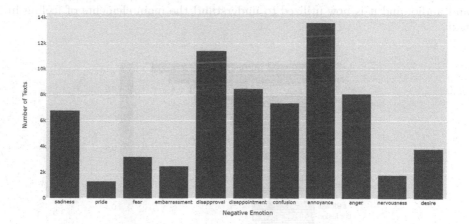

Fig. 4. Negative emotion histogram

The figure shows 12 types of emotions and they are admiration, amusement, approval, caring, curiosity, excitement, gratitude, joy, love, optimism, relief, surprise. Among these emotion approval was used maximum number of times in a text and relief was used minimum number of types.

Moreover, Fig. 4 portrays the negative emotions in the form of bar chart which shows a comparison of how many times these emotions where used in texts. The figure shows 11 types of emotions and they are sadness, pride, fear, embarrassment, disapproval, disappointment, confusion, annoyance, anger, nervousness, desire. Among these emotion annoyance was used maximum number of times in a text and pride was used minimum number of types.

We do not include neutral emotion since it is not considered part of the semantic space of emotion. Therefore, while carrying out the process we differentiated the types of positive emotion which are demonstrated right below in Fig. 3. Likewise, we differentiated the types of negative emotion which are demonstrated right below in Fig. 4.

5   Result Analysis

In the GoEmotions dataset, emotions are not distributed uniformly. The great frequency of good emotions, in particular, supports our need for a more diverse emotional scientific classification than the authoritative six core feelings. The principle preserved component analysis (PPCA) technique is used to examine two datasets by distinguishing linear combinations of emotion judgment that demonstrate the most elevated joint inconstancy across two arrangements of raters to ensure that the ordered decisions fit the basic data. As a result, it displays emotional elements that are well-understood by all raters. PPCA was already used to understand the main dimension of emotion recognition in video and audio, and it is now utilized to understand the main elements of feeling in text.

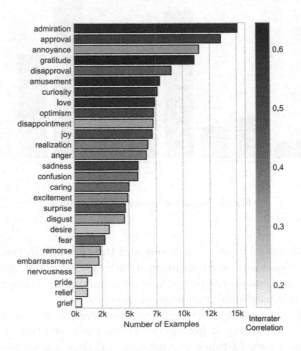

Fig. 5. Chart of frequency of emotions

Every part is observed to be critical (with p-values $< 1.5 \times 10^{-6}$ for all aspects), showing that every feeling catches a remarkable piece of information. This is not paltry, since in prior research on emotion recognition in conversation, just twelve out of thirty components of feeling were deemed critical. The frequency of correlation of the emotions has been shown through the chart in Fig. 5.

We look at the bunching of the characterized feelings dependent on relationships among rater decisions. With this methodology, two feelings will group together when they are as often as possible co-chose by raters.

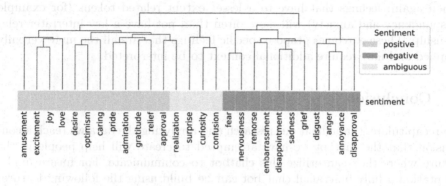

Fig. 6. Grouping of feelings

Despite the lack of a predetermined notion of feeling in our scientific categorization, we see that feelings that are associated in terms of their opinion (positive, negative, and ambiguous) group together, proving the quality and consistency of the evaluations. For instance, in the event that one rater decided "fervor" as a name for a given remark, almost certainly, another rater would pick a connected feeling, for example, "happiness", rather than, say, "dread". Maybe shockingly, all uncer tain feelings grouped together, and they bunched all the more intimately with positive feelings, shown in Fig. 6.

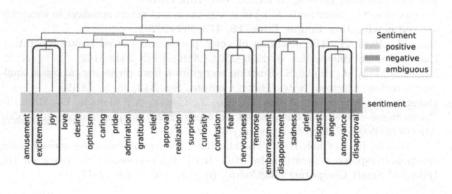

Fig. 7. Connection of sentiments

Essentially, feelings that are connected as far as their competence like joy and excitement, nervousness and fear, annoyance and anger, sadness and grief are likewise firmly related as seen in Fig. 7.

Working out the log odds ratio of all tokens for every feeling class differentiating to any remaining feelings, the lexical correspondences of each feeling were removed. As the log chances are z-scored, all qualities more prominent than three demonstrate a profoundly huge (>3 std) relationship with the comparing feeling. We observe that those feelings that are profoundly fundamentally connected with specific tokens (for example appreciation with "much obliged", entertainment with "haha") will generally have the most elevated interrater relationship. Then again, feelings that have to a lesser extent related tokens (for example despondency and anxiety) will more often than not have a low interrater relationship. These outcomes propose specific feelings that are all the more verbally implied and may require additional context to be interpreted.

6 Conclusion

To recapitulate, throughout our research and analyzing the data we reach a conclusion that the chatbot we have attempted to create will help people in the future where they can utilize this chatbot to communicate. For future implementation a fully functional chat bot can be build using the following features along with other features such as- neutral and ambiguous as mentioned earlier. According to GoEmotions dataset, people will input text-based sentences, where the chatbot comes in to detect emotions such as positive and negative emotions which will provide assistance in their daily lives to comprehend human emotions in a sophisticated manner.

References

1. Weizenbaum, J.: Eliza—a computer program for the study of natural language communication between man and machine. Commun. ACM **9**(1), 36–45 (1966)
2. Raj, S.: Building Chatbots with Python: Using Natural Language Processing and Machine Learning, 1st edn. APRESS, New York (2018)
3. Terzopoulos, G., Satratzemi, M.: Voice assistants and smart speakers in everyday life and in education. Inform. Educ. **19**, 473–490 (2020)
4. Jana, A., Patil, H.: The evolution of voice portal and virtual assistants. Int. J. Emerg. Technol. Innov. Res. **8**, d556–d564 (2021). ISSN 2349-5162
5. Egger, M., Ley, M., Hanke, S.: Emotion recognition from physiological signal analysis: a review. Electron. Notes Theor. Comput. Sci. **343**, 35–55 (2019)
6. Demszky, D., Movshovitz-Attias, D., Ko, J., Cowen, A., Nemade, G., Ravi, S.: Goemotions: a dataset of fine-grained emotions (2020). https://doi.org/10.48550/ARXIV.2005.00547. https://arxiv.org/abs/2005.00547
7. Lee, D., Oh, K.-J., Choi, H.-J.: The chatbot feels you-a counseling service using emotional response generation. In: 2017 IEEE International Conference on Big Data and Smart Computing (BigComp), pp. 437–440. IEEE (2017)

8. Kamita, T., Ito, T., Matsumoto, A., Munakata, T., Inoue, T.: A chatbot system for mental healthcare based on sat counseling method. Mob. Inf. Syst. **2019** (2019)
9. Munakata, T.: Reconstructing life and society with SAT therapy: foundations of the new generation CBT. Int. J. Struct. Assoc. Tech. **3**, 35–60 (2009)
10. Cook, B., Progovac, A., Chen, P., Mullin, B., Hou, S., Baca-Garcia, E.: Novel use of natural language processing (NLP) to predict suicidal ideation and psychiatric symptoms in a text-based mental health intervention in Madrid. Comput. Math. Methods Med. **2016** (2016). https://doi.org/10.1155/2016/8708434
11. Tewari, A., Chhabria, A., Khalsa, A.S., Chaudhary, S., Kanal, H.: A survey of mental health chatbots using NLP. Available at SSRN 3833914 (2021)
12. Kontoangelos, K., Economou, M., Papageorgiou, C.: Mental health effects of COVID-19 pandemia: a review of clinical and psychological traits. Psychiatry Investig. **17**(6), 491 (2020)
13. Rodrigo, M., et al.: Towards building mental health resilience through storytelling with a chatbot (2021)
14. Seligman, M.: PERMA and the building blocks of well-being. J. Posit. Psychol. **13**(4), 333–335 (2018)

Medical Cyber-Physical Systems Enabled with Permissioned Blockchain

Anupam Tiwari[✉] and Usha Batra

Department of CSE and IT, G D Goenka University, Gurgaon, India
anupamtiwari@protonmail.com, usha.batra@gdgu.org

Abstract. Global health security concerns have gained vast importance in recent times with outbreak of COVID-19. Today, the growing interdependence among countries and states has effected into accelerated growth of pandemics. A global need for rugged medical systems on a common platform is deemed today. Pandemics will not stop, they will resurrect again, they will happen irrespective till such times the medical world attains a disease less world in future. But till then, we can attempt to decelerate the pandemics growth enabled with new generation technologies. Medical cyber-physical systems are marred by a number of challenges and this paper proposes a model to negate these identified challenges enabled on multichain blockchain platform that imparts peculiar blockchain characteristics to the network of effected systems. The proposed model also enables to share encrypted data on select blockchain nodes granted defined access controls with proven encryption algorithms.

Keywords: Blockchain · Cyber-physical system · Electronic health records · Healthcare · Privacy · Security

1 Introduction

Blockchain technology mechanizing the widely known bitcoin cryptocurrency since 2009 found an extended role onwards 2014 after introduction of ethereum and smart contracts. Bitcoin cryptocurrency was followed up by a multitude of variants of cryptocurrencies which stand at around 5000 + today. Similarly after launch of ethereum blockchain, there has been a plethora of blockchain platforms variants like Hyperledger [1], R3 Corda, Quorum Ripple, Openchain, Multichain etc. These variants cater to various function requirements in multiple domains like supply chain, education, governance, cross border payments, personal identity, music royalties just to mention a few. Further to these associations, smart contracts [2] made way for the first time through ethereum in 2014 as unique transaction protocols that provision automatic execution of events through deterministic coding to render action deemed as per terms of an agreement. In addition to evolution of blockchain technology over last decade, cyber-physical systems (CPS) [3] is another domain which is evolving as a definite near future technology component.

M. Singh et al. (Eds.): ICACDS 2022, CCIS 1614, pp. 76–87, 2022.
https://doi.org/10.1007/978-3-031-12641-3_7

1.1 Medical Cyber-Physical Systems

Industry 4.0, Internet-of-Things (IoT) and Industrial IoT (IIoT) are few seemingly synonymous terms but have distinct functional differences. While IoT[4] focuses on connecting billions of devices on the internet and deriving critical information for data analytics, IIoT[5] is a subset variant of IoT that caters primarily to industrial network comprising multitude of devices and machine-to-machine connectivity duly synchronized with each other. Industry 4.0 [6] provisions to collect and derive data from connected machines and devices to expedite flexible and more efficient industry processes. Medical cyber-physical systems (MCPS) [7] are an emerging need in smart healthcare and specifically focus on integrating medical devices on a network. MCPS focuses on making the system act intelligently in real time with data derived, collated and analyzed. MCPS systems look forward to be employed in hospitals and medical institutions currently working in complex clinical scenarios. Context-aware MCPS intend to attain uninterrupted quality medical services, improved and expedited safety by rendering personalized services and treatment. The blockchain characteristics can facilitate intra and inter hospital interoperability with updated patient record sets in a trusted environment with transparent documentation. Besides these, with smart contracts these advantages can be extended further for negation of multiple middle men and agencies to bring a metamorphose change ever unthought-of. A true MCPS will be able to realise all these intentions subject to a rugged backend infrastructure enabled by strong cryptographically secure solutions. Peculiar to MCPS realisation, there are a horde of challenges that exist in the current technical capacities. These challenges are briefly discussed below:

1.1.1 Security and Privacy of Electronic-Health-Records (EHR)

EHR [8] refer to systemized collation of patient health information in a digital format and are bounded by many laws and complex fine tuned web of policies. Bare thread level grant of security permissions or access grants for each occurring event would be a design cephalalgia for IT administrators [9].

1.1.2 Permissions

Health Insurance Portability and Accountability Act of 1996 (HIPAA) [10] and similar variants of data protections laws are emerging globally. The millions of transaction events happening at any point of time globally would thus deem a complex architecture tuned with security and privacy balance metrics.

1.1.3 Trust

Another key challenge would be to gain trust of the system that would be handling healthcare data like has been seen never before.

1.1.4 Reliability of Data

It is estimated that global EHR data in 2020 is 25 exabytes which is likely to swell to 40 exabytes by 2025. The vulnerability to huge amount of EHR data to security threats and undesired modifications and tampering would be of crucial significance.

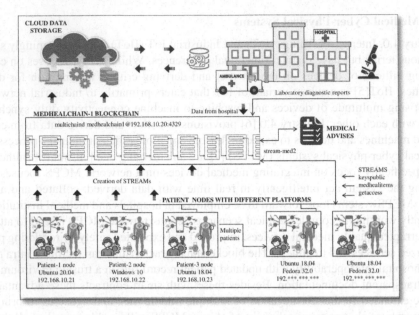

Fig. 1. Schematic of proposed blockchain enabled MCPS model

2   Current Works

While there has been a lot of work in associating CPS with blockchain, no specific limited scale simulated works associating MCPS with blockchain have been seen. Formal proofs of implementing blockchain characteristics using "Tamarin Prover" tool with proof-of-concept realizing a cyber physical trust systems driven by blockchain are seen [11]. Authors [12] discuss applying blockchain technology for CPS to focusing on challenges and future opportunities. Another work [13] provides a concise but systematic review through conceptual work on accrued advantages with association of blockchain and CPS concluded by identifying two research issues including excessive delay in reaching consensus and limited throughput. Works by [14] discuss in theoretical context the advantages that a traditional CPS can accrue if blockchain technology is associated with CPS. No models and works have been proposed in these works peculiar to MCPS. A holistic survey of various applications of CPS enabled with blockchain have been discussed by [15] in domains of smart grids, IIoT and MCPS etc. to realise a true CPS. The work in this paper proposes a model schematic which is peculiar to MCPS and follows it up with a proof of concept on the multichain enterprise blockchain platform.

3   Materials and Methods

3.1   Hardware

– Intel Xeon E3–1225 v5 3.3 G, 8 M cache, 16 GB UDIMM, 2400 MT/s, Single Rack with Ubuntu Server 20.04.1 LTS with Multichain blockchain admin console

– Quantity Two: Dell 2018 PowerEdge T30 Business Mini-Tower Server System with Intel Quad-Core Xeon E3–1225 v5 8 m Cache, Ubuntu Desktop 18.04.1 and Ubuntu 20.04 LTS
– 10th Generation Intel® Core™ i3-1005G1 Processor workstation with Windows 10 Home, 8 GB DDR4, 2666 MHz.

3.2 Software and Applications

MultiChain 2 Enterprise GPLv3- Open source fork of Bitcoin blockchain

3.3 Schema Setup

For the setup schematic as depicted in Fig. 1, the details assumed three patient nodes and one admin node with details as seen in Table 1.

Table 1. Nodes details in MCPS model

Node	OS simulating	IP configured	Description
Admin control node	Ubuntu Server 20.04.1 LTS	192.168.10.20 168.10.20.20	Address: 1Sype9YU5a6CfAeSgWuqbJHGb7Q1CZxfRWDuy and 1UwZVGo1NBykC1zLgBXUNgTErZA52QDRSa2Fm4
Patient 1	Ubuntu 20.04 LTS node	192.168.10.21 168.10.20.21	These nodes simulate many wearable's IoT medical devices vide which medical diagnostics data of patients is available in real time to systems. Address: 1MNCQL92NEru7UudC5T3n3Lm4nouoQwwmVqcbL and 17EKc8aRSqK6pUM26aKBjjkog1pAFFMe5SRQEc
Patient 2	Ubuntu 18.04 LTS node	192.168.10.22 168.10.20.22	Address: 1HYiMADQejGatd8tfr6fhG4pooJx5dsVmrYE3B
Patient 3	Windows 10 Home	192.168.10.23 168.10.20.23	Address: 1632aeetZ9SXB9VdEfV2168wzdiwA1Jb85L5v7

Multichain blockchain platform has been used to create a blockchain named *medhealchain1* with the important parameters as seen in Table 2.

3.4 Patient Identifiers

Patient data identifiers considered for the simulation include name, medical record number, date of birth, phone number, aadhar number (social security number), address.

Table 2. Parameters set in MCPS model

Parameter	Value	Description
Default-network-port	4329	Default TCP/IP port for peer-to-peer connection with other nodes
Default-rpc-port	4328	Default TCP/IP port for incoming JSON-RPC API requests
Chain-name	medhealchain1	Chain name, used as first argument for multichaind
Network-message-start	f6d9ddf7	Magic value first 4 bytes of every peer-to-peer message
Address-pubkeyhash-version	00e595f1	Version bytes used for *pay-to-pubkeyhash* addresses
Address-scripthash-version	0536ba7e	Version bytes used for pay-to-scripthash addresses
Address-checksum-value	83fea2bc	Bytes used for XOR in address checksum calculation
Genesis-pubkey	0333095fe01bc31a24f547e1e4375e4abfdb5a0b271b58234b14725fa2dc037f1e Genesis block coinbase output public key	
Genesis-timestamp	1605965150	Genesis block timestamp
Genesis-nbits	536936447	Genesis block difficulty (nBits)
Genesis-nonce	148	Genesis block nonce
Maximum-block-size	8388608	Maximum block size in bytes

4 Simulation Sequence

4.1 Creation of Streams and Appending Data

In the simulated scenario, persons at different geo-locations with device wearable's are transmitting data to the healthcare eco-system of things as depicted in Fig. 1. For the purpose of simulation, multichain streams have been used to enable *medhealchain1* blockchain to function as append-only database, with immutability, time-stamping, and notarization feature. While the *medhealchain1* multichain blockchain can have any number of streams, three streams [*stream-med-patient-iden, stream-test-report and stream-med2*] have been published referenced by a hash inside transactions. Further to this, other nodes as permitted by the admin node can subscribe to these nodes and access stream's data in real-time to enable efficient retrieval and access in multiple ways.

Stream *stream-med-patient-iden* was created with '{"restrict":"write"}' permissions and unique stream *stream-med-patient-iden* hash derived as per details in Table 3. The transaction IDs so generated with each of these streams created, assisted to search specific queries in the *medhealchain1* blockchain and serves as a unique identity for searching items inside specific streams.

5 Results

Patient identification data was published to *medhealchain1* blockchain with JavaScript Object Notation(JSON) and could be viewed through *multichain-cli* terminal as seen in

Table 3. Stream transactions ID generated

Stream name	Stream transaction id generated	Remarks
stream-med-patient-iden	e16c26b6160c1380aa6c27e323271ff94 65dcc761657eeb7b9c855bef3a7f2d6	This stream contains the patient identification details with unique credentials
stream-test-report	a5190c675c66a68655dc5fe638997003a 71fd1bb039ea301f0c9751163265474	This stream contains vital parameters of the patient retrieved online
stream-med2	19d630295d58ae25bf1da50a5d453f996 2de059e74d58738b876600941bd8138	This stream contains messages of medical advice nature

Fig. 2 along with blockchain *address, transactionid, confirmations* and *blocktime* details. It is pertinent to mention that the time stamping in the *medhealchain1* follows the epoch unix format alike the bitcoin blockchain and thus 1606899486 represents equivalent of GMT Wednesday, 2 December 2020 08:58:06.

The second stream *stream-test-report* indexes the parameters received from the wearable's of the respective persons on the *medhealchain1* blockchain. These streams were published from the patient node *192.168.10.21*. It was also seen that other nodes were not able to see any other data till such time specific nodes were granted subscription. Once subscription was granted to specific streams for specific nodes, the off chain data for these streams was retrieved by the other nodes *192.168.10.20* and *192.168.10.22* immediately. It was also seen that vital medical parameters of patient-1 published from patient IP address 192.168.10.21 were visible at admin address only after exclusive permissions were granted by patient-1 to admin. In third stream named *stream-med2*, a medical advice was published on the *medhealchain1* blockchain. The medical advice details were converted from plain text to hexadecimal as criteria to publishing in the blockchain. The sample advice and equivalent in hexadecimal are seen below:

Text Medical Advice
"One should maintain a minimum distance of ten feet while interacting with someone and masks should be always on. These masks should be as per the advised directions from WHO or specific country health department."

Hexadecimal Equivalent
"4f6e652073686f756c64206d61696e7461696e2061206d696e696d756d2064
697374616e6365206f662074656e206665657420776869c6520696e746572
616374696e67207769746820736f6d656f6e6520616e64206d61736b732073686f
756c6420626520616c77617973206f6e2e5468657365206d61736b732073686f7975
6c64206265206173206173207065722074686520616476697365642064697265637469 6f
6e732066726f6d2057484f206f7220737370656369666963206363f756e74727920

6865616c746820646465706172746d656e742e".

medhealchain1: getstreamitem stream-med-patient-iden
9e73e05100b3146c2366ba81055f7b1bd1beb68845274e170550f9bc51f7706a
{"method":"getstreamitem","params":["stream-med-patient-
iden","9e73e05100b3146c2366ba81055f7b1bd1beb68845274e170550f9bc51f7706a"],"id":"34902470-
1606567792","chain_name":"medhealchain1"}
{   → Query for Transaction hash
"publishers" : [
  "1Sype9YU5a6CfAeSgWuqbJHGb7Q1CZxfRWDuy" ¦
},   → Address of publisher
"keys" : [
  "key1"
],
"offchain" : false,
"available" : true,   → Patient identification fields
"data" : {
  "json" : {
    "name" : "Patient-1",
    "medical_record_number" : "medheal-1234",
    "date_of_birth" : "01-01-2000",
    "phone_no" :"987*******",
    "aadhar_number" : "98**-89**-98**-**98",
    "address" : "b14-sector9*9-noida-201301"
  }   → Total confirmations from node
},
"confirmations" : 22,   → Epoch & Unix Timestamp
"blocktime" : 1606550928,
"txid" : "9e73e05100b3146c2366ba81055f7b1bd1beb68845274e170550f9bc51f7706a"
}

Fig. 2. Extract from admin console query output of stream *stream-med-patient-iden*

Once published, the console query of retrieving stream items from *stream-med2* displayed the hexadecimal form of advice as published. Exclusive permissions were granted to all the nodes for access to such messages.

5.1 Data Confidentiality and Encrypted Storage on *Medhealchain1* Blockchain

The peculiar characteristics of bitcoin blockchain makes every transaction stored visible to every connected node and also to everyone exploring any bitcoin explorer sites. This might not be desired on a blockchain associated with storing medical data and private credentials. In a MCPS ecosystem, each bare thread data attributes like EHR, real-time patient vital parameters etc. would deem encryption by default before storage and decryption on permission basis. In the setup of network, permitted blockchain participants get access to encrypted data published on blockchain. The admin control node *192.168.10.20* will read the encrypted data which has been shared by node *192.168.10.21* with address *17EKc8aRSqK6pUM26aKBjjkog1pAFFMe5SRQEc*.

RSA public-private key set was generated at the admin control node terminal and the public key derived was published in the *keyspublic* stream with exclusive permissions granted to node *192.168.10.21* to *connect, send* and *read*. Three streams *keyspublic*, *medhealitems* and *getaccess* were created which allow publishing of data by any other

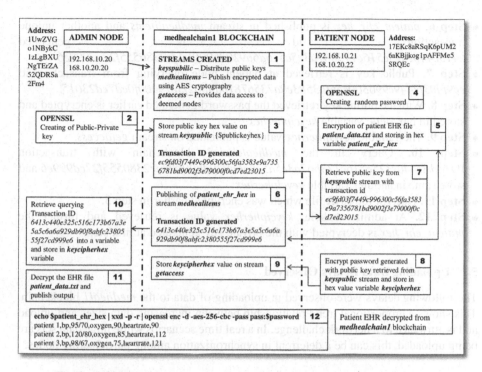

Fig. 3. Schematic of confidential data exchange on medhealchain1 blockchain

node with only exclusive global *send* permissions. Each transaction on *medhealchain1* blockchain including creation of three streams and publishing of public key or any event generates a 64 bit unique transaction id. These transaction id's were used for querying and filtering specific transactions.

Figure 3 depicts the exact schematic model which has been run on the patient machine nodes and admin nodes. The sequence is discussed below:

- **Step 1.** Creation of streams *keyspublic*, *medhealitems* and *getaccess*
- **Step 2.** Due permissions to *send, receive* and *connect* are granted to admin node(*192.168.10.20*) and machine 1 node generates a unique set of public-private keys
- **Step 3.** The public key hex value is stored in a variable *$publickeyhex* and a unique transaction id *ec9fd03f7449c996300c56fa3583e9a7356781bd9002f3e79000f0cd7ed23015* is generated and same is published on *medhealchain1* blockchain.
- **Step 4.** On patient node 1(*192.168.10.21*), a random password is generated with the *openssl*.
- **Step 5.** The random password generated further encrypts the EHR file i.e. *patient_data.txt* and stores in hex variable *patient_ehr_hex*.

- **Step 6.** *patient_ehr_hex* is published in stream *medhealitems* and another unique transaction   id   is   generated *6413c440e325c516c173b67a3e5a5c6a6a929db90f8abfc2380555f27cd999e6*
- **Step 7.** Public key is retrieved at patient node querying from transaction id *ec9fd03f7449c996300c56fa3583e9a7356781bd9002f3e79000f0cd7ed23015.*
- **Step 8.** With the public key retrieved the password generated earlier is encrypted and stored in hex value variable *keycipherhex.*
- **Step 9.** The variable *keycipherhex* value is published on stream *getaccess.*
- **Step 10.** Query in the *medhealchain1* blockchain with transaction ID *6413c440e325c516c173b67a3e5a5c6a6a929db90f8abfc2380555f27cd999e6* and store same in another variable *keycipherhex*
- **Step 11.** Decrypt the EHR file which was encrypted earlier in step 5.
- **Step 12.** At admin console, *keycipherhex* value is retrieved and further the *patient_ehr_hex* is decrypted with the private key generated at first step.

5.2 Uploading Data Delays Observed

The following delays were observed in uploading of data to the *medheal1* blockchain. This shows a proportional trend relationship between increasing data sizes and upload and is attributed to a network challenge. In a real time scenarios while huge data sets are being uploaded, this can be a deterrent in synchronization among participating nodes.

Table 4. Data upload delays observed on blockchain with increasing size

Size of Data	Time taken to upload to *medhealchain1* blockchain
1 MB	real 0m0.038 s
10 MB	real 0m0.148 s
100 MB	real 0m1.229 s
512 MB	real 0m6.648 s
1 GB	real 0m12.242 s
2 GB	real 0m23.877 s
4 GB	real 0m45.938 s
10 GB	real 1m35.623 s

6 Discussions

The schematic environment at Fig. 3 was realised as a model proof of concept. The work in this paper has been done in a limited environment which would not be the case in real time ground scenarios. The reality would witness millions of patients and their respective records being coordinated in the MCPS ecosystem. There will thus be a horde of challenges too, that need to be resolved which would primarily include the following:

6.1 Big Storage Issues

The collation and processing of billions of EHR records being generated globally in different formats with high velocity and from multitude of devices would be a challenge. Capturing among this collation the right data and cleaning it for the right analytics would be an extended challenge. While also the impending delay of data uploading as bought out in Table 4 may need a distributed storage solution like Inter planetary file systems storage (IPFS) [16].

6.2 Availability of True Data

A typical MCPS would consist of billions of networked medical devices which would be giving out real time EHR. These interconnected devices would be of different origins of manufacturers and mix consolidates of standard and non standard devices enabled with proprietary applications and software's which would have their own set of vulnerabilities. Vide [17], a security system model has been proposed that identifies unknown and suspicious devices with abnormal behavior for activating restricted or halted communications in an IoT ecosystem.

6.3 Scalability

MCPS ecosystem of healthcare with networked devices will scale up to billions by 2025. This large scale would surely come with its own set of challenges to include storage designs, IT infrastructure, communication bandwidth etc. [18].

Besides above, other challenges like byzantine brokers, identification of recognized networked medical devices and absence of standardisation of MCSP ecosystem devices would deem a forehand solution before going live and realizing a working MCPS ecosystem.

7 Conclusion

MCPS is an evolving necessity and is going to be an integral part of near future connected digital eco-system. While MCPS offers a plethora of advantages, at the same time with growing concerns of EHR, the threat to MCPS is imminent too. The growing digital penetration in human lives and the increasing dependency on MCPS with EHR data sets is going to be a matter of concern per se privacy and security issues. The paper proposes enabling MCPS with blockchain technology to exploit the built-in characteristics of a blockchain. Unlike the well known bitcoin ecosystem wherein all the transactions are transparent and visible to everyone on the network, the MCPS would deem a blockchain with exclusive permission access and encrypted data transfers. To realise this, the paper proposed a model based on multichain enterprise [19] blockchain which not only offers secure storage but also at the same time provisions bare thread level permission granting. Patients would be able to ensure which of their EHR are made accessible to whom under a secure environment. While in this paper, the blockchain enabled MCPS model has been proposed and a schematic proof of concept has been established, there remain a lot of other possibilities too. Storage of data sets in the current paper has been realised in blockchain and cloud services. However this storage can be further made distributed in nature with technologies like IPFS [20, 21]. The extension of decentralization further to storage in multiple blockchain enabled applications will realise the true potential of blockchain. Also artificial intelligence can enhance the blockchain capabilities to much higher scales [22]. Thus while this work has been able to establish a proof of concept on a limited scale model, a certain real time scenario is bound to face challenges as discussed in Sect. 6 which need to be resolved for realizing true potential of decentralization vide blockchain.

References

1. Benhamouda, F., Halevi, S., Halevi, T.: Supporting private data on hyperledger fabric with secure multiparty computation. In: 2018 IEEE International Conference on Cloud Engineering (IC2E), pp. 357–363 (2018). https://doi.org/10.1109/IC2E.2018.00069
2. Suvitha, M., Subha, R.: A survey on smart contract platforms and features. In: 2021 7th International Conference on Advanced Computing and Communication Systems (ICACCS), pp. 1536–1539 (2021). https://doi.org/10.1109/ICACCS51430.2021.9441970
3. Sabella, R., Shamsunder, S., Albrecht, S., Bovensiepen, D.: Guest editorial: networks for cyber-physical systems and industry 4.0. IEEE Commun. Mag. 59(8), 12 (2021). https://doi.org/10.1109/MCOM.2021.9530505
4. Seth, I., Panda, S.N., Guleria, K.: IoT based smart applications and recent research trends. In: 2021 6th International Conference on Signal Processing, Computing and Control (ISPCC), pp. 407–412 (2021). https://doi.org/10.1109/ISPCC53510.2021.9609484
5. Geng, H.: The industrial Internet of Things (IIoT). In: Internet of Things and Data Analytics Handbook, pp.41–81. Wiley (2017). https://doi.org/10.1002/9781119173601.ch3
6. El Hamdi, S., Abouabdellah, A., Oudani, M.: Industry 4.0: fundamentals and main challenges. In: 2019 International Colloquium on Logistics and Supply Chain Management (LOGISTIQUA), pp. 1–5 (2019). https://doi.org/10.1109/LOGISTIQUA.2019.8907280
7. Sokolsky, O.: Medical cyber-physical systems. In: 2011 18th IEEE International Conference and Workshops on Engineering of Computer-Based Systems, p. 2 (2011). https://doi.org/10.1109/ECBS.2011.40
8. Zaghloul, E., Li, T., Ren, J.: Security and privacy of electronic health records: decentralized and hierarchical data sharing using smart contracts. In: 2019 International Conference on Computing, Networking and Communications (ICNC), pp. 375–379 (2019). https://doi.org/10.1109/ICCNC.2019.8685552
9. Keshta, I., Odeh, A.: Security and privacy of electronic health records: concerns and challenges. Egypt. Inform. J. 22(2), 177–183 (2021). https://doi.org/10.1016/j.eij.2020.07.003
10. Shah, V., Peng, Z., Malla, G., Niu, N.: Towards norm classification: an initial analysis of HIPAA breaches. In: 2021 IEEE 29th International Requirements Engineering Conference Workshops (REW), pp. 415–420 (2021). https://doi.org/10.1109/REW53955.2021.00074
11. Milne, A.J., Beckmann, A., Kumar. P.: Cyber-physical trust systems driven by blockchain. IEEE Access 8, 66423–66437 (2020).https://doi.org/10.1109/ACCESS.2020.2984675
12. Dedeoglu, V., et al.: A journey in applying blockchain for cyberphysical systems. In: 2020 International Conference on COMmunication Systems \& NETworkS (COMSNETS), pp. 383–390 (2020). https://doi.org/10.1109/COMSNETS48256.2020.9027487

13. Zhao, W., Jiang, C., Gao, H., Yang, S., Luo, X.: Blockchain-enabled cyber–physical systems: a review. IEEE Internet Things J. **8**,(6), 4023–4034 (2021). https://doi.org/10.1109/JIOT.2020. 3014864
14. Braeken, A., Liyanage, M., Kanhere, S., Dixit, S.: Blockchain and cyberphysical systems. Computer **53** (2020). https://doi.org/10.1109/MC.2020.3005112
15. Rathore, H., Mohamed, A., Guizani, M.: A survey of blockchain enabled cyber-physical systems. Sensors **20**, 282 (2020).https://doi.org/10.3390/s20010282
16. Muralidharan, S., Ko, H.: An inter planetary file system (IPFS) based IoT framework. In: 2019 IEEE International Conference on Consumer Electronics (ICCE), pp. 1-22019). https:// doi.org/10.1109/ICCE.2019.8662002
17. Abdullah, A., Hamad, R., Abdulrahman, M., Moala, H., Elkhediri, S.: CyberSecurity: a review of Internet of Things (IoT) security issues, challenges and techniques. In: 2019 2nd International Conference on Computer Applications & Information Security (ICCAIS), pp. 1–6 (2019). https://doi.org/10.1109/CAIS.2019.8769560
18. Dabbagh, M., Kakavand, M., Tahir, M., Amphawan, A.: Performance analysis of blockchain platforms: empirical evaluation of hyperledger fabric and ethereum. In: 2020 IEEE 2nd International Conference on Artificial Intelligence in Engineering and Technology (IICAIET), pp. 1–6 (2020). https://doi.org/10.1109/IICAIET49801.2020.9257811
19. Ismailisufi, A., Popović, T., Gligorić, N., Radonjic, S., Šandi, S.: A private blockchain implementation using multichain open source platform. In: 2020 24th International Conference on Information Technology (IT), pp. 1–4 (2020). https://doi.org/10.1109/IT48810.2020.907 0689
20. Dwivedi, S.K., Amin, R., Vollala, S.: Blockchain-based secured IPFS-enable event storage technique with authentication protocol in VANET. IEEE/CAA J. Autom. Sin. **8**(12), 1913–1922 (2021). https://doi.org/10.1109/JAS.2021.1004225
21. Huang, H., Lin, J., Zheng, B., Zheng, Z., Bian, J.: When blockchain meets distributed file systems: an overview, challenges, and open issues. IEEE Access **8**, 50574–50586 (2020). https://doi.org/10.1109/ACCESS.2020.2979881
22. Singh, P.K.: Artificial intelligence with enhanced prospects by blockchain in the cyber domain. In: Singh, P.K., Singh, Y., Kolekar, M.H., Kar, A.K., Gonçalves, P.J.S. (eds) Recent Innovations in Computing. Lecture Notes in Electrical Engineering, vol. 832. Springer, Singapore (2022). https://doi.org/10.1007/978-981-16-8248-3_43

Multiclass Classification of Disease Using CNN and SVM of Medical Imaging

Pallavi Tiwari[✉], Deepak Upadhyay, Bhaskar Pant, and Noor Mohd

Department of Computer Science and Engineering, Graphic Era Deemed to be University, Dehradun, India
{Pallavitiwari_20061041.cse,deepakupadhyay.ece}@geu.ac.in

Abstract. This paper proposed a model that deals with automatic prediction of the disease given the medical imaging. While most of the existing models deals with predicting disease in one part of the body either brain, heart or lungs, this paper focuses on three different organs brain, chest, and knee for better understanding the real word challenge where problems do not include crisp classification but the multiclass classification. For simplicity this paper focuses on just determining whether that organ is affected with the disease or not and future work can be done by further expanding the model for multiple disease detection of that organ. We have used CNN for multiclass image classification to determine the input medical image is brain, chest or knee and then SVM is used for binary classification to determine whether that input image is detected with the disease or not. Three different datasets from Kaggle are used: Brain Tumor MRI Dataset, COVID-19 Chest X-ray Image Dataset and Knee Osteoarthritis Dataset with KL Grading. Images from these datasets are used to make fourth datasets for training and testing the CNN for the prediction of the three different organs and after that output will be the input of respective SVM classifier based on the output result and predict the weather it is diagnostic with the disease or not. The proposed model can be employed as an effective and efficient method to detect different human diseases associated with different parts of the body without explicitly giving the input that it belongs to that part. For the transparency this model displays the accuracy of prediction made for the input image.

Keywords: Coronavirus · Brain tumor · Osteoarthritis · CNN · SVM · Medical images

1 Introduction

1.1 Multiclass Classification

Classification is one of the crucial tasks in the field of machine and deep learning. In this era where there are data available is in millions carries with them some hidden information or patterns that are needed to be recognized for intelligent decision making. The two important forms of data analysis are classification and prediction which are being used to build a model for describing important class labels or to predict future data

M. Singh et al. (Eds.): ICACDS 2022, CCIS 1614, pp. 88–99, 2022.
https://doi.org/10.1007/978-3-031-12641-3_8

trends [1]. In this paper these two important concept classification and prediction were used. Classification used was multiclass classification which classify whether the given medical imaging is brain, chest, or knee and then prediction concept was used to predict whether the organ identified had disease or is normal.

Multiclass classification refers to the problem of classifying instances into more than two classes. In this problem encountered was to identify the disease of the patient given medical imaging in the form of MRI, X-ray, endoscopy, ultrasound, etc. There are three ways to solve multiclass classification, first is to extend binary classifier, second is to break the multiclass classification problem into several binary classification problem and third is Hierarchical classification methods [1].

Extending the binary classifier to solve multiclass classification problem is known as extensible algorithms. Some of the algorithms that can be naturally extends are neural networks, decision trees, k-Nearest Neighbor, Naive Bayes, and Support Vector Machines.

The method of breaking the multiclass classification problem into several binary classification problem is known as decomposing into binary classification. There are several methods for doing so which include OVA (one versus all), AVA (All versus all), ECOC (Error correcting output coding) and Generalized coding [2]. In OVA approach, the problem of classifying among k classes is reduced to k binary problems, where each problem is distinguishes a given class from the remaining k-1 classes. In the n = k classifier is required where every classifier is trained with positive examples belonging to class k and negative examples belonging to other classes k-1. In OVO each class is being compared with every other class and a classifier is built to discriminate between each class discriminating all other and voting is performed among the classifier, and whichever gets the maximum votes wins. In ECOC, the idea is to train n binary classifier to distinguish among k classes and every class is given a codeword of length n according to binary matrix m where m denotes a unique number representing a unique class. Generalized coding is the extension of ECOC where coding matrix can take values − 1, 0, and +1.

Hierarchical classification solves the problem of multiclass classification problem by diving the output space i.e., into the tree. The nodes are divided into multiple child nodes till each child nodes represents only one class.

In this paper, CNN was used to perform the multiclass classification problem and the detail regarding the classification will be discussed further in the paper.

1.2 Coronavirus

The Coronavirus disease 2019 (COVID-19) that has been originated from China has now been most frightening and death threatening disease that has spread all over the world. It not only affected the health of the public, but all affected the economy of the countries. It is a respiratory disease caused by SARS-COV-2 virus that seems like viral pneumonia but can result in death if not treated immediately [3]. According to WHO (World Health Organization), COVID-19 has affected over 280,119,931 people all over the world including 5,403,662 deaths as of 28th December 2021. It is challenging the medical system and needs to be taken care of as if diagnosed earlier people may be able to save by the treatment provided. Presently there are three ways for diagnosis of the

COVID-19 RT-PCR, CT-scan and CXR (chest Radiography) [4–7]. In this paper, CXR was used to perform the diagnosis of the COVID-19 as it provides reliable results [4]. SVM algorithm was to determine whether the CXR image is affected with COVID-19, or it is normal.

1.3 Brain Tumor

Brain tumor is formed by the excessive production and proliferation of the cells in the skull. Since the brain is the body's command center, tumors can impose strain on the head and can have severe impact on the human health [8]. As the people dying due to brain tumors are rising day by day hence the early diagnosis is important. There is several medical imaging that can be useful for the diagnosis available presently namely CT (Computed Tomography), SPECT (Single-Photon Emission Computed Tomography), PET (Positron Emission Tomography), MRS (Magnetic Resonance Spectroscopy) and MRI (Medical Resonance Imaging) [9]. In this paper for the diagnosis of the Brain tumors MRI images were used, where CNN was used for identifying whether it is the image of brain MRI and SVM for predicting whether it has brain tumor or is normal.

1.4 Knee Osteoarthritis

A Knee osteoarthritis (OS) is a serious disease that is the world's fourth greatest cause of disability. It is commonly known as degenerative joint disease and is caused by fatigue failure as well as the gradual loss of articular cartilage. It is a disease that usually worsens with time and can lead to impairment. The severity of clinical symptoms varies from person to person [10]. Despite this, risk analysis models for predicting pain progression in individuals with knee OA are still scarce. Current risk analysis models generally rely on demographic, clinical, and radiographic risk variables, or thorough MRI examination analysis. As a result, new and improved methodologies are required to develop ubiquitous, cost-effective, and easily obtained risk assessment models for forecasting pain development in knee OA patients [11]. In this paper X-Ray of the knee was used for the diagnosis and the model was build using CNN and SVM to predict whether a person is suffering from Knee OS or not.

2 Literature Survey

2.1 Deep Learning Based Medical Image Multiclass Classification

In [4] authors propose a DGCAN-based CNN model that makes use of CRX image for the classification of the images into the three categories which were normal, pneumonia and COVID-19 and confirmed that the performance of their proposed model is much better that the existing model as the proposed model makes use of DGCAN which is a multiple neural network that make use of random noise of the images and first extract the local features on several levels and then focuses on extracting the global images. Aishwarya et al. [3] in their paper does a review on the detection of COVID-19 using various dataset. The focus of their work includes studying detection using deep learning and machine learning technique on various datasets available. In [8] authors propose a web- based software that classify the brain tumors into three categories glioma, meningioma and pituitary using the CNN model and produces accuracy of more than 98%. Francisco et al. [9] proposes a CNN model for brain tumor into same categories as [8] that make use of multi-scale approach in which input images are processed in three different spatial scales and produces and accuracy of 97% which according to authors was higher than the existing models with the same datasets. G.wan. et al. [11] develops a deep based risk assessment model and compares with the traditional methods and concluded that deep learning model produces better diagnostic performances using baseline knee radiographs.

2.2 SVM Classifier for Image Classification

In [12] their suggested research used deep learning approach for the classification of the sugarcane as diseased or not, authors make used of three different feature extraction technique and output is given as input to seven different machine earning classifier and based on the result using RC and AUC used the VGG-16 as feature extractor and SVM as the classifier as it produces maximum AUC of 90.2%. Reem Alrais [13] proposes SVM model for classifying the medical image into two types of tumors. Author before classifying the images uses DWT (Discrete Wavelet Transform) to remove noise from the images and PCA (Principal Component Analysis) for reducing the dimensional feature to get the only required storage and computational space. In [14] authors presented a detailed study on the application of the SVM in image classification and how the enhancement of the image may lead to better accuracy in the classifier. Renukadevi et al. [15] in their study concluded that SVM accuracy changes with the value of gamma and produces better performance when the gamma value is increased. They use Coiflet wavelets for the feature extraction.

3 Proposed Methodology

The Objective of this proposed methodology is to use CNN-SVM classifier for the classification of medical images into three classes: Brain, chest and Knee and further predict whether the classified image is suffering from brain tumor or not, COVID-19 or not and Knee OS or not respectively. A flow diagram of the proposed methodology is depicted in Fig. 1.

3.1 Dataset

The Three different datasets, which are available in Kaggle publicly were used in this study. The first dataset is called Brain Tumor Classification (MRI) to classify the MRI images into four classes, but in this study these four classes are converted into one class for the first classifier namely CNN and is labelled as Brain and into two classes for the 2nd classifier namely SVM labelled as Brain tumor or normal to serve the purpose of the proposed methodology. The total number of images in this dataset is 3264. The second dataset is called COVID-19 Chest X-ray Image Dataset to classify the images into normal and Covid-19 affected. For the 2nd classifier the classes remain the same and for the first the images are merged into one folder or class called the chest for CNN classifier to be trained and classify the image as chest. The total number of images in this dataset is less which is 94. Lastly, the third dataset is called Knee Osteoarthritis Dataset with KL Grading- 2018 which is to predict the severity grading of Knee OS. In this dataset there are total 5 classes having labels as 0, 1, 2, 3, 4 in which 0 represent the healthy knee and other classes represent the levels of the person affected by the knee OS. For the first classifier this dataset is merged into one and labelled as Knee and for the 2nd it is merged into two 0 as the normal representing person having healthy knee and 1, 2, 3 and 4 are merged into one and labelled as Knee OS showing being affected by the Knee OS. The total number of images in this dataset is 9786. Figure 2 shows the sample of the dataset.

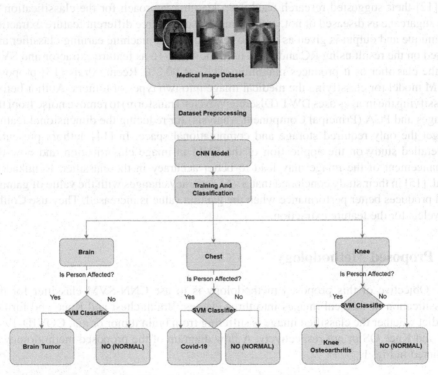

Fig. 1. Flowchart of proposed CNN-SVM model

3.2   Data Preprocessing

Getting Dataset

As discussed in above section that the dataset used is from Kaggle and for the purpose of processing the dataset they are arranged and kept in a folder. Firstly, this folder was downloaded into the google drive and with the help of function drive.mount the drive content is being loaded to our google collab environment, which platform was used to perform this study.

Importing Libraries

The programming language used to perform this study was Python and it is rich when comes to libraries. Hence all the necessary libraries were imported for preprocessing the data.

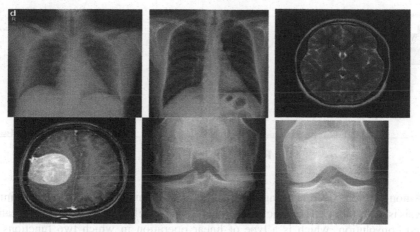

Fig. 2. Sample of medical images from dataset

Importing Dataset

For importing the dataset os library was used. Some of the function used were os.listdir(), os.walk() and os.path.join().

Data Augmentation

Data Augmentation refers to making slight changes to the current data to expand its uniqueness without having to collect new information. It prevents a neural network from learning features that are not useful [16].

Normalization

Since the dataset was collected from three different dataset, so the format used by them were also different some files were of.png, some were.jpeg. To apply same algorithm to these datasets, dataset needed to be formatted in one format and this is known as normalization.

Image standardization

Since the images used have different heights and width, Image Standardization technique was used to make all the image of same height and width.

Splitting

The last step of data preprocessing includes splitting the data into training and testing (in this case validation) data. Split () function was used. Since the dataset was imbalanced so was the training and testing dataset after splitting.

3.3 Convolutional Neural Networks (CNN)

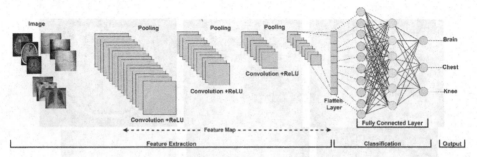

Fig. 3. The CNN model

Among various deep learning models, the most commonly used model for image analysis is CNN model. The basic idea of what CNN model does is based on the meaning behind convolution, which is a type of linear operation in which two functions are multiplied to produce a third function that expresses how one function's shape is changed by the other. Simply put, two matrices are multiplied to provide an output that is used to extract features from the image [17]. CNN model typically has two main parts: Feature extraction and Classification as shown in the Fig. 3. CNN model which is designed for image classification normally consist of four layers: Input layer, Convolution layer, Pooling layer, and Dense layer. Input layer, Convolution layer and Pooling layer are the part of feature extraction and Dense layer which is basically fully connected layers that receive inputs from convolution layer's output and classify the image accordingly is the part of the classification.

In the proposed model, CNN classifier classifies the medical image among three classes. The classifier is sequential and consist of 1 input layer, 8 convolution layer, 4 max pooling, 2 fully connected layer, 1 dropout, 1 softmax and 1 classification layer. Since this classifier was built to classify given into three classes, the output layer has three neurons. The dropout layer in between the dense layers was added to reduce the problem of overfitting as it switches off some of the neurons that are focused on forcing the data finding the new paths. The activation function used was ReLU. From the last fully connected layer, which was a three-dimensional feature vector, as given as input to softmax classifier which is the most recommended when comes to the multiclassification, makes the prediction about organ type. The CNN classifier is then compiled with the Adam optimizer, categorical cross entropy loss function, and an accuracy metric. The model is then trained with the the dataset having 10 epochs and producing accuracy of 0.99 and loss of 0.012. The output of the Classifier is shown in Fig. 4.

The classifier was then saved using model. Save () function as "model_detection.h5" so that we can later use the model in the algorithm of the proposed model.

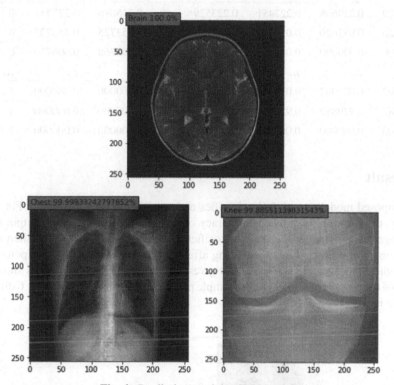

Fig. 4. Prediction result by CNN model

3.4 Support Vector Machine (SVM)

SVM is an efficient learning method and a powerful supervised classifier. The core principle behind an SVM classifier is to select a distinguishing hyper-plane for the two classes that are separated by a threshold. It was developed for range of problems based on learning, prediction, and classification. It uses the concept of learning and kernel functions [18]. When it comes to pattern categorization, problems are rarely linearly separable therefore, nonlinear functions are used to translate input patterns to a higher dimensional space [19–21]. In the proposed work, we had made three SVM binary classifier for classifying whether the organ is affected with the respective disease. SVM model is trained using the images by converting the image into array form as shown the sample in Table 1 and then was being saved using pickle. Dump () function. The output is in the form of 0 and 1 showing 0 as not affected and 1 as being affected.

Table 1. Sample of array of different images

0	1	2	3	...	67498	67499	Target
0.223529	0.219608	0.227451	0.223529	...	0.215686	0.227451	0
0.032620	0.032620	0.032620	0.305796	...	0.353725	0.353725	0
0.000000	0.000000	0.000000	0.000000	...	0.606760	0.606760	0
...	...	...	...	...	...	...	...
0.050902	0.050902	0.050902	0.005150	...	0.000000	0.000000	1
0.929882	0.929882	0.929882	0.992157	...	0.000000	0.000000	1
0.000000	0.000000	0.000000	0.000000	...	0.000000	0.000000	1

4 Result

The proposed model successfully classifies and predict the medical image. For transparency, the output also shows the accuracy of the predicted image. The Output shows the image, name whether the image classifies as brain, chest or knee and then shows the accuracy of predicted image as being affected with the disease and the percentage it is predicted as normal and then at last gives the Output. Following figures shows the output of the proposed model. For the sample purpose 6 outputs are shown for 6 different possible inputs (Figs. 5, 6, 7, 8, 9, 10).

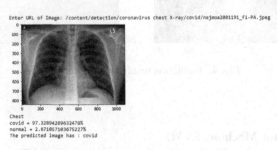

Fig. 5. Prediction result of covid medical imaging

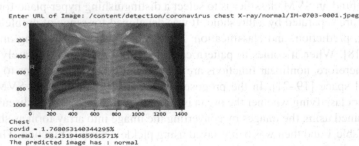

Fig. 6. Prediction result of chest normal medical imaging

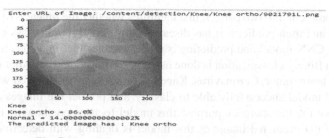

Fig. 7. Prediction result of knee OS medical imaging

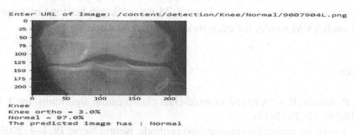

Fig. 8. Prediction result of knee normal medical imaging

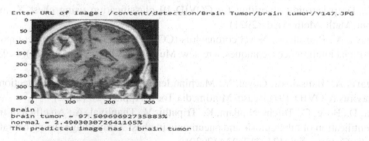

Fig. 9. Prediction result of brain tumor medical imaging

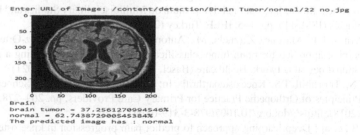

Fig. 10. Prediction result of brain normal medical imaging

5 Conclusion

This paper proposed a novel hybrid model CNN-SVM for the multiclass classification of three different disease brain tumor, Coronavirus and Knee OS using medical images.

It breaks the problem into two parts, it first identifies whether the image is of which body part or organ and then predict if it has disease or not. Firstly, multiclass classification is done using CNN model and prediction is made whether the image is chest, brain or knee and then Binary classification is done using SVM classifier to predict whether it is affected with brain tumor, Coronavirus, Knee OS or normal (not affected with disease). The proposed model successfully able to classify and predict the images as shown in the Sect. 4. For the transparency purpose this model focuses on showing the accuracy percentage of the predicted image as this model is dealing with detection of disease where wrong prediction can have consequences. For the future work, this model can further be extended to classify the different types or severity grading of disease and not just classifying as organ being affected or not. More work can be done by improving the accuracy of each SVM model for even more better prediction.

References

1. Chaitra, P., Kumar, R.S.: A review of multi-class classification algorithms. Int. J. Pure Appl. Math **118**(14), 17–26 (2018)
2. Aly, M.: Survey on multiclass classification methods. Neural Netw. **19**, 1–9 (2005)
3. Aishwarya, T., Ravi Kumar, V.: Machine learning and deep learning approaches to analyze and detect COVID-19: a review. SN Comput. Sci. **2**, 226 (2021)
4. VJ, S.: Deep learning algorithm for COVID-19 classification using chest X-ray images. Comput. Math. Meth. Med. (2021)
5. Bhargava, A., Bansal, A.: Novel coronavirus (COVID-19) diagnosis using computer vision and artificial intelligence techniques: a review. Multimedia Tools Appl. **80**(13), 19931–19946 (2021)
6. Bhargava, A., Bansal, A., Goyal, V.: Machine learning-based automatic detection of novel coronavirus (COVID-19) disease. Multimedia Tools Appl. 1–20 (2022)
7. Verma, D., Bose, C., Tufchi, N., Pant, K., Tripathi, V., Thapliyal, A.: An efficient framework for identification of tuberculosis and pneumonia in chest X-ray images using neural network. Procedia Comput. Sci. **171**, 217–224 (2020)
8. Ucuzal, H., YAŞAR, Ş., Çolak, C.: Classification of brain tumor types by deep learning with convolutional neural network on magnetic resonance images using a developed web-based interface. In: 3rd International Symposium on Multidisciplinary Studies and Innovative Technologies (ISMSIT), pp. 1–5. IEEE, Turkey (2019)
9. Díaz-Pernas, F.J., Martínez-Zarzuela, M., Antón-Rodríguez, M., González-Ortega, D.: A deep learning approach for brain tumor classification and segmentation using a multiscale convolutional neural network. Healthcare (Basel, Switzerland) **9**(2), 153 (2021)
10. Katz, J.N., Thornhill, T.S.: Knee osteoarthritis. In: Katz, J.N., Blauwet, C.A., Schoenfeld, A.J. (eds.) Principles of Orthopedic Practice for Primary Care Providers, pp. 269–278. Springer, Cham (2018). https://doi.org/10.1007/978-3-319-68661-5_16
11. Guan, B., et al.: Deep learning approach to predict pain progression in knee osteoarthritis. Skeletal Radiol. **51**(2), 363–373 (2021). https://doi.org/10.1007/s00256-021-03773-0
12. Srivastava, S., Kumar, P., Mohd, N., Singh, A., Gill, F.S.: A novel deep learning framework approach for sugarcane disease detection. SN Comput. Sci. **1**(2), 1–7 (2020)
13. Alrais, R., Elfadil, N.: Support vector machine (SVM) for medical image classification of tumorous. Int. J. Comput. Sci. Mob. Comput. **9**(6) 37–45 (2020)
14. Chandra, M.A., Bedi, S.S.: Survey on SVM and their application in image classification. Int. J. Inf. Technol. **13**(5), 1–11 (2018). https://doi.org/10.1007/s41870-017-0080-1

15. Renukadevi, N.T., Thangaraj, P.: Performance evaluation of SVM–RBF kernel for medical image classification. Glob. J. Comput. Sci. Technol. (2013)
16. Getting Started with Image Preprocessing in Python. https://www.section.io/engineer-ing-education/image-preprocessing-in-python/. Accessed 20 Jan 2022
17. Basic CNN Architecture. https://www.upgrad.com/blog/basic-cnn-architecture/. Accessed 20 Jan 2022
18. Mohd, N., Singh, A., Bhadauria, H.S.: A Novel SVM based IDS for distributed denial of sleep strike in wireless sensor networks. Wireless Pers. Commun. 111(3), 1999–2022 (2019). https://doi.org/10.1007/s11277-019-06969-9
19. Wahyuningrum, R.T., Anifah, L., Purnama, I.K.E., Purnomo, M.H.: A novel hybrid of S2DPCA and SVM for knee osteoarthritis classification. In: 2016 IEEE International Conference on Computational Intelligence and Virtual Environments for Measurement Systems and Applications (CIVEMSA), pp. 1–5. IEEE, Hungary (2016)
20. Katal, A., Wazid, M., Goudar, R.H.: Big data: issues, challenges, tools, and good practices. In: Sixth International Conference on Contemporary Computing, pp. 404–409. IEEE (2013)
21. Chauhan, H., Kumar, V., Pundir, S., Pilli, E.S.: A comparative study of classification techniques for intrusion detection. In: International Symposium on Computational and Business Intelligence, pp. 40–43. IEEE (2013)

Using KullBack-Liebler Divergence Based Meta-learning Algorithm for Few-Shot Skin Cancer Image Classification: Literature Review and a Conceptual Framework

Olusoji B. Akinrinade[1]([✉]), Chunglin Du[1], and Samuel Ajila[2]

[1] Computer System Engineering Department, Tshwane University of Technology, Pretoria, South Africa
sojiakinrinade2014@gmail.com, duc@tut.ac.za
[2] System and Computer Engineering Department, Carleton University, Ottawa, Canada
ajila@sce.carleton.ca

Abstract. Meta-learning often termed "learning to learn," seeks to construct models that can swiftly learn new information or adapt to new situations using only a few training data samples. Meta-learning is distinct from traditional supervised learning. The model is required to recognise training data before generalising to unknown test data in traditional supervised learning. Meta-learning, on the other hand, has only one goal: to learn. Few-shot makes predictions based on a limited amount of data using a prototypical network that it learns using a metric space in which classification is done by computing distances between prototype representations of each class. The Kullback-Leibler Divergence (KLD) a non-symmetric distance measure between two probability distributions, $p_{(x)}$ and $q_{(x)}$ is used in the meta-learning process of few-shot leaning to compute the distance between the prototype class to query cancerous image. This study seeks to investigate the usefulness of KLD in cancer image classification to assist medical practitioner in early diagnosis of the dreadful disease.

Keywords: Machine learning · Kullback-Leibler · Artificial intelligence · Data · Skin cancer · Images · Few-shot meta learning · Computing

1 Introduction

Image classification is the process of identifying and labelling groups of pixels or vectors inside an image using a set of rules. The classification law can be developed using one or more spectral or textural characteristics. There are several types of classification techniques: Supervised, Unsupervised, Logistic Regression, Decision Tree and Reinforcement. The supervised classification technique is a method of selecting training samples in images and allocating them to predetermined classes to generate statistical measures that can be used on the entire image. The numerical values of multiple visual features are evaluated in image classification, and data is grouped into distinct categories.

© The Author(s), under exclusive license to Springer Nature Switzerland AG 2022
M. Singh et al. (Eds.): ICACDS 2022, CCIS 1614, pp. 100–111, 2022.
https://doi.org/10.1007/978-3-031-12641-3_9

There are two stages in many classification systems – training and testing. During the training stage, characteristic qualities of conventional image features are determined, and a discrete representation of each category, i.e. the training category, is built based on them. In the testing stage, these feature-space partitions are used to classify image characteristics.

Few-shot classification is based on the premise that information may be easily extracted from baseline (seen) categories and transferred to novel (unseen) categories. A pre-trained core convolutional neural network (CNN) computes the input data of each image and is useful in detecting new image classes that were not in the baseline classes [1].

In this approach, meta-learning algorithms have a meta-training phase and a meta-testing phase. During the meta-training phase, the technique chooses N categories at random, then samples tiny base support set S_b and a base query set Q_b from datasets inside these categories. The goal is to train a classification model M that minimises the N-way classification loss LNway of the query set Q_b. To condition the classifier M, the specified support set S_b is used as in [1]. By training on a variety of tasks and producing predictions based just on the provided support set (known as "episodes"), a meta-learning algorithm can learn from little annotated images. In the meta-testing stage, all of the novel category images X_n was used as the support set for novel categories S_n, and the classifiers M can be adjusted to estimate novel categories using the new support set S_n [1].

The Kullback-Leibler divergence, as a distance metric, is a non-symmetric measure of the distance between two probability distributions, $p_{(x)}$ and $q_{(x)}$. The Kullback-Leibler (KL) divergence of $q_{(x)}$ from $p_{(x)}$, denoted D KL($p_{(x)}$, $q_{(x)}$), is a measurement of how much information is lost when $q_{(x)}$ is used to represent $p_{(x)}$.

Cancer is a medical disease caused by uncontrolled cell growth in body tissues Skin cancer is triggered by damaged deoxyribonucleic acid (DNA) in skin cells, which results in genetic abnormalities or changes in the skin [2]. Skin cancer can be defined as melanoma or non-melanoma. [3]. Melanoma is a rare and deadly kind of skin cancer. The American Cancer Society says melanoma skin cancer is rare, but it has a higher fatality rate [4]. In 2021 the United Nations estimated the world population at 7.9 billion, while the WHO projected overall new cancer cases at 1.8 million in 2021, with melanoma skin cancer accounting for only 4% of all cases [7]. The skin is the biggest organ and covers the entire body. Skin cancer can spread to nearby and distant organs rapidly, so, it's best to catch it early to ensure curability and cost-effectiveness. According to the research work in [5], the United States sees over 5.4 million new cases of skin cancer each year, and the global rate of spread is equally worrisome. Faced with limitations such as limited datasets researchers have developed several skin cancer detection techniques that are fast, accurate, and specific. Self-examination is typically the initial step toward skin cancer detection. Scaling, size, shape, colour, texture, and bleeding are all characteristics to inspect. A biopsy is recommended for any area showing any of the above-mentioned symptoms or developments. Prior to performing a biopsy, a dermatologist may need to examine the lesion with microscope or magnifying glass. Apart from self-examination and biopsy, numerous approaches for detecting and diagnosing skin cancer have been developed. Many of these techniques use imaging modalities such as CT scans, X-rays, and MRI in conjunction with machine learning algorithms for analysis of cancerous cells.

These imaging techniques are non-invasive, quick, and painless. Researchers developed skin cancer classification algorithms utilizing machine learning and computer vision techniques (ANN, SVM, logistic regression, and decision tree). While these investigations revealed promising results, the usage of computer vision and traditional machine learning has an impact on classification performance depending on the skin lesion segmentation result and the method's properties.

Due to the difficulties, as well as additional inherent limitations in machine learning algorithms for skin cancer diagnosis, deep learning make use of convolutional neural network to learn and detect highly discriminative features from a training image set. In CNN, each node is connected to a small amount of other nodes in the subsequent layer. With the use of appropriate filters, CNN can successfully capture a picture's temporal and spatial qualities [6]. This type of classifier normalises input vectors to a probability distribution using the modified softmax function [6], and the output values of the feature extractor are typically input to a fully connected network.

The purpose of this research is to develop a novel machine learning model for skin cancer diagnosis using few-shot met-learning and convolutional neural networks to enable quick detection of skin cancer

2   Related Works or Literature Review

The study in [8] investigated how to detect and analyse asymmetrical strips in dermoscopic images of skin lesions. The goal of the research was to see if skin lesions had streaks or not. This was achieved by analyzing the appearance of identified streak lines and conducting a three-way classification of streaks in a pigmented skin lesion, i.e. regular, irregular, and absent. In this approach, the directional pattern of detected lines was investigated in order to extract their directional characteristics and identify the key features. To model the geometric pattern of actual streaks and the spread and coverage of the structure, the method used a visual depiction. The accuracy of the three-class and two-class models was 76.1% with a weighted average area under the ROC curve (AUC) of 85%, 78.3% with an AUC of 83.2% (absent or present), and 83.6% with an AUC of 88.9% (regular or irregular), respectively. The AUC increased to 93.2%, 91.8%, and 90.9% respectively for the absent/present, absent/regular/irregular, and regular/irregular cases when the approach was applied to a cleaned subdivision of 300 images arbitrarily selected from the 945 images.

The work in [9] identified two major challenges in identifying dermatological condition from images. Real-world dermatological data distributions are often long tailed, and intra-class variation is large. They categorized the first problem as low-shot learning, where a base learner must adapt quickly to detect unfamiliar situations given a small number of labelled samples. A PCN trains a mixture of "prototypes" for every class to effectively depict intra-class variation [10]. Prototypes for each class were created using clustering and refined online. A weighted mixture of prototypes was used to categorize them. The weights represent the predicted cluster tasks. Also, they used ImageNet pre-training to train a modern convolutional neural network for image categorization,

ResNet-v2. They demonstrated the proposed technique's diagnostic capabilities on a real dataset of dermatological illnesses. For two alternative low shot configurations with train shots of 5 and 10 and test shot of 5, they presented the MCA of the test set on the top 200 classes available throughout the test time. Because the strategy uses episodic training to generate discriminative feature representations that may be extended to novel classes with a small number of samples, it performs much better (nine percent (or nine percent) absolute improvements over the existing approach).

The research work in [11] proposed a deep learning approach for automated melanoma lesion identification and segmentation. For efficient feature extraction and learning, improved encoder-decoder networks with encoder and decoder sub-networks connected by skip pathways. The method uses softmax classifiers to categorise melanoma tumours in pixels. They developed a lesion classifier that differentiates between melanoma and non-melanoma skin lesions. The suggested approach outperforms several state-of-the-art algorithms on the ISBI 2017 and Hospital Pedro Hispano (PH2) datasets, respectively. The approach obtained 95% ISIC 2017 data set and 92% exactness, 95% dice and 93% PH2 data set. Dice coefficient 95% and 93%.

The main challenge for few-shot learning is to reduce the error to limited data while maintaining sufficient adaptation to new classes [12]. Changing the tag restriction from new categories to base categories, accessible studies use huge quantities of annotated datasets [12]. Annotating datasets takes time and money, therefore reducing the requirement for annotation is essential [12]. proposed a more rigorous few-shot setting where no label access is permitted throughout training or testing. Utilizing self-supervision for training image representations and image similarities for classification at test time, they generated improved baselines with zero labels that had at least 10,000 times less labels than current approaches. The researchers hope the study will help design few-shot learning algorithms that don't rely on annotated data. Instead of using labels, they used Simple Contrastive Learning (SimCLR) a basic framework for contrastive learning of visual representations and Momentum Contrast (MoCo) for unsupervised visual representation learning. They employ image similarity to classify images since it is a good pre-training exercise for few-shot. Rather than anticipating image labels, students select a key image that is most similar to the input. During testing, images can be classified without using labels. They evaluated their algorithms on three commonly used few-shot image classification benchmark datasets: CIFAR-100 subset, Mini-Imagenet, and CIFAR-100 Few-Shot. Both SimpleShot's 1-Nearest Neighbor (1NN) and Matching Networks' soft cosine attention kernel (ATTN) were used as test time classifiers and used in multi-shot mode, the ATTN classifier improves accuracy by 2–10%, while the 1NN classifier achieves outstanding baselines (in line with known techniques). The ATTN classifier has the highest average accuracy of 30.0 2.8 in the 1-shot for the miniImagenet dataset, 53.0 0.2 in the 1-shot and 65.8 0.2 in the 5-shot for the CIFAR100 dataset, and 47.2 0.2 in the 5-shot for the FC100 dataset.

The problem of copious data requirements and labor-intensive training data preparation was solved by [13]. The researchers proposed a new few-shot object detection approach that focuses on finding new classes of objects using only a few labelled data samples Attention-RPN, Contrastive Training Strategy, and Multi-Relation Detector are the methodology's focal points. The similarity between the few-shot support set and query set is leveraged in this technique to identify novel items while suppressing false detection in the background. The attention- RPN uses support information to filter out most background boxes and those in non-matching classes, whereas the two-way contrastive training strategy was used to match the same category objects while also distinguishing different categories, with the multi-relation detector efficiently quantifying the similarity between query and support object proposal boxes. According to the findings, the proposed Functional Status and Outcome Database (FSOD) is capable of detecting objects of novel categories without the need for pre-training or network adaption. On the FSOD dataset, the proposed model generated AP, AP50, and AP75 values of 16.6, 31.3, and 16.1, respectively, whereas on the Common Object in Context (COCO) dataset, it generated AP (Adversarial Paraphrase), AP50, and AP75 values of 11.1, 20.4, and 10.6.

According to [14] discussed Deep Learning Solutions for Skin Cancer Diagnosis and Detection of Skin Cancer, in this study, a skin cancer detection Convolutional Neutral Network (CNN) model was developed to identify different skin cancer types and aid in early detection. Python with Keras was used to develop CNN classification model and Tensorflow on the backend. The model was built and analyzed with several network topologies and layer types, including Convolutional layers, Dropout layers, Pooling layers, and Dense layers, to mention very few. Transfer Learning methodologies are used to achieve fast convergence. The dataset was taken from the International Skin Imaging Collaboration (ISIC) competition archives and used to evaluate and train the model. The combined ISIC 2018 and ISIC 2019 datasets generated a new dataset. The dataset was cleaned up and the seven most common types of skin lesions were preserved. The models were able to learn more quickly as a result of increased number of examples provided per class. Melanoma (label 0), Melanocytic Nevus (label 1), Basal Cell Carcinoma (label 2), Actinic Keratosis (label 3), Benign Keratosis (label 4), Dermatofibroma (label 5), Vascular Lesion (label 6) and Squamous Cell Carcinoma (label 7) are among the seven groups (label 7). The input images were enhanced to make the model more resilient to unknown data, which improved testing accuracy dramatically. The Image Normalization technique was used to normalise the image's pixel values into a comparable distribution. Using the pre-trained weight of ImageNet classification, transfer learning was employed to train the CNN. The achieve the seven-class segmentation of the skin images, the network is trained and authenticated on state-of-the-art CNNs, including Inception V3, ResNet50, VGG16, MobileNet, and InceptionResnet. The Inception V3 and InceptionResnet CNN models obtained high ratings, with accuracies of 90 and 91 per cent, respectively. It is sufficiently accurate to classify lesion images into one of the seven categories [15] proposed an autonomous computer vision technique for identifying melanoma skin cancer. The approach is divided into three stages: first, adaptive principal curvature is used to pre-process skin lesions images, then colour normalisation is used to segment skin lesions, and last, the Asymmetry, Border, Color, Diameter (ABCD) rule is used to extract features. Because the efficiency of skin lesions segmentation, as well as the efficiency of feature extraction, is negatively influenced by hair, the pre-most processing's important duty is to locate and remove hairs. The hairs were identified using adaptive principal curvature, and they were removed using the inpainting approach. [16] came up with a way to improve the thresholding segmentation value because the colours of skin lesions can change and can be the same as the underlying skin tone. They used a publicly available dataset from the International Skin Imaging Collaboration (ISIC) skin lesions database to show the result of the proposed approach. The results of the melanoma skin cancer detection revealed that the proposed scheme has good accuracy and cumulative strong performance; the accuracy, Dice, and Jaccard ratings for the segmentation phase are 96.6%, 93.9%, and 88.7%, respectively. The melanoma detection phase of the ISIC dataset exhibits a near-perfect accuracy rate for a small portion of the dataset. They compared their technique to four other approaches and discovered that theirs outperformed the others [17] proposed the use of a fine-grained classification approach to develop a classification model to address the issues associated with autonomous detection of dermoscopy image lesions. The model used MobileNet and DenseNet as well as a feature discrimination network. The suggested methodology fed two types of training data into the recognition model's feature extraction unit. To train the recognition model's feature discrimination networks, two sets of feature maps were created. This method extracts more discriminative lesion features and improves the model's classification performance with a small number of model parameters. The proposed technique improved segmentation accuracy by 96.2% with only a few model parameters.

According to [18] they proposed the TIP to increase the generalisation capabilities of few shot object detection models. Between fully supervised and few-shot scenarios, the proposed TIP technique imposes invariant consistency among altered pictures to overcome the generalisation gap. To increase the generalisation capabilities of Few-Shot Object Detection (FSOD) models by sample expansion, the methodology was applied to semi-supervised learning. Using the PASCAL VOC and MSCOCO datasets, the technique was implemented. The experimental setup included two scenarios: Semi-supervised FSOD uses a large number of unannotated data samples for model training. The proposed technique performed well in both fully supervised and semi-supervised FSOD cases. The study found average precision of 25.8% for 10 shots and 29.2% for 30 shots. For 10 and 30 shots, average recall was 43.8 and 45.1%.

Contrary to popular belief, the intrinsic tension between innovative class representation and classification is typically disregarded. In order to attain high accuracy in few-shot object detection, the distributions of two base classes must be far apart (max-margin). Moreover, to effectively characterise new classes, the distributions of base classes should be close together (min-margin). The solution to the aforementioned issue is the focus of [18]. The authors used a CME technique to improve both novel class reconstruction and feature space division. When learning the feature characterisation, CME preserves appropriate margin space for innovative classes. With novel classes, CME enhances feature space division by tracking margin equilibrium in an argumentative min-max approach. The proposed CME detector outperforms its competitors in class margin equilibrium by up to 5%.

According to [19] creating algorithms that can generalise to a specific task with only a few labelled samples is a crucial issue in closing the performance gap. Human cognition is based on structured, reusable concepts that allow humans to easily adapt to new activities and make judgments. Contrarily, standard meta-learning algorithms allegedly learn complex representations from previously labelled events without applying any pattern. COMET is a meta-learning approach designed to educate humans how to learn across human-interpretable notion dimensions. COMET trains semi-structured metric spaces rather than a single unstructured metric space, and successfully aggregates the results of different concept classifiers. They evaluated the proposed model on few-shot tasks from several domains, including fine-grained image classification, document categorization, and cell type annotation. The COMET has an average accuracy of 67.9 0.9 for the CUB dataset and 85.3 0.5 for the Tabula Muris dataset. It also outperformed baseline meta-learning techniques with 1-shot and 5-shot accuracy of 71.5 0.7 and 89.8 0.3 for the Reuters dataset. Leaning and meta-algorithm.

Table 1 below summarize the above section

Table 1. .

Study	Dataset description	Proposition and metrics of evaluation	Results
Prabhu et al. (2019)	–	Application of prototypical clustering networks (PCN) for dermatological diseases detection MCA	MCA of 30.0 ± 2.8 for 5 shots MCA of 49.6 ± 2.8 for 10 shots
Thanh et al. (2020)	International skin imaging collaboration (ISIC)	Melanoma skin cancer detection using autonomous computer vision techniques Accuracy, Dice, and Jaccard ratings	Accuracy: 100%
Nahata and Singh (2020)	International skin imaging collaboration (ISIC	CNN model for skin cancer detection Accuracy	Inception V3 CNN: Accuracy = 90% Resnet CNN: Accuracy = 91%

Based on related works summarized in Table 1 from different researchers and to the best knowledge of the authors no work has be done using meta learning algorithm to solve a classification problem in skin cancer detection when a limited number of samples are available, while achieving dimensionality reduction and cut in overall computational cost.

Furthermore, this research work will use Kullback-Leibler distance learning metric based on few-shot meta-learning technique using convolutional neural networks and prototypical networks to address skin cancer detention when a limited number of samples are available

3   The Fundamentals of the Few-Shot Learning Concept, as Well as Ongoing Research into the Application of Few-Shot Learning to Skin Cancer Detection

As aforementioned in this research work, one of the most significant results indicated by deep learning in medical applications has been the shortage of labelled data in the target domain. This is mostly related to the excessively high expense of paying a professional clinician to examine the patient's health status [20]. Data augmentation is one way to get around a restriction. It means making imaginary things from the original data [21, 22]. Transfer learning is another often-used technique. This technique entails training the network using evidence from a relevant domain, then transferring the weights and biases to a new network and fine-tuning it for the target domain [20]. The disadvantage of this algorithm is that it behaves poorly when the volume of the test set is minimal or when the pattern of the test set is not adequately adjusted [23].

Therefore, few-shot learning models have been described in the literature to facilitate learning from tiny amounts of labelled data sets in order to simulate humans' ability to generalise new knowledge based on a tiny range of instances [24]. With just a few training data points, few-shot techniques create a model that is applicable to different tasks and conditions. The fundamental principle is to learn the model's key parameters in such a manner that everything performs effectively on an inventive task when the model's settings are updated using many gradient steps computed using a bit of data gathered from the inventive challenge [30].

Few-shot learning's aim is to obtain the best possible show models from small training data [25, 31, 32].

Numerous approaches to few-shot learning are outlined in [31], particularly Matching networks, which generate a learning algorithm over support set labels to manage query set labels for emerging classes. By merging embedding and centroid representations (as class prototypes), prototypical networks [25] categorise unique data using Euclidean distance. Also, as established in the works of [31,25], embeddings are developed end-to-end using an episodic research sample [26].

Research work [27] used an improved relational network for metric learning to identify skin ailments. They used a small amount of documented skin lesion image data for this study, "Few-shot learning for skin lesion image classification," which was called "Few-shot learning for skin lesion image classification." The system comprised a relative position network (RPN) and a relative mapping network (RMN), with the RPN accumulating and collecting classification models via an attention mechanism and the RMN generating picture categorizing matching using a weighted sum of concentration mapping distance. Using the public ISIC melanoma dataset, the average accuracy of the classification is 85%, which shows that the technique is both effective and practical [28] suggested a few-shot recognition model for image segmentation that requires minimum pixel-level tagging. To begin, we collected the co-occurrence area between the support and query images so that it could be used as a prior mask to eliminate unwanted background areas. Then, the outcomes are combined and provided to the inference module, which predicts the division of a query image. Furthermore, by inverting the support and query roles, the recommended network was retrained, utilising the symmetrical structure.

Thorough testing on ISIC-2017, ISIC-2019, and PH2 showed that the method provides a promising framework for dividing skin lesions in a few shots.

Finally, they designed a few-shot prototype network based on an internet-based medical database [20] to resolve the shortage of documented samples. To begin with, a comparative learning branch was developed to enhance the capabilities of the feature extractor. Second, a new procedure for producing positive and negative sample pairs was developed for comparison learning, and it was reported to alleviate the obligation to explicitly maintain a sample queue. Finally, the contrast learning branch was used to solve data corruption and establish a category prototype. Finally, the hybrid loss, which combines prototype and contrastive losses, has been used to increase classification accuracy and convergence speed. Their method was said to perform exceptionally well on mini-ISIC-2i and mini-ImageNet datasets [29].

4   Kullback-Leibler Divergence Framework

This research utilizes Convolutional Neural Network architecture (Resnet-50 or VGG-19) to extract features from skin cancer image support and query sets, as well as investigations to evaluate the Kullback–Leibler divergence measure's suitability. Skin Cancer Detection Using a Few-shot Meta-Learning Technique Based on Convolutional Neural Networks and Prototypical Networks. This study compares the performance of the various results obtained from the experiments using a publicly available skin cancer dataset to improve skin cancer detection in a limited dataset setting utilizing a few shot meta-leaning technique.

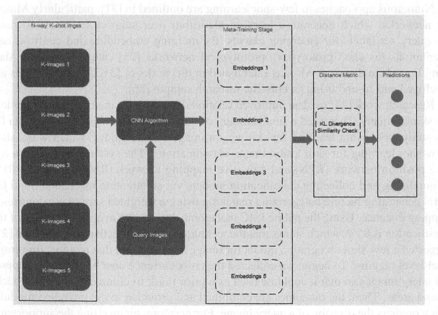

Fig. 1. Proposed framework for the Kullback-Leibler approach above.

N-way K-shot images were inputed from the support set into a convolutional neural network (CNN) algorithm that first computes the feature embedding representation of each of the K images in each of the N classes in a meta-learning fashion, as shown in Fig. 1. The images in the query set of unseen images go through the same feature embedding process. A prototype of the class is computed for each class in the support set to typically represent that class of images from the feature vector representations of all the images. The Kullback-Leibler distance learning metric is then used to evaluate how similar each image in the query set is to each of the prototype representations of each class in the support set. Each image in the query set is classified into the appropriate class based on the computed distance. The mathematical expression for the Kullback-Leibler divergence measure can be seen in Eq. 1.

$$D_{KL}(p_{(x)} \| q_{(x)}) = \text{Summation } p_{(x)} \text{in} p_{(x)}/q_{(x)}. \tag{1}$$

where $p_{(x)}$ and $q_{(x)}$. (x) be two discrete random variable x probability distributions. That is, for any x in X, both $p_{(x)}$. (x) and $q_{(x)}$. (x) aggregate to 1, and $p_{(x)}$. (x) > 0 and $q_{(x)}$. (x) > 0.

5 Conclusion

In this research work, the authors proposed the new Meta Learning Algorithm for few-shot learning while a distance learning metric Kullback-Liebler plays the role of computing the distance between the protype class and the query class in which a classification problem can be solved when a limited number of samples is available. Another area of concern is that a prototypical network is capable of learning to learn a metric space in which classification can be performed by computing distances to prototype representations of each class. With our proposed new algorithm to measure the distance for an accurate prediction, future researchers can improve some of the limitations noted in this current study by improving the area of study.

Acknowledgments. Special thanks and appreciation go out to my supervisors, Prof. C. Du and Prof. S. Ajila, for their professional guidance during this research study, as well as my colleague Olusola Salami, who assisted in the review of this paper and provided advice on the flow that was used.

References

1. Chen, W.-Y., et al.: A closer look at few-shot classification. ICLR, 1–16 (2019). https://git hub.com/wyharveychen/CloserLookFewShot. Accessed 09 Jan 2022.
2. Ashraf, R., et al.: Region-of-interest based transfer learning assisted framework for skin cancer detection. IEEE Access **8**, 147858–147871 (2020). https://doi.org/10.1109/ACCESS.2020.3014701
3. Elgamal, M.: Automatic skin cancer images classification. Int. J. Adv. Comput. Sci. Appl. **4**(3) (2013). https://doi.org/10.14569/IJACSA.2013.040342
4. Dildar, M., et al.: Skin cancer detection: a review using deep learning techniques. Int. J. Environ. Res. Public Heal. **18**(10), 5479 (2021). https://doi.org/10.3390/IJERPH18105479

5. Esteva, A., et al.: Dermatologist-level classification of skin cancer with deep neural networks. Nature **542**(7639), 115–118 (2017). https://doi.org/10.1038/NATURE21056
6. Szegedy, C., et al.: Going deeper with convolutions. In: CVPR, pp. 1–9 (2015)
7. Siegel, R.L., Miller, K.D., Fuchs, H.E., Jemal, A.: Cancer statistics. CA Cancer J. Clin. **71**(1), 7–33 (2021). https://doi.org/10.3322/caac.21654
8. Sadeghi, M., Lee, T.K., McLean, D., Lui, H., Atkins, M.S.: Detection and analysis of irregular streaks in dermoscopic images of skin lesions. IEEE Trans. Med. Imaging **32**(5), 849–861 (2013). https://doi.org/10.1109/TMI.2013.2239307
9. Prabhu, V., et al.: Few-Shot learning for dermatological disease diagnosis. Proc. Mach. Learn. Res. **106**, 1–15 (2019). http://www.dermnet.com/. Accessed 09 Jan 2022
10. Snell, J., Swersky, K., Zemel, R.: Prototypical networks for few-shot learning. Adv. Neural Inf. Process. Syst. **30**, 4078–4088 https://arxiv.org/abs/1703.05175v2. Accessed 09 Jan 2022
11. Adegun, A.A., Viriri, S.: Deep learning-based system for automatic melanoma detection. IEEE Access **8**, 7160–7172 (2020). https://doi.org/10.1109/ACCESS.2019.2962812
12. Bharti, A., Balasubramanian, V., Jawahar, C.V.: Few shot learning with no labels, undefined (2020). https://anonymous. Accessed 09 Jan 2022
13. Fan, Q., Zhuo, W., Tang, C.K., Tai, Y.W.: Few-shot object detection with attention-RPN and multi-relation detector. In: Proceedings of the IEEE/CVF Conference on Computer Vision and Pattern Recognition, pp. 4012–4021 (2019). https://doi.org/10.1109/CVPR42600.2020.00407
14. Nahata, H., Singh, S.P.: Deep learning solutions for skin cancer detection and diagnosis. In: Jain, V., Chatterjee, J.M. (eds.) Machine Learning with Health Care Perspective. LAIS, vol. 13, pp. 159–182. Springer, Cham (2020). https://doi.org/10.1007/978-3-030-40850-3_8
15. Thanh, D.N.H., Prasath, V.B.S., Hieu, L.M., Hien, N.N.: Melanoma skin cancer detection method based on adaptive principal curvature, colour normalisation and feature extraction with the ABCD rule. J. Digit. Imaging **33**(3), 574–585 (2019). https://doi.org/10.1007/s10278-019-00316-x
16. Kulkarni, N.: Color thresholding method for image segmentation of natural images. Int. J. Imag, Graph. Signal Process. **4**(1), 28–34 (2012). https://doi.org/10.5815/ijigsp.2012.01.04
17. Wei, L., Ding, K., Hu, H.: Automatic skin cancer detection in dermoscopy images based on ensemble lightweight deep learning network. IEEE Access **8**, 99633–99647 (2020). https://doi.org/10.1109/ACCESS.2020.2997710
18. Li, A., Li, Z.: Transformation invariant few-shot object detection. In: Proceedings of the IEEE/CVF Conference on Computer Vision and Pattern Recognition, pp. 3094–3102 (2021)
19. Cao, K., Brbićbrbić, M., Leskovec, J.: Concept Learners for Few-Shotlearning (2020)
20. Garcia, S.I.: Meta-learning for skin cancer detection using Deep Learning Techniques, (NeurIPS), 1–7. http://arxiv.org/abs/2104.10775 (2021)
21. Wong, S.C., Gatt, A., Stamatescu, V., McDonnell, M.D.: Understanding data augmentation for classification: when to warp? In: 2016 International Conference on Digital Image Computing: Techniques and Applications, DICTA, pp. 1–6 (2016)
22. Mikołajczyk, A., Grochowski, M.: Data augmentation for improving deep learning in image classification problem. In: 2018 International Interdisciplinary PhD Workshop, IIPhDW, August 2019, pp. 117–22 (2018)
23. Kumar, V., Glaude, H., de Lichy, C., Campbell, W.: A closer look at feature space data augmentation for few-shot intent classification. In: DeepLo@EMNLP-IJCNLP 2019 - Proceedings of the 2nd Workshop on Deep Learning Approaches for Low-Resource Natural Language Processing – Proceedings, pp. 1–10 (2021)
24. Duan, R., Li, D., Tong, Q., Yang, T., Liu, X., Liu, X.: A survey of few-shot learning: an effective method for intrusion detection. Secur. Commun. Netw. (2021)
25. Snell, J., Swersky, K., Zemel, R.: Prototypical networks for few-shot learning. Adv. Neural Inf. Process. Syst. (Nips) **30**, 4078–4088 (2017)

26. Prabhu, V., Kannan, A., Ravuri, M., Chablani, M., Sontag, D., Amatriain, X., et al.: Few-shot learning for dermatological disease diagnosis. In: Proceedings of Machine Learning Research [Internet], vol. 106(Icd), pp. 1–15. http://www.dermnet.com/

27. Liu, X.J., Li, K., Luan, H., Wang, W., Chen, Z.: Few-shot learning for skin lesion image classification. Multi. Tools Appl. 1–12 (2022).https://doi.org/10.1007/s11042-021-11472-0

28. Xiao, J., Xu, H., Zhao, W., Cheng, C., Gao, H.: A Prior-mask-guided few-shot learning for skin lesion segmentation. Computing, 1–23 (2021)

29. Xiao, J., Xu, H., Fang, D., Cheng, C., Gao, H.: Boosting and rectifying few-shot learning prototype network for skin lesion classification based on the internet of medical things. Wireless Netw. **0123456789**, 1–15 (2021). https://doi.org/10.1007/s11276-021-02713-z

30. Mahajan, K., Sharma, M., Vig, L.: Meta-dermdiagnosis: few-shot skin disease identification using meta-learning. In: IEEE Computer Society Conference on Computer Vision and Pattern Recognition Workshops, pp. 3142–3151 (2020)

31. Vinyals, O., Blundell, C., Lillicrap, T., Kavukcuoglu, K.: Matching networks for one shot learning. Adv. Neural Inf. Process. Syst. **29** (2020)

32. Santoro, A., Botvinick, M., Lillicrap, T., Deepmind G., Com, C.G.: One-shot Learning with Memory-Augmented Neural Network (2016)

WE-Net: An Ensemble Deep Learning Model for Covid-19 Detection in Chest X-ray Images Using Segmentation and Classification

Rupanjali Chaudhuri, Divya Nagpal[✉], Abhinav Azad, and Suman Pal

Cerner Corporation, Bangalore 560045, India
{rupanjali.chaudhuri,divya.nagpal,abhinav.azad,suman.pal}@cerner.com
https://www.cerner.com/

Abstract. Amidst the increasing surge of Covid-19 infections world-wide, chest X-ray (CXR) imaging data have been found incredibly helpful for the fast screening of COVID-19 patients. This has been particularly helpful in resolving the overcapacity situation in the urgent care center and emergency department. An accurate Covid-19 detection algorithm can further aid this effort to reduce the disease burden. As part of this study, we put forward WE-Net, an ensemble deep learning (DL) framework for detecting pulmonary manifestations of COVID-19 from CXRs. We incorporated lung segmentation using U-Net to identify the thoracic Region of Interest (RoI), which was further utilized to train DL models to learn from relevant features. ImageNet based pre-trained DL models were fine-tuned, trained, and evaluated on the publicly available CXR collections. Ensemble methods like stacked generalization, voting, averaging, and the weighted average were used to combine predictions from best-performing models. The purpose of incorporating ensemble techniques is to overcome some of the challenges, such as generalization errors encountered due to noise and training on a small number of data sets. Experimental evaluations concluded on significant improvement in performance using the deep fusion neural network, i.e., the WE-Net model, which led to 99.02% accuracy and 0.989 area under the curve (AUC) in detecting COVID-19 from CXRs. The combined use of image segmentation, pre-trained DL models, and ensemble learning (EL) boosted the prediction results.

Keywords: COVID-19 · Deep learning · Segmentation · Classification · Ensemble

1 Introduction

The new Severe Acute Respiratory Syndrome - Coronavirus - 2 (SARS-CoV-2) led to the novel Coronavirus (COVID-19), an extremely contagious disease that surfaced in late 2019. The world witnessed a massive surge in COVID-19

© The Author(s), under exclusive license to Springer Nature Switzerland AG 2022
M. Singh et al. (Eds.): ICACDS 2022, CCIS 1614, pp. 112–123, 2022.
https://doi.org/10.1007/978-3-031-12641-3_10

cases; consequently, on 11th March 2020, World Health Organisation (WHO) announced the disease a pandemic [1]. Globally, as of 8th April 2022, there have been 496,507,539 verified cases of COVID-19, including 6,177,354 mortalities, reported to WHO.

The Reverse Transcription-Polymerase Chain Reaction (RT-PCR), a COVID-19 detection test, is frequently used and is considered the gold standard. However, RT-PCR has shortcomings concerning high variability in results, time taken for the laboratory analysis along with logistics, testing kit availability, and testing capacity constraints [2].

Radiography examination provides an effective alternative to RT-PCR for fast screening of this disease, including screening CXRs and Computed Tomography (CT) images by radiologists. Portable CXRs, although not as sensitive as CT scans, provide a faster alternative. Being a readily available and economical option, CXRs can be used to fight the disease in a quick and cheap way, thereby aiding rapid containerization of the spread of this disease.

To further fasten the COVID screening process, the use of artificial intelligence (AI) enabled automated computer-aided diagnostic (CADx) tools can be of great help to assist radiologists.

This study illustrates a novel ensemble DL-based CADx for COVID-19 detection from the CXR images collected from different data sources. The pipeline consists of CXR image pre-processing, data augmentation, lung segmentation, and RoI extraction. The extracted RoI images were further used to train the transfer learned DL models. These fine-tuned models were further leveraged to create an ensemble model using various strategies. A comparison of these different ensemble models is performed and evaluated.

We have detailed the core methodology and model pipeline in Sect. 3. The subsections elaborate on dataset and methodology details. The conclusion is clearly stated in Sect. 5, preceded by the results and discussion section captured in Sect. 4.

2   Literature Survey

AI studies have demonstrated promising results using convolutional neural networks (CNNs) for COVID-19 classification [3,4]. The researchers of [5] have developed a tailored CNN and pre-trained AlexNet model to classify CXRs into normal or COVID-19 pneumonia categories with an accuracy of 94.1% and 98%, respectively. The authors of [6] have used a pre-trained ResNet-50 [7] to classify CXRs into normal, pneumonia, and COVID-19 viral pneumonia categories and achieved an accuracy of 98.18% and F-score of 98.19%. Analysis of CXRs through CNNs can also aid the diagnosis and help to differentiate amongst other types of pneumonia-like viral pneumonia not induced by COVID-19 and bacterial pneumonia [8]. In work by Singh et al. [9], CT scan images have been analyzed by the CNN model to detect COVID-19 patients. Ng et al. [10] along with Huang et al. [11] have discussed the superiority of CXR image analysis for COVID detection over other techniques, owing to promising screening results, the ready availability of CXR machines, and their low maintenance cost. Bagging ensemble or bootstrap aggregating of transfer learned models ResNet34, Inception V3, and DenseNet201 named ET-Net was used by authors of [12] for Covid19 detection using CXR and achieved 97.81 ± 0.53% accuracy on 5-fold cross-validation.

In work by Rajaraman et al., [13], modality-specific features are learned by a selection of ImageNet pre-trained models. The trained models are fine-tuned to improve model performance and generalization. They have iteratively pruned the best-performing models. Pruning helped to reduce the complexity of the model and thereby improved the memory efficiency. Overall classification performance was improved by combining the prediction results from the pruned models, which had the best performances through ensemble strategies. Quan et al. [14] came up with DenseCapsNet that integrates a CNN, i.e., DenseNet, and a capsule network, i.e., CapsNet. They also proposed TernausNet for lung segmentation, where only the lung contour region was used for training the DenseCapsNet. The results obtained had 90.7% accuracy along with a 90.9% F1 score. In context of AI-based lung detection, Mohammad Y. et al. [15] also proposed a DL CNN model based on Unet [25] for automatic lung segmentation. They achieved 0.976 and 0.979 dice score on Japanese Society of Radiological Technology (JSRT) [17] and Montgomery [18] dataset respectively.

In the recent past, a massive surge in COVID-19 research led to the increasing availability of publicly available COVID-19 CXR image datasets and increased computational power. The research studies continue with an effort toward building an accurate classification algorithm for COVID detection.

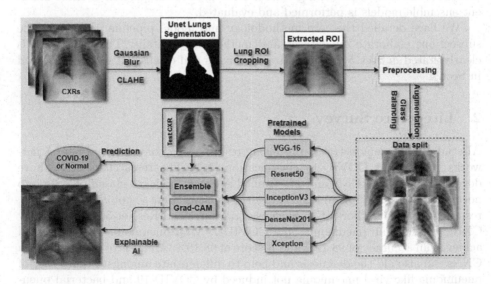

Fig. 1. Workflow diagram

3   Methodology and Model Pipeline

This study incorporates image pre-processing, segmentation, transfer learning (TL), and EL for training the COVID-19 detection model. The study highlights the benefits of segmentation and ensemble of transfer learned DL models to train an accurate and generalized model for COVID-19 classification. Figure 1 summarises the model pipeline.

3.1   Dataset Summary

Table 1 summarizes the data distribution for CXRs used for developing segmentation and classification models. Listed below are the publicly available CXRs collection used for the study. The first three datasets were used to develop the segmentation model, followed by the dataset used to develop the classification model.

Table 1. Data summary

Type	Sources	Data Samples	
Segmentation	Shenzhen	566	
	Montgomery	138	
	NIH	247	
Classification	Covid Radiographic Database	Covid	3616
		Normal	10192

Segmentation in Chest Radiographs (SCR) database: This database contains Posterior-Anterior (PA) radiographs from the JSRT database collected from one in the United States and 13 institutions in Japan [16,17]. Scanned images from films are saved to the resolution of 2048 by 2048 pixels. Manual segmentation masks for lungs, heart, and clavicles have been provided for each image.

Shenzhen CXR set: The National Library of Medicine - Maryland, the USA, created a CXR database in association with Shenzhen No.3 People's Hospital - Shenzhen, China. The CXRs are provided in PNG format sized to approximately 3K × 3K pixels [18].

Montgomery County CXR set: The Department of Health and Human Services - Montgomery County, Maryland, USA, curated this dataset [18]. The Eureka X-ray machine was leveraged to capture 12-bit gray-level CXRs in PNG format. The size of CXRs is either 4,020 × 4,892 or 4,892 × 4,020 pixels.

Kaggle Radiography database for COVID-19: This dataset has been curated with the collaboration of experts from Dhaka, Qatar, Pakistan, and Malaysia [19,20]. It consists of CXRs with four classes of diseases, namely Covid-19, Viral Pneumonia, Lung opacity, and Normal. The number of images in this database is increasing over time, with more patient CXRs screens made publicly available.

3.2   Data Preparation

We prepared data for model development by breaking the combined data set into three parts, i.e., sixty percent for training and twenty percent each for validation and test sets, respectively. The Ground Truth (GT) for the segmentation model is the lung masks, and the GT for the COVID-19 classification model are the labels made available in the above-discussed datasets.

To balance the dataset and reduce overfitting, augmentation was done on the training data [21], which was achieved by adding random variations in the following attributes of the images: rotations, translational shifts, zooming, shear shifts, horizontal flips, and brightness intensity changes. Augmentations in the CXR - lung mask pairs followed the same template. After the augmentation, additional pre-processing was performed on the CXR masks, thresholding the diluted pixel back to 0 values (black). Further, the CXRs used for the segmentation model were pre-processed using Contrast Limited Adaptive Histogram Equalization (CLAHE) followed by Gaussian blurring to mimic the COVID positive CXRs, thus enabling the segmentation model to learn on abnormal CXRs [22].

Finally, all the images were resized to 256×256 for input to segmentation and classification models. We performed up-scaling using inter-cubic interpolation and down-scaling using inter-area interpolation types respectively [23,24]. The pre-trained TL-based DL models are employed for creating the COVID-19 detection model and are trained to work with color RGB images as input, but our x-rays are grayscale. To fill this gap, we converted our 1-channel grayscale X-rays into 3-channel images as the last step of our data pre-processing.

3.3   Lung Segmentation and Region of Interest Extraction

In this study, we followed RoI extraction by segmentation and image cropping. These images were used to train our DL classification models with relevant features, thereby aiding reliable decision-making. We performed semantic segmentation of lungs using U-Net, trained on CXRs with lung masks. U-Net is a specially designed network utilizing CNN for biomedical image segmentation tasks [25]. We experimented and trained the U-Net model with different values of epochs along with early stopping functionality using the Adam optimizer. For regularization, dropout layers are added for the first convolution in every step with increasing order from 0.1 to 0.3 in the contracting paths and 0.2 to 0.1 in the expansive path. Checkpoints were used to analyze model performance and save the best model weights, which are further leveraged to generate the final lung masks.

We then performed contour detection over the lung masks. A union of the fit bounding boxes around the contours was cropped, providing us ROI approximation of the thoracic cavity consisting of the lung lobes. The cropped image was resized to a resolution of 256×256 pixels. Figure 2 summarizes the segmentation and ROI extraction process.

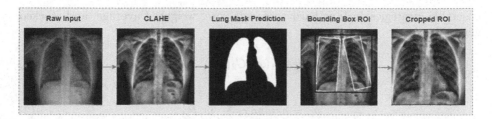

Fig. 2. ROI extraction

3.4 Classification Using Transfer Learning

TL is a Machine Learning (ML) method where a model developed for a particular problem is employed as the commencing point for creating another model to solve another problem. TL models have the benefit of learning from a small dataset, reduced training time, and computational complexity [26].

Using a pre-trained model is one of the best approaches for applying the TL strategy. We can use off-the-shelf features from state-of-the-art ImageNet based CNN pre-trained TL models for solving the task [27]. The core idea for extracting features on the new data is to consume the weighted layers of the pre-trained TL models while training.

We experimented with several pre-trained models, finally retaining the best five ImageNet pre-trained CNN architectures: VGG-16, ResNet50, DenseNet201 (DNet201), InceptionV3 (InceptV3), and Xception in our study.

Truncating the rearmost convolution layer and replacing the classifier layers with custom build layers, fine-tuning is done for the pre-trained TL-based models. We added an average pooling layer to reduce the dimension of the feature map. Batch normalization has also been included in the architecture of all pre-trained models except for VGG-16. After the convolutional and pooling layers, a classification network is placed at the model culmination consisting of Fully Connected Layers (FCLs). We have used FCL with Rectified Linear Unit (ReLU) as the activation function along with 256 neurons which are followed by a dropout layer.

The FCLs combine high-level features to get the final result. In the end, a loss layer using softmax activation function and loss function as binary cross-entropy was incorporated.

3.5 Ensemble Model

DL models may hit generalization issues with statistical noise in training data, training on a limited dataset, and random weight initialization. Using EL, we can overcome these issues. EL is a hybrid learning paradigm that has the art of mixing several base learners and improvises on the generalization and predictive power of the model [28, 29].

Our proposed work has studied different ensemble techniques for COVID-19 detection by aggregating predictions from multiple DL models through various ensemble strategies such as hard vote, soft vote, weighted average, and stacked generalization.

$$\hat{y}_h = mode\{C_1(x), C_2(x), ..., C_n(x)\} \tag{1}$$

$$\hat{y}_w = argmax_i \sum_{j=1}^{m} w_j P_{ij} \tag{2}$$

Hard voting $(\hat{y}_h)$ (as shown in Eq. 1) is the simple ensembling technique wherein the prediction from each classifier C is a vote, and the class with the maximum votes is certified as the final prediction. For the simple average ensemble technique, we calculate the average prediction probabilities from every model and use it as the final prediction. An extension of this technique is a weighted average $(\hat{y}_w)$, wherein weights are assigned to every model such that the base model with higher predictive power is given more importance [30]. In Eq. 2, we have p_{ij} and w_j as the probabilities of the predictions and the weights assigned respectively for the jth classifier and i^{th} class label. Additionally, by substituting uniform weights, we compute average probabilities. Stacked generalization has been implemented using XGBoost and SVM classifiers as meta-learners that are used to combine the predictions of the base models optimally. A stacked generalization approach has been described in Algorithm 1.

3.6   Explainable AI

While using DL models in medical image prediction problems, it's important to understand their learned behavior, explain the predictions from these black-box models, and understand the clinical decision making.

Gradient-weighted Class Activation Mapping (Grad-CAM) was implemented to tackle this problem. It is an algorithm that can visualize class activation maps and debug the CNN, ensuring that the network is "looking" at the correct locations in an image. Grad-CAM works by finding the last convolutional layer in the network and evaluating the gradient information flowing into that layer. The output of Grad-CAM is a heat-map visualization for a given class label that can be used to visually verify where the CNN is looking in the image. A Grad-CAM heat-map is mathematically defined as a combination of weighted feature maps, followed by a ReLU activation [31].

For an image class i, the class differentiating localization map L^i can be defined as follows:

$$L_{x,y}^i = ReLU(\sum_n w_n^i M_{x,y}^i) \tag{3}$$

where the activation map for the n-th kernel is denoted by $M_{x,y}^n$ at (x, y) index, and ReLU activation function. The weights for the target class at the n-th kernel are computed as shown in Eq. 4:

Algorithm 1. Stacked Generalization With K-fold Cross Validation

---

Input: $Data(D) = [(x_i, y_i)|x_i \in X, y_i \in Y]$;
N = No of base learners;
Base Classifiers: $C = [C_1, C_2, ...C_N]$
Output: Ensemble Classifier (EC)

1: **Step 1:** Split data into 2 parts: $D = [D_1, D_2]$
2: **Step 2:** Train base classifiers on D1 and store the weights
3: **for** $i \leftarrow 1$ to N **do**
4:      Learn C_i based on D_1
5:      Store trained weights
6: **end for**
7: **Step 3:** Stack predictions
8: **for** $j \leftarrow 1$ to N **do**
9:      Get base classifier predictions(p) on D_2
10:      Stack preds($\hat{p}$) $\leftarrow$ dstack(p) from numpy package
11: **end for**
12: **Step 4:** Use $(\hat{p}_i, y_i)$ as an input to meta learner
13: **Step 5:** Stratified split $(\hat{p}_i, y_i)$ into K equal folds:-
14:      $\hat{p} = [\hat{p}_1, \hat{p}_2, \hat{p}_3, ...\hat{p}_K], y = [y_1, y_2, y_3, ...y_K]$
15: **Step 6:** Perform cross validation
16: **for** $i \leftarrow 1$ to K **do**
17:      Train meta learner on $K - 1$ folds
18:      Get predictions on K^{th} fold
19: **end for**

$$w_n^i = \frac{1}{Z} \sum_x \sum_y \frac{\partial Y^i}{\partial M_{x,y}^n} \tag{4}$$

where the number of pixels is denoted as Z and the probability score by Y^i.

4 Results and Discussion

We first report the segmentation model results obtained after cropping the lung ROI in CXRs. The performance metrics are summarized in Table 2.

Table 2. Segmentation results

Dataset	IoU Score	Accuracy	Precision	Sensitivity	Specificity	Support
Train	0.925	0.978	0.940	0.983	0.976	2294
Val	0.937	0.982	0.966	0.968	0.987	95
Test	0.935	0.983	0.971	0.961	0.990	90

Table 3 summarizes the performance of pre-trained TL-based models and the ensemble models. Results indicate accurate classification performance with high

AUC values and narrow Confidence Interval (CI) for pre-trained models. Apart from F1 and accuracy, we also have Matthews Correlation Coefficient (MCC), which is a more reliable score and is invariant towards class swapping, unlike F1-Score.

Table 3. Performance metrics of classification models

Models	Accuracy	MCC	Precision	F1-Score	ROC-AUC	ROC-AUC CI
VGG-16	0.927	0.828	0.983	0.949	0.937	[0.927-0.947]
ResNet50	0.976	0.939	0.990	0.984	0.975	[0.969-0.982]
DNet201	0.974	0.935	0.992	0.982	0.976	[0.970-0.982]
InceptV3	0.973	0.932	0.990	0.982	0.973	[0.966-0.980]
Xception	0.976	0.938	0.984	0.984	0.970	[0.962-0.978]
Ensembled Models						
Hard Voting	0.987	0.966	0.992	0.991	0.984	[0.979-0.990]
Average	0.988	0.968	0.993	0.992	0.985	[0.980-0.991]
WE-Net	0.990	0.975	0.995	0.993	0.989	[0.985-0.994]
XGBoost	0.989	0.969	0.991	0.992	0.983	[0.975-0.995]
SVM	0.990	0.973	0.991	0.993	0.985	[0.974-0.995]

Ensemble results of the pre-trained models in Table 3 show improved performance with the weighted-average ensemble (WE-Net) technique out-performing hard voting, averaging, and stacked generalization techniques. The weights assigned are 0.1, 0.2, 0.1, 0.4, and 0.2 to VGG16, ResNet50, InceptionV3, DenseNet-201, and Xception model respectively. This combination performed remarkably well by accomplishing 99% overall accuracy, 0.989 as the area under the ROC curve, and 0.993 as the F1-score. Further, it was noticed that the stacked generalization ensemble with XGBoost and SVM as meta learners provided comparable results to the WE-Net ensemble model considering accuracy, F1-score, and MCC metrics with a marginal difference of [0.0–0.001] and [0.002–0.006] for accuracy and MCC respectively. A shorter CI range by the WE-Net ensemble model [0.985-0.994] signified a minimal margin of error, thus leading to a high precision score compared to other ensemble techniques for binary classification of CXRs into COVID vs. normal.

To locate the salient areas resulting in the classification decision, we have used Grad-CAM implementation. A heat map was generated based on the feature weights and superimposed on the input image. In Fig. 3, a raw COVID positive X-ray image consists of Ground Glass Opacity (GGO) and some consolidation that leads to the concealment of vascular markings of lungs due to the replacement of air spaces by fluid. The figure depicts that our models were able to recognize these patterns with the help of red color corresponding to the pixel with higher feature importance, thereby highlighting the areas in the lung that led to the classification of CXR as Covid-19.

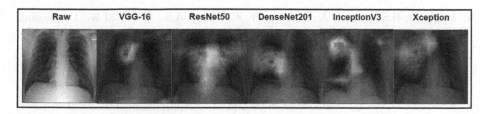

Fig. 3. Visual representations via GRAD-CAM

5 Conclusion

We have built WE-Net, a weighted ensemble DL model to aid the fast screening of COVID-positive patients from CXR images as part of this study. Feeding the ROI to the DL models helps to learn from relevant features. The framework presented in this study uses an ensemble of five state-of-the-art pre-trained models that helped achieve a classification accuracy of 99.02% with an AUC score of 0.989. Also, utilizing the ensemble technique helped provide a generalized model agnostic to noise and over-fitting issues. The results obtained by our framework outperformed the individual pre-trained models and some of the state-of-the-art models presented in the literature. Color visualization using Grad-Cam can further aid radiologists in clinical decision-making for detecting positive COVID-19 cases.

As future work, this study can be extended to multi-class classification problems with an increase in the dataset and supporting classes. Also, we plan to expand our work for patient risk-stratification to help care planning of hospitalized COVID patients.

References

1. WHO Director-General's opening remarks at the media briefing on COVID-19, 11 March 2020. https://www.who.int/director-general/speeches/detail/who-director-general-s-opening-remarks-at-the-media-briefing-on-covid-19---11-march-2020. Accessed Jan 2022
2. Teymouri, M., et al.: Recent advances and challenges of RT-PCR tests for the diagnosis of COVID-19. Pathol. Res. Pract. **221**, 153443 (2021). https://doi.org/10.1016/j.prp.2021.153443
3. Gozes, O., et al.: Rapid AI Development Cycle for the Coronavirus (COVID-19) Pandemic: Initial Results for Automated Detection & Patient Monitoring using Deep Learning CT Image Analysis. arXiv abs/2003.05037 (2020)
4. Li, L., et al.: Using artificial intelligence to detect COVID-19 and community-acquired pneumonia based on pulmonary CT: evaluation of the diagnostic accuracy. Radiology **296**(2), E65–E71 (2020). https://doi.org/10.1148/radiol.2020200905
5. Maghdid, H.S., et al.: Diagnosing COVID-19 pneumonia from x-ray and CT images using deep learning and transfer learning algorithms. In: Proceedings of SPIE 11734, Multimodal Image Exploitation and Learning 2021, p. 117340E (2021). https://doi.org/10.1117/12.2588672

6. Maghdid, H.S., Asaad, A., Ghafoor, K.Z., Sadiq, A.S., Khan, M.K.: Diagnosing COVID-19 pneumonia from x-ray and CT images using deep learning and transfer learning algorithms. Defense + Commercial Sensing (2021)
7. He, K., Zhang, X., Ren, S., Sun, J.: Deep residual learning for image recognition. In: IEEE Conference on Computer Vision and Pattern Recognition (CVPR), pp. 770–778 (2016). https://doi.org/10.1109/CVPR.2016.90
8. Kermany, D.S., Goldbaum, M., Cai, W., et al.: Identifying medical diagnoses and treatable diseases by image-based deep learning. Cell **172**(5), 1122–1131.e9 (2018). https://doi.org/10.1016/j.cell.2018.02.010
9. Singh, D., Kumar, V., Vaishali, et al.: Classification of COVID-19 patients from chest CT images using multi-objective differential evolution-based convolutional neural networks. Eur. J. Clin. Microbiol. Infect Dis. **39**, 1379–1389 (2020). https://doi.org/10.1007/s10096-020-03901-z
10. Ng, M.Y., Lee, E.Y.P., Yang, J., et al.: Imaging profile of the COVID-19 infection: radiologic findings and literature review. Radiol. Cardiothorac. Imaging **2**(1), e200034 (2020). https://doi.org/10.1148/ryct.2020200034
11. Huang, C., Wang, Y., Li, X., et al.: Clinical features of patients infected with 2019 novel coronavirus in Wuhan, China. Lancet **395**(10223), 497–506 (2020). https://doi.org/10.1016/S0140-6736(20)30183-5. Epub 2020 Jan 24. Erratum. In: Lancet. 2020 Jan 30; PMID: 31986264; PMCID: PMC7159299
12. Kundu, R., et al.: ET-NET: an ensemble of transfer learning models for prediction of COVID-19 infection through chest CT-scan images. Multimed. Tools Appl. **81**, 31–50 (2022). https://doi.org/10.1007/s11042-021-11319-8
13. Rajaraman, S., et al.: Iteratively Pruned Deep Learning Ensembles for COVID-19 Detection in Chest X-rays (2020)
14. Quan, H., Xu, X., Zheng, T., Li, Z., Zhao, M., Cui, X.: DenseCapsNet: detection of COVID-19 from X-ray images using a capsule neural network. Comput. Biol. Med. **133**, 104399 (2021). https://doi.org/10.1016/j.compbiomed.2021.104399
15. Yahyatabar, M., Jouvet, P., Cheriet, F.: Dense-Unet: a light model for lung fields segmentation in Chest X-Ray images. In: Annual International Conference of the IEEE Engineering in Medicine and Biology Society, pp. 1242–1245 (2020). https://doi.org/10.1109/EMBC44109.2020.9176033. PMID: 33018212
16. van Ginneken, B., Stegmann, M.B., Loog, M.: Segmentation of anatomical structures in chest radiographs using supervised methods: a comparative study on a public database. Med. Image Anal. **10**(1), 19–40 (2006). https://doi.org/10.1016/j.media.2005.02.002. PMID: 15919232
17. Shiraishi, J., et al.: Development of a digital image database for chest radiographs with and without a lung nodule: receiver operating characteristic analysis of radiologists' detection of pulmonary nodules. AJR Am. J. Roentgenol. **174**(1), 71–74 (2000). https://doi.org/10.2214/ajr.174.1.1740071
18. Jaeger, S., et al.: Two public chest X-ray datasets for computer-aided screening of pulmonary diseases. Quant. Imaging Med. Surg. **4**(6), 475–477 (2014). https://doi.org/10.3978/j.issn.2223-4292.2014.11.20
19. Chowdhury, M.E.H., et al.: Can AI help in screening viral and COVID-19 pneumonia? IEEE Access **8**, 132665–132676 (2020). https://doi.org/10.1109/ACCESS.2020.3010287
20. Rahman, T., Khandakar, A., Qiblawey, Y., et al.: Exploring the effect of image enhancement techniques on COVID-19 detection using chest X-ray images. Comput. Biol. Med. **132**, 104319 (2021). https://doi.org/10.1016/j.compbiomed.2021.104319

21. Mikołajczyk, A., Grochowski, M.: Data augmentation for improving deep learning in image classification problem. In: 2018 International Interdisciplinary PhD Workshop (IIPhDW), pp. 117–122 (2018). https://doi.org/10.1109/IIPHDW.2018.8388338
22. Yadav, G., Maheshwari, S., Agarwal, A.: Contrast limited adaptive histogram equalization based enhancement for real time video system. In: 2014 International Conference on Advances in Computing, Communications and Informatics (ICACCI), pp. 2392–2397 (2014). https://doi.org/10.1109/ICACCI.2014.6968381
23. Keys, R.: Cubic convolution interpolation for digital image processing. IEEE Trans. Acoust. Speech Signal Process. **29**(6), 1153–1160 (1981). https://doi.org/10.1109/TASSP.1981.1163711
24. Lehmann, T.M., Gonner, C., Spitzer, K.: Survey: interpolation methods in medical image processing. IEEE Trans. Med. Imaging **18**(11), 1049–1075 (1999). https://doi.org/10.1109/42.816070
25. Ronneberger, O., Fischer, P., Brox, T.: U-Net: convolutional networks for biomedical image segmentation. In: Navab, N., Hornegger, J., Wells, W.M., Frangi, A.F. (eds.) MICCAI 2015. LNCS, vol. 9351, pp. 234–241. Springer, Cham (2015). https://doi.org/10.1007/978-3-319-24574-4_28
26. Best, N., Ott, J., Linstead, E.J.: Exploring the efficacy of transfer learning in mining image-based software artifacts. J. Big Data **7**, 59 (2020). https://doi.org/10.1186/s40537-020-00335-4
27. Deng, J., Dong, W., Socher, R., Li, L.J., Li, K., Fei-Fei, L.: ImageNet: a large-scale hierarchical image database. In: IEEE Conference on Computer Vision and Pattern Recognition, pp. 248–255 (2009). https://doi.org/10.1109/CVPR.2009.5206848
28. Tang, S., et al.: EDL COVID: ensemble deep learning for COVID-19 case detection from chest X-ray images. IEEE Trans. Industr. Inf. **17**(9), 6539–6549 (2021). https://doi.org/10.1109/TII.2021.3057683
29. Das, A.K., Ghosh, S., Thunder, S., et al.: Automatic COVID-19 detection from X-ray images using ensemble learning with convolutional neural network. Pattern. Anal. Applic. **24**, 1111–1124 (2021). https://doi.org/10.1007/s10044-021-00970-4
30. Frazão, X., Alexandre, L.A.: Weighted convolutional neural network ensemble. In: Bayro-Corrochano, E., Hancock, E. (eds.) CIARP 2014. LNCS, vol. 8827, pp. 674–681. Springer, Cham (2014). https://doi.org/10.1007/978-3-319-12568-8_82
31. Selvaraju, R.R., Cogswell, M., Das, A., Vedantam, R., Parikh, D., Batra, D.: Grad-CAM: visual explanations from deep networks via gradient-based localization. In: IEEE International Conference on Computer Vision (ICCV), pp. 618–626 (2017). https://doi.org/10.1109/ICCV.2017.74

Time to Response Prediction for Following up on Account Receivables in Healthcare Revenue Cycle Management

Rupanjali Chaudhuri[✉], Sai Phani Krishna Parsa, Divya Nagpal, and Kalaivanan R

Cerner Corporation, Bangalore 560045, India
{rupanjali.chaudhuri,saiphanikrishna.parsa,divya.nagpal,
kalaivanan.r}@cerner.com
https://www.cerner.com/

Abstract. Healthcare systems find it difficult to decide on how long they should wait between submitting insurance claims and following up on the Accounts Receivables (AR) for the submitted claims. The solution to this can be the study of payer-specific historical data to understand payment trends of the submitted claims. This data may include censored data points where the responses were not received from the payers and hence is appropriate to be analyzed with Survival Analysis. To aid the follow-up process on submitted claims, we developed and validated a Time to Response (TTR) survival models using the machine learning method Random Survival Forests (RSF) for Medicare, Medicaid, and Commercial payers. TTR models aim to streamline the insurance claim follow-up process for smoother functioning of healthcare systems. These models capture the previous response time patterns based on the covariates and predict the probability of getting a response for each claim filed, from the day of claim file submission. We studied the effect of demographic, geographic, date-time related field, diagnoses, procedural and clean claim covariates on TTR. The model performance was assessed in terms of Concordance Index (C-index) and Integrated Brier score (IBS). Train and test C-index of Medicare was 0.68 and 0.67, Medicaid 0.74 and 0.74, Commercial 0.76 and 0.75 respectively. Test IBS was reported as 0.02, 0.01 and 0.06 for Medicare, Medicaid and Commercial payers respectively.

Keywords: Survival analysis · Time to response · Medicare · Revenue cycle management · Healthcare

1 Introduction

Managing Accounts Receivables becomes challenging for healthcare systems due to day by day increasing number of healthcare services being offered to patients. The payment entry initiates with healthcare systems generating a bill for the medical care rendered to patients. In cases where the patient is covered with insurance, the generated invoices are sent out to insurance companies for payment. The outstanding bills form the healthcare system's Accounts Receivables (AR).

M. Singh et al. (Eds.): ICACDS 2022, CCIS 1614, pp. 124–137, 2022.
https://doi.org/10.1007/978-3-031-12641-3_11

Following up on the AR on a timely basis holds immense importance for healthcare systems to ensure uninterrupted cash flow. Timely payments will further help hospitals in planning the recovery of the patient's owed amount. Any delay in payments from the insurance payers will hamper the entire health care system's functioning as they have to bear operational and maintenance costs [1].

Staff members of healthcare systems should keep a tab on the AR and see if the payments reach on time, for eliminating aged AR. An efficient AR management team will keep track of all insurance claims that have been filed. Further, it will execute an action plan for claims whose payment is delayed beyond a set time-limit [2–4].

However, healthcare systems usually find it difficult to understand the timing of collecting AR thereby causing inefficient deployment of collection resources and difficulty estimating cash flow. Analyzing payer-specific (insurance companies) historical claims to understand the factors affecting the payment response behavior and developing payer-specific AR follow-up prediction models can help overcome this difficulty.

Studying the payer-specific historical data for identifying the factors affecting the corresponding payer's response behavior is challenging due to non-uniformity in response patterns. There may be claims which saw rejections by payers for which response was never received. Also, there may be claims for which response was not received during the time of the study. This leads to censored data in the dataset which can be appropriately modeled using Survival Analysis.

For efficient decision-making in AR follow-up of outstanding claims, we propose "Time to Response" models based on Survival Analysis. "Time to Response" (TTR) means the duration between the claim submission time until a response was heard from the insurance companies. We have developed the TTR models with the aim to recover maximum amounts of AR in as short a time as possible, thereby reducing AR days and maximizing the hospital revenues. These models capture the factors affecting the payer-specific response time patterns and predict the probability of getting a response for each claim filed, from the day of claim file submission. TTR model results along with the bill amount can aid prioritize claims to be followed up [3].

The remainder of this paper is organized as follows. The related work is detailed in Section 2. Section 3 provides a data summary, Sects. 4 and 5 summarizes the features considered and the method followed for data preparation and model development. The results from the implementation and discussion are summarized in Section 6. Section 7 provides information on how the TTR model can be integrated within the AR workflow. Finally, concluding remarks and future scope are summarized in Section 8.

2 Related Work

2.1 Traditional Approach

This approach involves a fixed follow-up time ranging from 20–40 days after the initial claim submission based on the practice. This fixed follow-up time does not account for the fact that different payers have different time to response patterns [2].

2.2 Data-Driven Rule-Based Approach

This approach involves prioritizing claim for follow-up using descriptive data analysis on historical data set to understand the general time taken by different payers in combination with binning of claims based on charge amount. Payers can be grouped in different high-level bins based on historical TTR. However, this approach does not account for other covariates influencing TTR [2].

3 Data Summary

3.1 Key Terminologies

Institutional Claims are the claims, that are generated for services rendered by healthcare institutions such as hospitals, skilled nursing facilities, and any other outpatient and inpatient services. Institutional services include any use of equipment and supplies, laboratory and radiology services, etc., and others. The paper and electronic forms of the institutional claim are "UB-04" and "837I" respectively [5].

Professional claims are the claims, that are generated for services rendered by healthcare professionals and partners such as physicians, suppliers, and other non-institutional providers for both outpatient and inpatient services. The paper and electronic forms of professional claim are "CMS-1500" and "837P" respectively [6].

A Subscriber is a person, who signs for a healthcare insurance plan with insurance company. Subscriber should be uniquely identified to an information source by a Member Identification Number. Subscriber can also enroll his/her dependents in the insurance plan. Hence, it is not always necessary that the subscriber is the patient.

Payment floor is defined as the minimum amount of time a claim must be held by insurance companies before payment can be released to the healthcare institutions [7].

3.2 Data Source

TTR models were developed using data from "Electronic Data Interchange" (EDI) files named 837I for Institutional claims and 837P for Professional claims and the corresponding remits named 835 [8, 9]. The information available in the EDI file is summarized in Table 1. All the data files used for training our model are in x12 data (version 5010) format, which is the American National Standards Institute (ANSI) standard transaction set [10]. The use of this standard data format as an input source gives us the benefit of developing a vendor-agnostic model which can be deployed across different platforms.

3.3 Data Creation

The claims and remits data provided in x12 format was converted into XML format. Further, a sequence of data transformation steps were performed to convert the data in XML format to the pandas data frame. This pandas data frame was the input for the next steps towards model development.

The data is created in such a way that the service line items within a claim are in different rows and the claim-related information is repeated for all the line items. Line

Table 1. EDI file summary

Segment	Description
Transaction header	Contains information about the type of claim (whether it is a professional, institutional, or a dental claim), transaction set control numbers,etc
Billing provider detail	Contains information about provider's name, demography, identification, and currency details
Subscriber detail	Contains information about the subscriber's name, identification, gender, demography, etc
Patient detail	Contains information on the patient's name,gender, demography, and relationship with the subscriber
Claim information loop	Contains information about claim submitter's identifier, claim level charge amount, facility code, claim frequency code, date fields, etc
Service line loop	Contains line level information like line-item control number, Current Procedural Terminology (CPT) codes, date of service, service facility, etc

item, here refers to the entry made for each service provided to the patient insured under a claim. This gives us the flexibility to create claim-level aggregated features. Further, as our, goal is to predict TTR for a given claim, the line level information was converted to claim level information after creating the aggregate features based on the line items data.

3.4 Ground Truth Creation

For model development, the ground truth required is the "time to response" for a claim submitted. Below are the fields considered from claim-remit pair for ground truth creation:

1. BHT04 Data Element in both 837I and 837P files contain the Transaction Set Creation Date - "the date that the original submitter created the claim file from their business application system."
2. BPR16 Data Element in 835 file contain the Cheque Issue or EFT Effective Date (Remit File) - "the date the originating company intends for the transaction to be settled (i.e., Payment Effective Date)."
3. The difference in days between "BPR16" and "BHT04" was taken to derive the label of "time to response".

3.5 Dataset Summary

The dataset prepared for model development after mapping claims to the corresponding remits primarily constituted the claims submitted to Medicare, Medicaid, and Commercial payers. Table 2 below lists down the dataset summary for each of the Health Care Plan.

Table 2. Description of dataset statistics.

Health care plan	Samples count	Event ratio	Time to response		
			Minimum	Median	Maximum
Medicare	24037	98%	1	14	26
Medicaid	14239	99%	3	5	56
Commercial	2781	78%	1	9	77

4 Features Considered

A wide range of features were considered for inclusion in this model. Broadly speaking, they fit into the categories as mentioned in Figure 1.

1. Patient particular details such as gender, age, weight, pregnancy indicator, death date, length of stay in hospital etc., are considered.
2. Geographic features such as patient's city and state, insurance company's city & state, billing provider's city & state are considered.
3. Diagnoses and Procedural features capture the Diagnoses and Procedural information related to the patient. The diagnoses and procedural codes were converted into their respective hierarchical groups using Clinical Classifications Software Refined (CCSR) to reduce cardinality of these features.
4. Date Time Related Features such as hour of the day, weekday, day of the month etc., are derived from the claim creation date and claim creation time data on the claim files. The underlying idea beneath creating these features is to capture any seasonal patterns in the way claim files are adjudicated by the insurance companies. Further, "count of holidays" feature was created specifically for Medicare payers as they follow a 13-day payment floor. However, in the event of a holiday on 14th day, the ideal response date gets shifted based on the number of holidays. Hence, this feature was created to account for this pattern.
5. Code Value Features - There are a rich variety of code value segments such as Diagnosis Related Group (DRG), Condition Information Codes and etc. available in the claim files. All these code values were included into the features set.
6. Line Item Features are created by aggregating the information across all the line items for each claim file. Features such as count of line items per claim and sum, mean, and median values of line item quantities per each claim were created. All

the code values data such as Revenue codes, HCPCS codes and NDC codes were translated as Boolean features at a claim level, to capture whether a line item with a particular code value is available for a given claim or not.

Features Considered

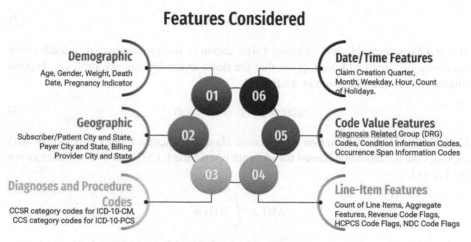

Fig. 1. Features considered

5 Method

5.1 Survival Analysis

Survival analysis is a branch of statistics, that is focused primarily on the time until the occurrence of an event. Survival analysis techniques have the capability to handle Censoring (censored data). Censoring is a form of missing data problem in which some of subjects in experiment do not have the time to event value [11, 12]. Censoring comes in different variants, while the most common being "Right Censoring". Right Censoring is said to occur when either the subjects choose not to continue with the study due to several reasons such as dropout, lost to follow-up, unavailable information, or the study ends before the event of interest occurs before the trial end date [13, 14].

5.2 Survival Analysis Key Terminologies

Survival Function: Let T - a possibly infinite, but always non-negative random lifetime taken from the population under study. The survival function $S(t)$ defines the probability of surviving past time t, as represented in Eq. (1).

$$S(t) = Pr(T > t) \tag{1}$$

1. $0 \leq S(t) \leq 1$

2. $S(t)$ is a non-increasing function of t.

Cumulative Density Function: Cumulative Density Function (CDF) captures the probability of an event happening before time t. Given the survival probabilities $S(t)$, CDF can obtained as per the Eq. (2).

$$CDF = 1 - S(t) \tag{2}$$

Hazard Function: Hazard Function ($\lambda(t)$) captures the probability of a death event occurring occurring at time t, given that the death event has not occurred yet. Hazard function is computed as seen in the Eq. (3).

$$\lambda(t) = Pr(T = t|T \geq t) \tag{3}$$

Cumulative Hazard Function Cumulative Hazard Function (CHF) denoted by $\Lambda(t)$ captures the accumulated hazard for an event up to time t. CHF can be computed as per the Eq. (4).

$$\Lambda(t) = \int_0^t \lambda(i)\, di \tag{4}$$

Further information on survival terminologies and how each of the functions are inter-related can be found at Lifelines python package documentation [15].

5.3 Random Survival Forest

Random Survival Forest (RSF) algorithm is a survival tree based ensemble method for analyzing right-censored survival data [16]. RSF can handle both categorical and numerical features for detecting feature interaction. RSF also supports a large number of predictors and works with small sample size. Also, as demonstrated by Breiman et al. [17] forests based algorithms are robust to noise variables.

Below are the steps involved in the RSF algorithm [16, 18]:

1. Randomly select B bootstrap samples from the original dataset. On an average 37% of the data from the original dataset is excluded in each bootstrap sample, which is called out-of-bag data (OOB data).
2. Grow B survival trees, one from each bootstrap sample. While growing the tree, at each node, p candidate variables (covariates) are selected at random to identify the best candidate variable that maximizes the survival difference between daughter nodes.
3. Grow each tree to full size under the constraint that a terminal node should have no less than a specified number of unique response time values.
4. Compute an ensemble cumulative hazard by taking the average of cumulative hazard information from the B trees.
5. Using OOB data, calculate prediction error for the ensemble cumulative hazard function (CHF).

5.4 Feature Selection

When dealing with a huge set of variables, it is easy to pick up noisy variables that do not generalize well. Our goal is to extract a parsimonious set of features which captures the most meaningful information about "time to response". Redundant, rare, and irrelevant features were dropped by using missingness, variance and correlation analysis conditions. Prior to running RSF feature selection, stratified 10-fold cross validation was used to split the data into train and test sets. For each fold in the 10-fold groups, a "noise" feature was added from Poisson distribution and Gaussian distribution. Later feature importance scores were computed using permutation-combination approach [16, 19, 20]. Using RSF model, train and test set Concordance Index (C-Index) was further computed. Across 10-folds, aggregated feature importance, the number of times a feature was ranked above the "noise" and the number of times a feature was given a positive score were tracked to subset the important features. Further, 10-fold grid search was performed to identify the best RSF parameters using the finalized features. The final set of features and the best parameters selected for RSF model training is summarized in Table 3.

Table 3. RSF model parameters

Models	No. of features	No. of trees	Max depth of tree	Min samples at internal nodes	Min samples at leaf nodes
Medicare	6	200	10	20	2
Medicaid	4	200	5	20	2
Commercial	11	200	15	40	10

5.5 TTR Predictions

The central elements of the RSF algorithm are growing a survival tree and constructing the ensemble CHF [16].

For making TTR predictions, each claim file's features are dropped down each tree in the forest until it reaches a terminal node. Terminal nodes are the most extreme nodes in a saturated tree [16]. Let τ denote all the terminal nodes in a tree. Let $(T_{1,h}, \delta_{1,h}), \ldots,$ $(T_{n(h), h}, \delta_{n(h),h})$ be pairs of the survival times (response times) and censoring information (was there a response for a claim or not) for individuals (cases) in a terminal node h δ. A claim file is said to be right-censored at time $T_{i,h}$ if $\delta_{i,h} = 0$, meaning that there is no response for the claim file by the end of experiment time; otherwise, if $\delta_{i,h} = 1$, meaning that a response was heard for the claim file at $T_{i,h}$. Let $t_{1,h} < t_{2,h} < < t_{N(h),h}$ be the $N(h)$ distinct response times. Define $d_{l,h}$ and $Y_{l,h}$ to be the number of deaths and individuals at risk at time $t_{l,h}$.

$$\hat{H}_h(t) = \sum_{t_{l,h} \leq t} \frac{d_{l,h}}{Y_{l,h}} \tag{5}$$

The CHF estimate of h denoted as $\hat{H}_h(t)$ is obtained using the the Nelson-Aalen [21] estimator as given in equation (5). All claim files within h have the same CHF. Each claim file i has a d -dimensional covariate x_i. Let $H(t|x_i)$ be the CHF for i. To determine this value, drop x_i down the tree. Because of the binary nature of a survival tree, x_i will fall into a unique terminal node $h \in \tau$.

$$H(t|x_i) = \hat{H}_h(t), \; if \; x_i \in h \qquad (6)$$

Equation (6) defines the CHF for all cases and defines the CHF for the tree and is denoted as $H(t|x_i)$. This CHF is derived from a single tree and in order to compute an ensemble CHF, all the CHF values across B survival trees are averaged.

Since the problem statement discussed in this paper is aimed at interpreting the time to death as time until the occurrence of an event, for purpose of convenient interpretation the survival probabilities obtained for each claim file are translated into cumulative distribution function.

$$F(t) = P(T \le t) = 1 - S(t) \qquad (7)$$

Cumulative distribution values (denoted as $F(t)$) are obtained from survival values using equation (7), where S(t) is the survival function. The reason for CDF implementation instead of survival function is primarily because this problem statement demands probability of TTR event before time t for AR follow-up. The provider would be interested in following up with the payer if they are yet to receive response to the submitted claim. This time t of interest would be the time where predicted probability of response is high by the survival model indicating the provider should have ideally got a response for a claim submitted.

6   Results and Discussion

Payer-wise three RSF models were trained for Medicare, Medicaid, and Commercial payers with the final set of features and the best parameters using scikit-survival package [22]. The performance of the developed models was assessed using two measures- the concordance index (C-index) and the integrated brier score.

The C-index represents the global assessment of the survival model's discrimination power. In this use case C-index indicates model's ability to correctly provide a reliable ranking of the survival times based on the time to responses of claims. In general, when the C-index is close to 1, the model has an almost perfect discriminatory power; but if it is close to 0.5, it has no ability to discriminate between low and high time to response claims.

Brier score is a measure of the model's accuracy and evaluates the accuracy of a predicted survival function at a given time t. Integrated Brier Score (IBS) is an overall measure for the prediction performance of the model at all times. The lower the Brier score is the better is a model's performance.

6.1 Performance Evaluation

Stratified grouped 10-fold cross validation grouped by a patient was used to assess the model performance. This approach helps get a fair assessment of model performance ensuring minimal longitudinal correlation between train and test splits. The performance of the models is summarized in Table 4.

Table 4. Model performance

Model	Train C-index	Test C-index	OOB C-index	Train IBS	Test IBS
Medicare	0.68	0.67	0.67	0.05	0.02
Medicaid	0.74	0.74	0.74	0.01	0.01
Commercial	0.76	0.75	0.75	0.06	0.06

We further studied the discriminating capability of the selected features to model the time to response for the submitted claims from each of the payers. The feature importance of selected features is depicted in Fig. 2.

We found conclusive evidence of seasonal pattern driven time to response for Medicare and Medicaid payers. Medicare has a payment floor of 13 days, wherein most claims get cleared on the 14th day from the day it was filed [7]. However, if there are any public holidays in between the processing gets delayed by the number of holidays. We observed count of holidays as the feature with highest importance Fig. 3 (a) captures the variance in predictions depending on the value of "count of holidays" feature. Further, Medicare claims TTR is also affected by other date-time related features and claim charge amount.

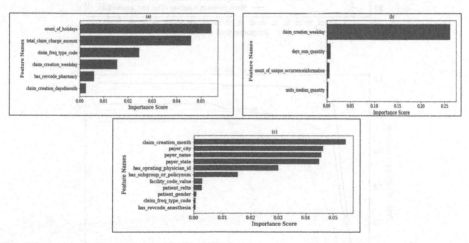

Fig. 2. Final model features. (a): Medicare feature importance; (b): Medicaid feature importance; (c) Commercial feature importance

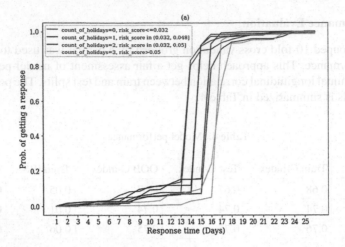

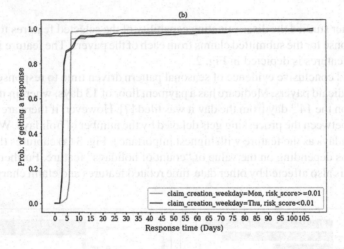

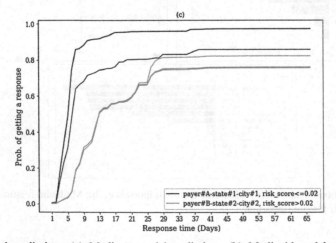

Fig. 3. Model predictions. (a): Medicare model predictions; (b): Medicaid model predictions; (c) Commercial model predictions

Medicaid claims showed a weekly seasonality pattern with claims having bulk clearance on Mondays. This seasonality may differ in different healthcare system's data based on location. This is primarily because Medicaid being a state- based government payer, the time to response and claim clearance patterns may be driven by state policies. Fig. 3 (b) captures the Monday clearance pattern, where in we can see that if the claim files were created on weekday's they have a quicker response time. However, Saturday's and Sunday's are exception as they will be cleared on the second Monday from claim creation date instead of the first Monday(as observed for other claim creation weekdays).

For Commercial payers geographic and demographic features were important as commercial model has several sub-payers showing mixed payer time to response behaviour. Fig. 3 (c) captures the variance in prediction plots given different values for payer name, payer state and payer city.

Please note that for the purpose of interpretation, inverse of actual risk score values are projected as risk scores in Fig. 3. As per survival analysis risk score will be high for subjects with a shorter time-to-event value. To ensure that a claim with higher response time value is appropriately aligned instead with high risk score values aiding to highlight claims at risk to follow up, inverse of risk score values are projected as risk scores.

7 TTR Model Integration in AR Workflow

Our proposed AR follow-up process includes payer specific TTR models with their corresponding features and is depicted in Fig. 4. Here is an outline of AR follow-up process with TTR models:

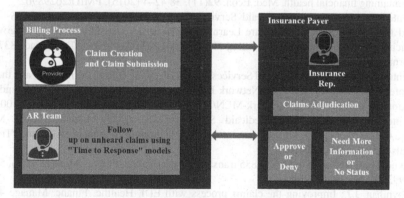

Fig. 4. Proposed RCM workflow

1. Billing Process: Healthcare providers or providers will create the claim file and submit it to the insurance companies for clearance.
2. Insurance Payer: Insurance companies can either choose to settle the claim provided the claim is clean and provides all the required information for settlement. Claims can also be denied, or insurance companies reach out to the healthcare providers in case they need more information.

3. AR Team: AR teams will leverage the payer specific TTR models directly or TTR models output in combination with the "total claim charge amount" to prioritize follow-up on outstanding claims for timely receiving the outstanding amounts.

8   Conclusion and Future Scope

We have developed and validated TTR models using the machine learning method RSF for Medicare, Medicaid, and Commercial payers based on data collected from a Healthcare system. RSF algorithm being simple and easy to make predictions with, is well suited for integration to production environment. However, in future, we would like to perform comparison study between RSF and other algorithms to conclude on research on better prediction performance for healthcare insurance claims.

References

1. Kamenetzky, S.A.: Cash flow management. Arch. Ophthalmol. **111**(6), 757–60 https://doi.org/10.1001/archopht.1993.01090060045021, PMID:8512475
2. Manley, R., Satiani, B.: Revenue cycle management. J. Vasc. Surg. **50**(5), 1232 (2009). ISSN:0741–5214 PMID: 19782521 https://doi.org/10.1016/j.jvs.2009.07.065
3. McDaniel, J.W., Baum, N.: Putting the receive in accounts receivable. J. Med. Pract. Manag. **22**(1), 31–33 (2006). PMID: 16986638
4. Beaulieu-volk, D.: Accounts receivable. Strategies for better management. From eligibility verification to patient engagement, managing accounts receivable is a vital process to maintaining financial health. Med. Econ. **92**(11), 38 42–4 (2015). PMID:26298962
5. Center for Medicare and Medicaid Services(CMS) Medicare Billing Form CMS- 1450 and the 837 Institutional Medicare Learning Network Booklet. https://www.cms.gov/Outreach-andEducation/Medicare-Learning-Network-MLN/MLNProducts/Downloads/837I-FormCMS-1450-ICN006926.pdf
6. Center for Medicare and Medicaid Services (CMS) Medicare Billing Form CMS- and the 837 Professional Medicare Learning Network Booklet: https://www.cms.gov/Outreach-and-Education/Medicare-Learning-Network-MLN/MLNProducts/Downloads/837P-CMS-1500.pdf
7. Center for Medicare and Medicaid Services(CMS) CMS Manual System Medicare Claims Processing. https://www.cms.gov/Regulations-and-Guidance/Guidance/Transmittals/Downloads/R114CP.pdf
8. Romo, E.: Adapting to the ANSI 835 transaction set. Healthc. Financ. Manage **47**(1), 54–56 (1993). PMID:10145738
9. Moynihan, J.J.: Improving the claims process with EDI. Healthc. Financ. Manage **47**(1), 48–49 (1993). PMID: 10145737
10. Clement, J.M., Marc, O.J., Paul, D., Blaine, T., Jeffrey, G.S.: What is done, what is needed and what is realistic to expect from medical informatics standards. Int. J. Med. Inform. 48(1–3), 5-12. ISSN 1386-5056 (1998). https://doi.org/10.1016/S1386-5056(97)00102-0
11. Prinja, S., Gupta, N., Verma, R.: Censoring in clinical trials: review of survival analysis techniques. Indian J. Comm. Med. Official Publ. Indian Assoc. Prev. Soc. Med. **35**(2), 217–221 (2010). https://doi.org/10.4103/0970-0218.66859
12. Wikipedia Survival Analysis. https://en.wikipedia.org/wiki/Survival_analysis
13. Rich, J.T., et al.: A practical guide to understanding Kaplan-Meier curves. Otolaryngol Head Neck Surg. **143**(3), 331–336 (2010). https://doi.org/10.1016/j.otohns.2010.05.007

14. Enrique, B., Laura, B.: Effect of right censoring bias on survival analysis. arXiv2012.08649 (2020)
15. Lifelines Documentation. https://lifelines.readthedocs.io/en/latest/
16. Ishwaran, H., Kogalur, U.B., Blackstone, E.H., Lauer, M.S.: Random survival forests. Ann. Appl. Stat. **2**, 841–860 (2008). https://doi.org/10.1214/08-AOAS169
17. Breiman, L.: Random forests. Mach. Learn. **45**, 5–32 (2001)
18. Janika, S.: Modelling Late Invoice Payment Times Using Survival Analysis and Random Forests Techniques. Financial and Actuarial Mathematics Curriculum Master's thesis, Institute of Mathematics and Statistics, University of Tartu
19. Eli5 Package Document. https://eli5.readthedocs.io/en/latest/blackbox/ permutation_importance.html
20. Lars, B., Gilles, L., et al.: API design for machine learning software: experiences from the scikit-learn project. In: ECML PKDD Workshop: Languages for Data Mining and Machine Learning, pp. 108–122 (2013)
21. Ørnulf, B.: Nelson-aalen estimator. Encycl. Biostatis. (2005). https://doi.org/10.1002/047001 1815.b2a11054
22. Pölsterl, S.: scikit-survival: a library for time-to-event analysis built on top of scikit-learn. J. Mach. Learn. Res. **21**(212), 1–6 (2020)

A Deep Learning Approach for Automated Detection and Classification of Alzheimer's Disease

Deepthi K. Oommen[✉] and J. Arunnehru

Department of Computer Science and Engineering, SRM Institute of Science and Technology,
Vadapalani, Chennai, Tamilnadu, India
do6523@srmist.edu.in

Abstract. Alzheimer's Disease is progressive dementia that begins with minor memory loss and develops into the complete loss of mental and physical abilities. Memory-related regions of the brain, such as the entorhinal cortex and hippocampus, are the first to be damaged. A person's mental stability is harmed as a result of the severity. Later on, it affects cortical regions that engage with language, logic, and social interaction. Subsequently, it spreads to other parts of the brain, resulting in a substantial reduction in brain volume. Although computer-aided algorithms have achieved significant advances in research, there is still room for improvement in the feasible diagnostic procedure accessible in clinical practice. Deep learning models have gone mainstream in the latest decades due to their superior performance. Compared to typical machine learning approaches, deep models are more accurate in detecting Alzheimer's Disease. To identify the labels as demented or non-demented, the researchers used the Open Access Series of Imaging Studies (OASIS) dataset. The novelty comes in doing extensive research to uncover crucial predictor factors and then selecting a Deep Neural Network (DNN) with five hidden layers and carefully tuned hyper-parameters to achieve exemplary performance. The assertions were supported by evidence of 90.10% correlation accuracy at various iterations and layers.

Keywords: Alzheimer's Disease · Dementia · Deep learning model · Hyper-parameters

1 Introduction

Alzheimer's disease (AD) is defined by disordered brain function caused by the deterioration of neurons. The condition originates in the temporal lobe of the brain hemisphere and extends to other areas of the brain. AD is one of the most critical public health conditions globally and a widespread neurodegenerative phenotype. Multiple cognitive impairments, such as memory, executive control, linguistics, and visuospatial skills, form Alzheimer's disease [1]. According to Alzheimer's Association research, 10% to 20% of adults aged 45 or older have Mild Cognitive Impairment (MCI), an introductory stage of AD. The Mental Health Gap Action Program designated dementia as a high-priority health problem that requires immediate attention [2].

M. Singh et al. (Eds.): ICACDS 2022, CCIS 1614, pp. 138–149, 2022.
https://doi.org/10.1007/978-3-031-12641-3_12

Dementia is not the same as the usual mental decline with age [3]. Dementia patients frequently lose track of time and place, and they may forget to eat regularly or maintain proper hygiene. It's an expected fact that the aetiology of AD and other brain disorders involves the destruction of brain cells, which can be seen through imaging The cerebral cortex, shrinkage is a factor for Alzheimer's disease. The formation of amyloid plaques and neurofibrillary tangles culminates in irreversible atrophy. Amyloid protein plaques adhere to and damage neuron synapses (the connections between neurons), causing memory lapses, which is an onset of Alzheimer's disease. When a protein strand is twisted, neurofibrillary tangles develop, causing damage to neurons and nerve synapses [4].

Furthermore, the WHO's prevalence rate projections indicate that the proportion of individuals living with dementia would continue to climb, primarily among the elderly. According to the Alzheimer's Association's 2019 global statistics, 47 million individuals worldwide have Alzheimer's Disease; As a result, it has become a substantial public health issue in contemporary society. Furthermore, the total number of cases is predicted to reach 76 million by 2030 [5]. The increasing number of AD cases is attributed to various factors, including population expansion, ageing, and changing social and economic development behaviours. At present, there is no treatment available that can halt or reverse this disease's progression. However, if detected early enough, the condition can be controlled. Diagnosis of AD patients usually with brain imaging techniques such as MRI scans combined with clinical exams that search for indications of memory impairment [6]. Many explorations in the writing have investigated the potential outcomes of Deep Learning (DL) approaches in the domain of neurodegenerative illnesses like Alzheimer's Disease. The use of information created from attractive reverberation imaging (MRI) or positron emanation tomography (PET) has been widely read up for this reason [7]. Brain shrinkage, particularly in the hippocampal regions, shows the loss of cognitive abilities associated with AD. The main goal of this research is to detect Alzheimer's Disease with a deep neural network model using brain image analysis and clinical examination.

The mini-mental state examination (MMSE), clinical dementia rating (CDR), and Alzheimer's disease assessment scale (ADAS) are all longitudinal clinical tests that have shown a significant association with Alzheimer's disease progression [8]. The mapping of this link is advantageous since it enables practitioners to act during the undiagnosed stage. While Alzheimer's disease has no therapy, it can be delayed if symptoms are handled early enough. Memory medications, behaviour mediation, sleep and occupation therapies, and supported living are common treatments [9, 10].

The current work incorporates the prediction of dementia, the OASIS dataset is utilized to perform a hierarchical analysis of all accessible data points. Experimentation with various deep neural network parameters is carried out for achieving the best parameter for the DNN model, and compare the results of the proposed model to those of previously implemented deep learning and machine learning classifiers to demonstrate its efficacy.

The following describes the structure of this paper: Sect. 2 discusses the related works, Sect. 3 depicts a summing-up of the proposed methodology and the explanation of the Deep neural network, Sect. 4 discloses the experiment results, which describes the dataset and predictor variables and experiment work performed with the deep neural network on the given dataset and results explains performance measure of the classifier and accuracy results with confusion matrix and ROC curve inculcated to depicts the effectiveness of proposed work. Finally, Sect. 5 inscribes the conclusion with future work.

2 Research Background

This division aims to evaluate the recent approaches in predicting and classifying Alzheimer's disease using predictor variables with the help of different types of computational models. To trace the progression of Alzheimer's Disease throughout all six phases, complex nonlinear multifactorial multimodal modelling is necessary. The reduction of preclinical MMSE scores in adults over 75 years old (n = 528) was simulated using growth mixture modelling (GMM) [11]. The study used a subset of the OASIS data to develop approaches for identifying binary-coded AD using Eigenbrain imaging It excluded persons under the age of sixty and data that was incomplete, defined Alzheimer's Disease, confined their analysis with a few numbers of a sliced image of the cortex, and performed a cross fold validations from 10–50 [12]. For 1,000 patients, the MMSE scores and the ADAS scale were used to predict AD symptom development classes over six years. To predict AD progression, the researchers employed MR Imaging, genetic proteins and clinical-based data, and a Siamese neural network with two identical equally weighted subnetworks [8].

In paper [13, 14] the electronic health record scores were utilized to detect dementia. A cross-sectional initial data collection, ADNI registry data sets, MRI measures, and artificial neural network for dementia prediction were also utilized by the researchers. Another analysis created discriminating maps using MRI brain images to implement perfusion technique in the brain tissues and its scores are utilized to properly identify AD and its stages.

The paper [15] developed a deep learning strategy based on sparse autoencoders and 3D convolutional neural networks. An MRI scan should predict if an individual has Alzheimer's disease or has a healthy brain. A workflow for extracting multivariate neuroimaging features for multiclass AD diagnosis was provided in the article. It developed a deep-learning model that maintains all relevant information stored in imaging data using a zero-masking methodology. It had an accuracy rate of 87% [16]. The research discussed in the paper, [17, 18] utilized profound learning-based pipelines and neuroimages to recognize Alzheimer's infection from sound controls records for a specific age range. It was almost difficult to recognize Alzheimer's patients from solid minds.According to the study in paper [19, 20], biomarkers for predicting progression of AD from Mild Cognitive Impairment (MCI) include volumetric MRI-based assessments of regional brain atrophy, notably medial temporal volume loss. For decades, the most extensively used screening test for identifying AD is MMSE. It is not capable of detecting MCI, but it can detect AD at a much higher rate. The hippocampus volume, entorhinal cortex thickness, middle temporal gyrus, and retrosplenial cortex weightings for each ROI obtained using a linear discrimination analysis are used to generate the atrophy score. the greatest way to tell the difference between Alzheimer's sufferers and healthy controls.

Deep learning (DL) has shown promise in clinical information systems for diseases such as diabetes, cancer, and Alzheimer's disease through statistics and imaging analysis. DL's key advantage over other shallow learning approaches is it's capable of learning the best forecasting characteristics directly from the raw information given a set of labelled cases.. In the case of incomplete data, deep learning approaches can help with training and prediction [21].

This work proposes a deep neural network-based automated AD recognition method. The goal is to develop deep neural network models that can accurately identify the existence of Alzheimer's Disease utilizing deep learning techniques combined with clinical evaluations and brain atrophy volume.

3   Proposed Methodology

Our work employed a Deep Neural Network (DNN) model for training and testing, Alzheimer disease prediction criteria called predictor variables to categorise demented and non-demented persons. The neural network model's hidden layers learn the categories incrementally. Figure 1 depicts the proposed deep neural network model for predicting Alzheimer's Disease by taking into account risk variables. The AD Predictor variables dataset contains a set of seven parameters. It can be considered as two sets; the first set typically analyses a subject's cognitive ability. The second is to provide brain volume statistics, as it plays a more significant role in identifying AD patients because the particular disease can shrink the brain. The dataset contains information about both genders. Inconsistent and improper data have been deleted with the benefits of data preprocessing. Feature Elimination was used to delete features that are no longer relevant.

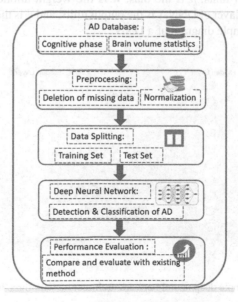

Fig. 1. The proposed model architecture.

A DNN-based prediction model was constructed to detect demented and non-demented people to locate the AD afflicted patients. The model's performance was assessed by comparing it to other best deep and machine learning models found in prior studies. The model has used a layout of 400 data, given a training-test ratio of 70:30 and used five hidden layers to acquire its best performance in accurately classifying the data.

3.1 Deep Neural Network

The current study suggested a DNN model predict AD using a dataset of predictor factors. A DNN is a type of artificial neural network (ANN) that contains numerous hidden layers between the input and output layers and has lately become popular for classification. Each unit in a DNN receives connections from all units in the preceding layer since it is ultimately linked. As a result, each unit has its own bias and weight in two consecutive layers for every pair of units. The net input was computed by multiplying each input by its weight and adding the results. Each unit applied the activation function to the net input in the hidden layer. As a result, the computation of a network with hidden layers and its output layer may be described as in Eqs. 1 & 2.

$$h_i = \phi \left(\sum_j W_{ij} x_j + b_i \right) \tag{1}$$

$$y_i = \phi \left(\sum_j W_{ij} h_j + b_i \right) \tag{2}$$

The x is the input units, b is the bias, w is the weight allotted to the network, h represents the hidden layer of DNN, y is the output units, and ϕ is the activation function. Figure 2 depicts a sample diagram of a DNN.

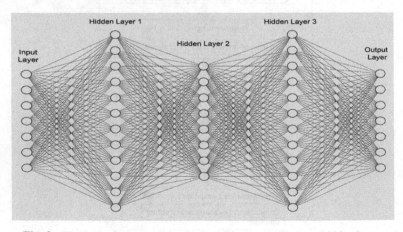

Fig. 2. A sample deep neural network architecture with three hidden layers

4 Experimental Setup

The section describes the set of parameters explored for the proposed work. The experiments are conducted under Windows 10 with an Intel i7 processor of 2.10 GHz with 8 GB of RAM. The software requirements used for the model is set with Python (3.7.6) through the Keras (2.4.3) framework and as Tensorflow (2.3.1) the backend with the libraries of the deep learning model. Many investigations have investigated the capability of Deep Learning (DL) procedures in the field of neurodegenerative problems, for example, Alzheimer's DiseasOne of the most difficult parts of building profound brain networks models is picking the mixes of hyper-boundaries utilizing a circling methodology to expand the exactness and productivity of the model. Profound Learning with matrix search helps in choosing the reasonable boundaries to acquire the best expectation model that keeps away from overfittinge.. Our Neural Network has explored various ranges of hidden layers with a different set of nodes, learning rates, optimizers, activation functions, batch size, epochs, dropouts and output functions to build the network to obtain a network with fine-tuned parameters. The model with different boundary ranges is recorded in Table 1.

Table 1. Hyperparameter setting for the proposed network.

Hyper-parameters	Values explored
Hidden layers	3, 5, 7, 10, 15
Optimizers	SGD, Adam, Adamax, Adagrad
Activation functions	Sigmoid, Relu, Leaky Relu, Tanh
Number of nodes	12, 16, 20
Number of epochs	100, 150, 200, 250, 300, 350, 400
Learning rate	.0001–.01
Batch size	16, 32, 64, 128
Output function	Softmax

4.1 Dataset

The proposed model is tested using data from the Open Access Series of Imaging Studies (OASIS) [22]. The data set consists of 400 observations from 142 participants ranging in age from 58 to 96. There are both men and women in the dataset, and all are right-handed. Medical statistics, such as intracranial volumes (eTIV) and brain volumes (nWBV) of the participants, are included in addition to the Electronic Health Records(HER) such as Clinical Dementia Rate (CDR) and Mini-Mental State Examination (MMSE). The CDR scale has four levels: 0, 0.5, 1, and 2 respectively. A CDR of 0 indicates that the patient is not mentally ill. Very mild dementia, mild dementia, and severe dementia are all represented by CDR values of 0.5, 1 and 2. This medical test is significant in determining whether or not a person is an Alzheimer's patient. The MMSE is a 30-question questionnaire designed to assess people's cognitive abilities in arithmetic, memory, and orientation. Demented and non-demented controls are the two categories considered in the current work for predicting Alzheimer's Disease [23].

4.2 Predictor Variables

- MMSE (Mini-mental-state-examination): The MMSE is a 30-point valid and reliable questionnaire for detecting dementia.It asses the mental stability based on language skills, visual-spatial skills, orientation skills, memory and attention
- The Clinical Dementia Rating (CDR): It is an ordinarily involved clinical method for rating dementia seriousness, with values going from 0 (no impedance) to 3 (extreme weakness). It checks a singular's judgment ability, critical thinking abilities, local area undertakings, individual consideration and leisure activities to figure out mental dependability.
- Estimated Total Intracranial Volume (eTIV) aka Intracranial volume (ICV): It is frequently employed in volumetric brain research as a measure of pre-morbid brain size as well as to account for inter-subject variability in head size.
- Normalized whole brain volume (nWBV): This element represents the total mass of the brain. Individual subject data benefit from brain normalization since it enables reporting observed active areas in a standard spatial coordinate system.
- The Atlas scaling factor (ASF): It's a parameter scaling factor that enables eTIV comparisons based on human anatomical differences. Table 2 summarizes the dataset's characteristics.

Table 2. Features and the associated range of values in the OASIS database for AD.

Factors	Description	Value ranges
Subject ID	Each subject is assigned a unique identifier	1–142 (Number of subjects)
Age	Represents human growth factor	60–96
Group	For labelling the class	Non-demented controls or demented controls
Gender	Represents the main category of the human being	Male or Female
MRI ID	Each test is identified by a unique identity. For a single person, there might be many MRI IDs	1–354
CDR	Clinical dementia rating	0–2
MMSE	Mini-mental state examination	0–30
eTIV	Estimated total intracranial volume	1488 ± 176.13

(*continued*)

Table 2. (*continued*)

Factors	Description	Value ranges
nWBV	Whole-brain volume normalised as a percentage of all voxels ("constant" for an estimate of total intracranial volume)	0.730 ± 0.037
ASF	For the size of the brain, there is a volume scaling factor:- Atlas Scale Factor	1.195 ± 0.138

4.3 Performance Measures

The proposed method's performance was assessed using four metrics for quantitative evaluation and comparison as Accuracy, Sensitivity, Recall or Specificity, Precision and F1-Score. The following Table 3 describes the definitions for each of these parameters.

Table 3. Performance measures.

Parameter	Equation
Accuracy	$A = \left[\frac{tp+tn}{tp+fp+tn+fn}\right]$
Precision	$P = \left[\frac{TP}{TP+FP}\right]$
Recall	$R = \left[\frac{TP}{TP+FN}\right]$
F1 Score	$F1 - score = 2\left[\frac{P \times R}{P+R}\right]$

The TP, TN, FP, FN in the table indicates True Positive shows the number of subjects categorized correctly. True Negatives represents the number of subjects who were incorrectly classified. False Positives denotes the number of subjects classified as non-demented controls and False Negatives means the number of subjects who were non-demented is classified as demented controls.

5 Results and Discussion

Our approach used the predictor variables from the OASIS dataset and train the features in a deep neural network model with five hidden layers, with distinct neurons in each. The network optimizes its weight with a learning pace of .001 and a batch size of 32 to lower the loss function. The predictor variables are iterated for 400 epochs using the categorical cross-entropy and Adam optimizer for compiling the Keras framework. The network used Rectified Linear Unit (ReLu) as the activations for all the units in the hidden layer and considered the softmax for the output function to optimize the network performance. Table 4 Illustrates the specifics of the proposed DNN for AD's improved hyper-parameters. The performance of our model is assessed by training the deep network with 15 layers and 10 layers but has shown a significant decline of accuracy rate. As a result, we have chosen our architecture with few layers and it reciprocated with an accuracy of 90.10%. The acquired features of the OASIS dataset focused on two main criteria the Electronic Health Records (EHR) of the cognitive abilities and the brain volumes of each person for predicting AD. The proposed model has exhibited a remarkable inference in classifying the demented and non-demented controls. F1-Score.

Table 4. Optimized hyper-parameter applied in the DNN Model

Hyperparameter	Optimized value
Hidden layers	5
Activation function	ReLu
Learning rate	.001
Maximum iteration	400
Optimization algorithm	Adam
Batch size	32
Output function	Softmax

The performance metrics for the binary classification of AD detection with the deep neural network model is depicted in Table 5. The study has concentrated a dataset of size 400 from OASIS with seven parameters to obtain the demented and non-demented subjects, with a training and test ratio of 70 and 30. The chosen parameters were optimum enough to do binary classification with the proposed network model. As shown in Table 5, the model acquired an accuracy of 90.10%, precision with 89.5%, Recall of 91.5% and F1score of 90.5%.

Table 5. Model performance acquired for the AD detection

Class	Accuracy	Precision	Recall	F1 score
Non-demented	.92	.89	.92	.91
Demented	.88	.90	.91	.90

In addition to the model performance, the complete Confusion Matrix (CM) for the OASIS dataset is shown in Fig. 3. It can derive a conclusion about the performance of the classification algorithm. The model displays better accomplishment in analyzing the non-demented compared to the demented group. As present in CM, the model predicted true negative values and true positive values as .92 and .88. It describes that the non-demented class is predicted well, whereas 12% of the demented class is misplaced as non-demented, which needs more focus in further work. Figure 4 Depicts the ROC curve for the binary classification AD detection. The Receiver Operating Characteristics (ROC) defines a trade-off between the two-parameter metrics sensitivity (True Positive Rate) and specificity (1-False Positive Rate). Its purpose is to identify the accuracy rate of given classifiers. In this method, the ROC curve depicted an analysis based on a TPR and FPR by considering AD parameters for determining demented and non-demented controls. The correlation of area under the curve (AUC = 1) classifies the ROC curve in visualizing a high assurance in accurately organizing healthy controls and affected subjects.

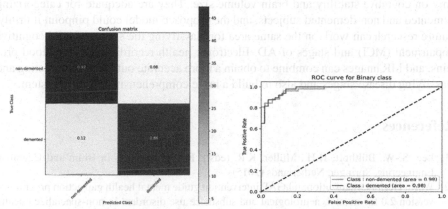

Fig. 3. Confusion matrix: the classification of demented Vs non-demented controls

Fig. 4. ROC curve for AD classification

6 State of the Art Comparison

The outcomes acquired by the DNN with indicator factors for AD discovery are quantitatively contrasted and cutting edge outcomes with the latest discoveries and examinations. The near outcomes are introduced in Table 6. The examination grasps that the proposed strategy shows the best outcome.

Table 6. State of the art recognition accuracy for OASIS dataset.

Method	Accuracy	Methodology	Dataset
Proposed	90.10%	DNN	OASIS
Battineni et al. [15]	68.75%	SVM	OASIS
Priyanka et al. [16]	87%	SVM	OASIS
Hamzah et al. [17]	78%	ANN & ML	OASIS

7 Conclusion

The proposed work has exhibited a productive way to deal with Alzheimer's Disease finding utilizing a bunch of indicator factors that are clinically supported. The work examined the viability of the profound DNN model in grouping AD and sound control subjects.The current DNN model of five hidden layers and the fine-tuned hyperparameters could achieve 90.10% accuracy in the binary classification for diagnosing AD. The model has undergone various iterations with distinct parameters to finalize the best-tuned parameters. The predictor variables stated in the current paper focused only on cognitive stability and brain volume size. They are adequate for categorizing demented and non-demented subjects, and the proposed model could pinpoint it firmly. Future research can work on the same area for classifying the stages of mild cognitive impairment (MCI) and stages of AD. Electronic health records, gene and blood proteins, and MR images can combine to obtain a more accurate outcome in predicting and classifying disease stages and also to build a more comprehensive e-health system.

References

1. Lee, S.-W., Bülthoff, H.H., Müller, K.R. (eds.): Recent Progress in Brain and Cognitive Engineering. Springer, Netherlands (2015)
2. World Health Organization: mhGAP intervention guide mental health gap action programme version 2.0 for mental, neurological and substance use disorders in non-specialized health settings. World Heal Organ **2016**, 1–173 (2016)
3. Alzheimer's Facts and Figures Report | Alzheimer's Association https://www.alz.org/alzheimers-dementia/facts-figures. Accessed 21 Oct 2021
4. What Happens to the Brain in Alzheimer's Disease? | National Institute on Aging. https://www.nia.nih.gov/health/what-happens-brain-alzheimers-disease. Accessed 21 Oct 2021
5. Almustafa, K.M.: Classification of epileptic seizure dataset using different machine learning algorithms. Inform. Med. Unlocked **21**, 100444 (2020)
6. Cabral, C., Margarida, S.: Classification of Alzheimer's disease from FDG-PET images using favourite class ensembles. In: 2013 35th Annual International Conference of the IEEE Engineering in Medicine and Biology Society (EMBC), pp. 2477–2480. IEEE (2013)
7. Oommen, D.K., Arunnehru, J.: A comprehensive study on early detection of Alzheimer disease using convolutional neural network. In: Proceedings of AIP Conference, vol. 2385, no. 1. AIP Publishing LLC (2022)
8. Bhagwat, N., Viviano, J.D., Voineskos, A.N., Chakravarty, M.M., Initiative, A.D.N.: Modeling and prediction of clinical symptom trajectories in Alzheimer's disease using longitudinal data. PLoS Comput. Biol. **14**(9), e1006376 (2018)
9. Huang, L., Gao, Y., Jin, Y., Thung, K.-H., Shen, D.: Soft-split sparse regression based random forest for predicting future clinical scores of alzheimer's disease. In: Zhou, L., Wang, Li., Wang, Q., Shi, Y. (eds.) MLMI 2015. LNCS, vol. 9352, pp. 246–254. Springer, Cham (2015). https://doi.org/10.1007/978-3-319-24888-2_30
10. Arunnehru, J., Geetha, M.K.: Motion intensity code for action recognition in video using PCA and SVM. In: Prasath, R., Kathirvalavakumar, T. (eds.) MIKE 2013. LNCS (LNAI), vol. 8284, pp. 70–81. Springer, Cham (2013). https://doi.org/10.1007/978-3-319-03844-5_8
11. Bansal, D., Chhikara, R., Khanna, K., Gupta, P.: Comparative analysis of various machine learning algorithms for detecting dementia. Procedia Comput. Sci. **132**, 1497–1502 (2018)

12. Collij, L.E., et al.: Application of machine learning to arterial spin labelling in mild cognitive impairment and Alzheimer disease. Radiology **281**(3), 865–875 (2016)
13. Bhagwat, N., et al.: An artificial neural network model for clinical score predictionin Alzheimer disease using structural neuroimaging measures. J. Psychiatry Neurosci. JPN **44**(4), 246 (2019)
14. Arunnehru, J., Chamundeeswari, G., Prasanna Bharathi, S.: Human action recognition using 3D convolutional neural networks with 3D motion cuboids in surveillance videos. Procedia Comput. Sci. **133**, 471–477 (2018)
15. Battineni, G., Chintalapudi, N., Amenta, F.: Machine learning in medicine: Performance calculation of dementia prediction by support vector machines (SVM). Inform. Med. Unlocked **16**, 100200 (2019)
16. Priyanka, G., Priya, R.T., Vasunthra, S.: An effective dementia diagnosis system using machine learning techniques. J. Phys. Conf. Ser. **1916**(1). IOP Publishing (2021)
17. Hamzah, H.A., Diazura, A.M., Alfatah, A.E.: Design of Artificial Neural Networks for Early Detection of Dementia Risk Using Mini-Mental State of Examination (MMSE) (2020)
18. Vaijayanthi, S., Arunnehru, J.: Synthesis Approach for Emotion Recognition from Cepstral and Pitch Coefficients Using Machine Learning, pp. 515–528. Springer, Singapore (2021)
19. Sarraf, S., Tofighi, G.: Alzheimer's disease neuroimaging initiative. DeepAD: alzheimer's disease classification via deep convolutional neural networks using MRI and fMRI. BioRxiv, 070441 (2016)
20. Arunnehru, J., Kalaiselvi Geetha, M.: Automatic human emotion recognition in surveillance video. In: Dey, N., Santhi, V. (eds.) Intelligent Techniques in Signal Processing for Multimedia Security. SCI, vol. 660, pp. 321–342. Springer, Cham (2017). https://doi.org/10.1007/978-3-319-44790-2_15
21. Duc, Nguyen T., et al.: 3D-deep learning based automatic diagnosis of Alzheimer's disease with joint MMSE prediction using resting-state fMRI. Neuroinformatics **18**(1), 71–86 (2020)
22. Venugopalan, J., et al.: Multimodal deep learning models for early detection of Alzheimer's disease stage. Sci. Rep. **11**(1), 1–13 (2021)
23. LaMontagne, P.J., et al.: OASIS-3: longitudinal neuroimaging clinical, and cognitive dataset for normal aging and alzheimer disease. medRxiv 2013 (2019)

A Real-Time Driver Assistance System Using Object Detection and Tracking

Jamuna S. Murthy[✉], Sanjeeva S. Chitlapalli, U. N. Anirudha, and Varsha Subramanya

Department of ISE, B N M Institute of Technology, Bangalore, India
jamunamurthy.s@gmail.com

Abstract. ADAS (Advanced Driver Assistance System) has become a vital part of the driving experience. In recent years, there have been several advancements in ADAS technology such as parking assistance and lane detection. The proposed work presents a real-time Driver assistance framework by implementing the state-of-the-art object detection algorithm YOLOv4. This paper provides a comparison between and other state-of-the-art object detectors. Comparison is done based on mean average precision (mAP) and frames per second (FPS) on three different datasets and one standard dataset. YOLOv4 proves to be faster and more accurate than the other object detection algorithms in the comparison. This framework is used to build an application which helps users make better decisions on the road. This application consists of a simple user interface that displays alerts and warnings.

Keywords: Object detection · ADAS · LIDAR sensor · CNN · YOLOv4

1 Introduction

According to WHO, approximately 1.3 million people die each year due to road traffic crashes. With a rise in accidents and with the increase in the number of vehicles, ADAS (Advanced Driver Assistance System) has become a vital part of the driving experience. Prior warnings seconds before an incident can help the driver handle the situation in a better manner. ADAS has emerged as an extremely vital tool with respect to safety in the automobile industry. Notable automotive giants have stepped in to integrate ADAS into their models. Existing ADAS technologies operate on sensors such as LIDAR for the object detection module.

Realizing the importance of timing information, the proposed frame was introduced. Our proposed framework is applied to build an interactive application. The proposed framework assists the user by notifying them with unique alerts and warnings. The proposed framework aims towards providing Alerts and warnings a few seconds prior based on the Real-time data. In order to build such a system, YOLOv4 was implemented by analyzing multiple Object Detection Systems based on their speed and accuracy. Our proposed framework includes a system that will be able to assist drivers in compromising situations by giving a heads up with significant speed and accuracy.

M. Singh et al. (Eds.): ICACDS 2022, CCIS 1614, pp. 150–159, 2022.
https://doi.org/10.1007/978-3-031-12641-3_13

2 Literature Review

Numerous researches are done on different aspects of ADAS and Autonomous vehicles. The Simultaneous Localization And Mapping (SLAM) techniques along with the Detection And Tracking Of Moving Objects (DATMO) techniques were used in autonomous vehicles to solve the problem of object detection by Wang C C et al. (2003), by using a laser scanner [1]. The different observed shapes on each laser scan made it difficult to identify the object. Wang C C et al. (2005) provided a night-vision-based driver assistance system by using their proposed system that performs vehicle recognition as well as lane detection [2]. ADAS also includes Driver Monitoring Systems. Driver Monitoring System (DMS) helps in keeping track of various facial features of the driver like eyelid and mouth movement. One such system was proposed by Shaily S et al. (2021) [3].

Lane detection, being a basic problem in ADAS, has many challenges such as for instance-level discrimination and detection of lane lines with complex topologies. Liu, L et al. (2021) proposed CondLaneNet. The CondLaneNet framework first determines the lane instances and thereafter generates the line shape for every instance dynamically is predicted. A conditional lane detection strategy was introduced by them based on row-wise formulation and conditional convolution to solve instance-level discrimination and in order to tackle detection of lane lines having complex topologies, including fork lines and dense lines, they designed the Recurrent Instance Module (RIM) [4].

In the field of autonomous vehicles, Manoharan S (2019) proposed a better safety algorithm for artificial intelligence [5]. There is a lot of research done in Object Detection since it plays a crucial role in many of the technologies. To get a better understanding of state-of-the-art object detection techniques and models, cloud-based. Liu L et al. (2016) conducted a survey of most of the research that provides a clear picture of these techniques. The main goal of this survey was to recognize the impact of deep learning techniques in the field of object detection that has led to many groundbreaking achievements. This survey covers many features of object detection ranging from detection frameworks to evaluation metrics [6, 7].

For many region-based detectors, like Fast R-CNN [8], a costly per-region sub-network is applied several times. In order to address this, Dai J et al. (2016) introduced R-FCN by proposing location-sensitive score maps to address a dilemma between translation-invariance in image classification and translation-variance in object detection [9]. One of the major challenges of object detection was to detect and localize multiple objects across a large spectrum of scales and locations, due to which the pyramidal feature representations were introduced. In this, an image is represented with multiscale feature layers. Feature Pyramid Network (FPN), one such model to generate pyramidal feature representations for object detection, presents no difficulty and as well as effective but may not be the optimal architecture design. For image classification in a vast search space, the Neural Architecture Search (NAS) algorithm demonstrates favorable results on the productive discovery of outstanding architectures. Hence, inspired by the modularized architecture proposed by Zoph et al. (2018), Ghiasi G et al. (2019) proposed the search space of scalable architecture that generates pyramidal representations. They proposed an architecture, called NAS-FPN, that provides a lot of flexibility in building object detection architecture and is adaptable to a variety of backbone models, on a wide range of accuracy and speed tradeoffs [10].

Various detection systems repurpose classifiers by taking a classifier for an object and then evaluating it again at multiple locations scales and locations in a test image. For example, R-CNN uses region proposal methods to first produce bounding boxes that are likely to appear in an image and then, on these suggested boxes, run a classifier. These intricate pipelines were slow and hard to optimize. Hence Redmon, J. et al. (2016) proposed You Only Look Once (YOLO), an algorithm that is a single convolutional network simultaneously that predicts multiple bounding boxes and class probabilities for those boxes. Unlike R-CNN and other similar algorithms, YOLO is found to be extremely fast, sees the entire image during training and testing hence making fewer background errors. When trained on natural images and tested on the artwork, YOLO outperforms other algorithms by a wide margin. However, YOLO was shown to fall short of state-of-the-art detection systems in terms of accuracy, and it struggled to precisely localize some objects [11]. Redmon, J. et al. (2017), by focusing mainly on improving recall and localization while maintaining classification accuracy, proposed YOLOv2. It was then found that detection methods are constrained to a small set of objects, hence they as well proposed a joint training algorithm that allows one to train object detectors on both detection and classification data, using which they trained the YOLO9000 algorithm which was built by modifying YOLOv2 [12].

The majority of the accurate CNN-based object detectors required high GPU power and training in order to achieve their optimal accuracy. High GPU power is essential for achieving accuracy and speed in real-time since it is vital in a car collision or obstacle warning model. Bochkovskiy A et al. (2020) proposed a modified version of the state-of-the-art object detection models, YOLOv4, with significant improvement in the speed and accuracy of the models. An impressive aspect of this model is that it can operate in real-time on a conventional GPU and training as well requires only a single GPU. Hence using conventional GPUs such as 1080Ti or 2080 Ti one can train an accurate and extremely fast object detector [13]. Since YOLOv4 outperforms other frameworks, our proposed framework is based on it.

Contributions:

- The proposed framework helps in providing an additional safety layer for the Driver (User) and the passengers of the vehicles.
- The Extraction and Detection modules present in the proposed framework enable the software to outperform its competitors.
- Due to the addition of the visualization module the user can easily interact with the framework. This is one of the key features of the framework.

Based on the previous work done in the field of ADAS and object detection as reviewed the proposed framework was formulated.

3 Proposed Work

The input video is processed as frames, each of which acts as input to the object recognition and detection algorithm (YOLOv4). Each frame is processed along three stages in the algorithm namely-Backbone, neck, and head as presented in Fig. 1.

- Backbone: CSPDarknet53,
- Neck: Concatenated Path Aggregation Networks with Spatial Pyramid Pooling (SPP) additional module
- Head: YOLOv3

Concerning the framework, object detection can be categorized into 3 major modules. These 3 modules are

1. Extraction (Backbone and Neck)
2. Detection (Head)
3. Visualization

3.1 Extraction

The backbone and neck take images (each of the frames) as input to extract the feature maps using CSPDarknet53 and SPP, PANet path-aggregation. Darknet53 comprises 53 Convolutional layers. For detection tasks, 53 layers stacked on to the original architecture of 53 layers give us 106 layers of architecture.

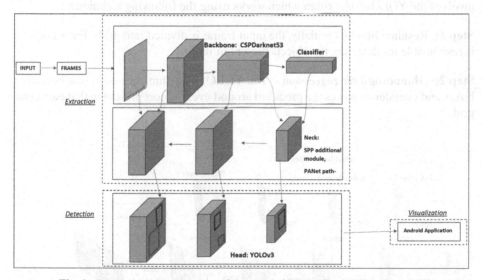

Fig. 1. Proposed object detection framework for ADAS using deep learning

Step 1: Input: the video input is processed frame by frame.

Step 2: CSPDarknet53 - Cross-Stage-Partial-connections are Concerning used to eliminate duplicate gradient information that occurs while using conventional DenseNet [14].

- In CSPDenseNet the base layer is divided into 2 parts; here part A and part B.
- One part will go into the original Dense Block and is processed accordingly; here part B is processed in the Dense block.
- The other part will directly skip to the transition stage.

As a result of this, there is no duplicate gradient information; it also reduces a lot of computations. As shown in Fig. 2.

Step 3: Additional Layers are added between the backbone and the head using the neck. To aggregate the information, the YOLOv4 algorithm applies a modified Path aggregation network [15] with a modified spatial attention module and a modified SPP (Spatial Pyramid Pooling) [16]. Concatenated Path Aggregation Networks [15] with Spatial Pyramid Pooling (SPP) additional modules [16] is used to increase the accuracy of the detector.

3.2 Detection

Each frame processed in the backbone and neck is then transferred to the head which involves the YOLOv3 algorithm which works using the following techniques:

Step 1: Residual **blocks** - initially, the input frame is divided into grids. Each grid cell is responsible for detecting the objects present in its cell.

Step 2: **Bounding box regression** - The YOLO algorithm runs such that bounding boxes and confidence scores are predicted around every object present in that particular grid.

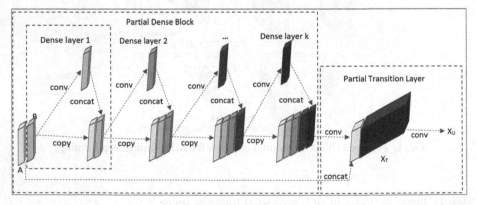

Fig. 2. CSPDenseNet

Every bounding box consists of these attributes: width (bw), height (bh), bounding box center (x, y) and confidence score (c). The confidence score represents how confident, accurate the algorithm is of a particular object in that bounding box. Together with these attributes, YOLO uses a single bounding box regression to predict the probability of an object appearing in the bounding box. Figure 1 shows the YOLOv4 algorithm being run in real-time on a webcam. The algorithm detected objects in the frames by indicating the classes they belong to and the confidence scores representing how sure it is of the objects.

Step 3: Intersection over Union (IoU) - If there is no object in a grid cell, the confidence score is 0; otherwise, the confidence score should be equal to the intersection over union (IoU) between the ground truth and predicted box. Here, the ground truth boxes are manually pre-defined by the user, hence greater IoU means greater confidence score, which means higher accuracy of prediction by the algorithm. This expressed in the equation below (Fig. 3). Filtration of those boxes with no objects is done based on the probability of objects in that box. Non-max suppression processes eliminate the unwanted bounding boxes and the box with the highest probability or confidence score will remain [17, 18].

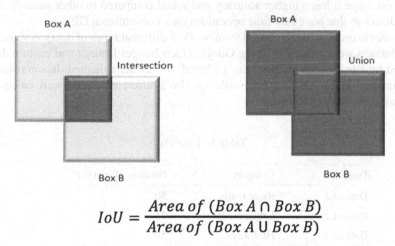

$$IoU = \frac{Area\ of\ (Box\ A \cap Box\ B)}{Area\ of\ (Box\ A \cup Box\ B)}$$

Fig. 3. Equation for intersection-over-union (IoU)

IoU calculation is used to measure the overlap between two proposals.

Step 4: Final detection - The algorithm detects the object and class probabilities.

3.3 Visualization

The final module of the proposed system involves an android based application. The application inputs a real-time video stream from the device camera runs an object detection algorithm on it and notifies the user under any case of any condition that requires to be brought to the user's attention and needs to be acknowledged.

4 Evaluation Results

With the aim of creating a CNN for real-time operation on a conventional GPU, YOLOv4 was introduced. In the process of doing so, various training improvement methods on the accuracy of the classifier on the ImageNet dataset were tested and their influence was noted along with the accuracy of the detector on the MS COCO dataset.

While comparing YOLOv4 with other state-of-the-art object detectors, it was found that YOLOv4 improved YOLOv3's AP by 10% and FPS by 12% and within comparable performance, YOLOv4 ran twice as fast as EfficientDet. It was found that the classifiers accuracy was enhanced by proposing features such as CutMix and Mosaic data augmentation, Class label smoothing, and Mish activation. YOLOv4 was chosen in our proposed framework since it has a higher accuracy and speed compared to other state-of-the-art object detectors that have real-time operations on a conventional GPU.

In order to evaluate the proposed framework, 3 different types of datasets were used. The 3 datasets were created by using Google Open Images Dataset and custom dataset obtained by labeling the images were gathered. 8% of the collected data consisted of blurry images and images with low visibility. The 3 different datasets were categorized as explained in Table 1 below.

Table 1. Datasets

Datasets	Category	Number of images
Dataset 1	Rural roads	50
Dataset 2	Urban roads	75
Dataset 3	Highways	50

The mAP of a few state-of-the-art object Detectors such as YOLOv3 [19, 20], Faster-RCNN [21, 22], EfficientDet was compared using these datasets. The results of this comparison are represented in Fig. 4 (Fig. 4). With respect to mAP, it is clearly seen that YOLOv4 outperforms its competitors by a significant margin.

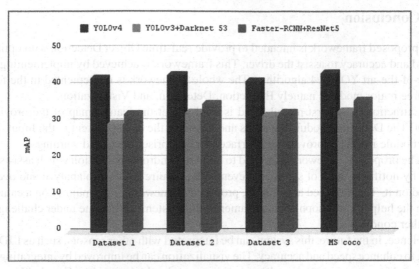

Fig. 4. Comparison of state-of-the-art object detectors on the custom dataset and standard MS coco dataset

The graph in Fig. 5 (Fig. 5) represents a comparative analysis of YOLOv4 with other state-of-the-art object detection algorithms regarding average precision (Y-axis) and frames per second (X-axis). It can be inferred that indeed YOLOv4 algorithm out-performs others in real-time detection. It achieves an average precision between 38 and 43, and frames per second between 65 and 124. The YOLOv3 algorithm, on the other hand, obtains an average precision (AP) of 31 to 33 and frames per second (FPS) of 73 to 120.

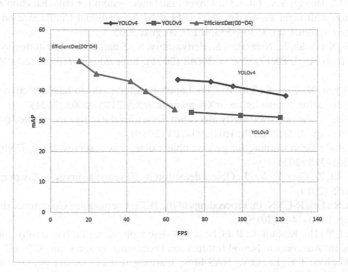

Fig. 5. Comparison of YOLO and EfficientDet(D0~-D4) on Standard MS-coco dataset

5 Conclusion

The proposed framework is intended to provide real-time Object Detection with optimal speed and accuracy to assist the driver. This framework is achieved by implementing the state-of-the-art YOLOv4 algorithm. The whole framework is implemented in the form of three major modules namely Extraction, Detection, and Visualization.

Extraction is the first module, and is used to get the feature map of the provided Input. The Detection module identifies and localizes the object present in the Input. The last module is used to provide an interface that comprises alerts and warnings.

The proposed framework is applied to build the android application which assists the user by notifying them of significant events that require the user to analyze and decide based on it. The proposed application, proposed framework relies majorly on a camera. With the help of some sophisticated cameras, this system can operate under challenging weather conditions.

Hence, in the future, this system can be integrated with other sensors, such as LIDAR [23], to enhance speed and accuracy. The visualization can be improved by integrating the proposed framework with other driver assistance technologies, such as Google Maps and voice assistant. In the future, with the help of a cloud-based approach [24], the processes can be recorded and analyzed. The cloud-based approach also helps in increasing the accessibility of the application. Raspberry pi can also be used in order to have a smooth flow in the processes and increased efficiency. In the future, the proposed framework can be integrated with the Electronic Control Unit (ECU) [25] present inside the vehicles.

References

1. Wang, C.C., Thorpe, C., Thrun, S., Hebert, M., DurrantWhyte, H.: Simultaneous *localization, mapping and moving object tracking. Int. J. Robot. Res. **26**(9), 889–916 (2007)
2. Wang, C.C., Huang, S.S., Fu, L.C.: Driver assistance system for lane detection and vehicle recognition with night vision. In: 2005 IEEE/RSJ International Conference on Intelligent Robots and Systems, pp. 3530–3535. IEEE, August 2005
3. Shaily, S., Krishnan, S., Natarajan, S., Periyasamy, S.: Smart driver monitoring system. Multimedia Tools Appl. **80**(17), 25633–25648 (2021). https://doi.org/10.1007/s11042-021-108 77-1
4. Liu, L., Chen, X., Zhu, S., Tan, P.: CondLaneNet: a top-to-down lane detection framework based on conditional convolution. arXiv preprint arXiv:2105.05003 (2021)
5. Manoharan, S.: An improved safety algorithm for artificial intelligence enabled processors in self driving cars. J. Artif. Intell. **1**(02), 95–104 (2019)
6. Liu, L., et al.: Deep learning for generic object detection: a survey. Int. J. Comput. Vision **128**(2), 261–318 (2020)
7. Zou, Z., Shi, Z., Guo, Y., Ye, J.: Object detection in 20 years: a survey. arXiv preprint arXiv: 1905.05055 (2019)
8. Girshick, R.: Fast R-CNN. In: Proceedings of the IEEE International Conference on Computer Vision, pp. 1440–1448 (2015)
9. Dai, J., Li, Y., He, K., Sun, J.: R-FCN: object detection via region-based fully convolutional networks. In: Advances in Neural Information Processing Systems, pp. 379–387 (2016)
10. Ghiasi, G., Lin, T.Y., Le, Q.V.: NAS-FPN: learning scalable feature pyramid architecture for object detection. In: Proceedings of the IEEE/CVF Conference on Computer Vision and Pattern Recognition, pp. 7036–7045 (2019)

11. Redmon, J., Divvala, S., Girshick, R., Farhadi, A.: You only look once: unified, real-time object detection. In: Proceedings of the IEEE Conference on Computer Vision and Pattern Recognition, pp. 779–788 (2016)
12. Redmon, J., Farhadi, A.: YOLO9000: better, faster, stronger. In: Proceedings of the IEEE Conference on Computer Vision and Pattern Recognition, pp. 7263–7271 (2017)
13. Bochkovskiy, A., Wang, C.Y., Liao, H.Y.M.: YOLOv4: optimal speed and accuracy of object detection. arXiv preprint arXiv:2004.10934 (2020)
14. Wang, C.Y., Liao, H.Y.M., Wu, Y.H., Chen, P.Y., Hsieh, J.W., Yeh, I.H.: CSPNet: a new backbone that can enhance learning capability of CNN. In: Proceedings of the IEEE/CVF Conference on Computer Vision and Pattern Recognition Workshops, pp. 390–391 (2020)
15. Liu, S., Qi, L., Qin, H., Shi, J., Jia, J.: Path aggregation network for instance segmentation. In: Proceedings of the IEEE Conference on Computer Vision and Pattern Recognition, pp. 8759–8768 (2018)
16. He, K., Zhang, X., Ren, S., Sun, J.: Spatial pyramid pooling in deep convolutional networks for visual recognition. IEEE Trans. Pattern Anal. Mach. Intell. 37(9), 1904–1916 (2015)
17. Rezatofighi, H., Tsoi, N., Gwak, J., Sadeghian, A., Reid, I., Savarese, S.: Generalized intersection over union: a metric and a loss for bounding box regression. In: Proceedings of the IEEE/CVF Conference on Computer Vision and Pattern Recognition, pp. 658–666 (2019)
18. Neubeck, A., Van Gool, L.: Efficient non-maximum suppression. In: 18th International Conference on Pattern Recognition (ICPR 2006), vol. 3, pp. 850–855. IEEE, August 2006
19. Redmon, J., Farhadi, A.: YOLOv3: anincremental improvement. University of Washington (2018)
20. Lee, Y.H., Kim, Y.: Comparison of CNN and YOLO for object detection. J. Semicond. Disp. Technol. 19(1), 85–92 (2020)
21. Ren, S., He, K., Girshick, R., Sun, J.: Faster R-CNN: towards real-time object detection with region proposal networks. Adv. Neural. Inf. Process. Syst. 28, 91–99 (2015)
22. Shine, L., Edison, A., Jiji, C.V.: A comparative study of faster R-CNN models for anomaly detection in 2019 AI city challenge. In: Proceedings of the IEEE/CVF Conference on Computer Vision and Pattern Recognition Workshops, pp. 306–314 (2019)
23. Beltrán, J., Guindel, C., Moreno, F.M., Cruzado, D., Garcia, F., De La Escalera, A.: Bird-Net: a 3D object detection framework from LiDAR information. In: 2018 21st International Conference on Intelligent Transportation Systems (ITSC), pp. 3517–3523. IEEE, November 2018
24. Cabanes, Q., Senouci, B.: Objects detection and recognition in smart vehicle applications: point cloud based approach. In: 2017 Ninth International Conference on Ubiquitous and Future Networks (ICUFN), pp. 287–289. IEEE, July 2017
25. Talavera, E., Díaz-Álvarez, A., Naranjo, J.E., Olaverri-Monreal, C.: Autonomous vehicles technological trends. Electronics 10(10), 1207 (2021)

Cyber Crime Prediction Using Machine Learning

Ruchi Verma[✉] and Shreya Jayant

Department of Computer Science and Engineering, Jaypee University of Information
Technology, Solan, India
ruchi.verma@juit.ac.in

Abstract. With the plethora of information exchange in online communication, occurrences of some unwanted vices like cybercrimes and cyber bullying have taken shape. It is imperative to develop a prediction model for cyber-crime prediction. The existing technical approaches to identify these unwarranted activities have not achieved the required success rate. In the proposed approach, methods of Machine Learning and Neural Network are used for analyzing and predicting the occurrence of cybercrime. We have considered four major categories of cybercrime, namely cyber bullying, cyber harassment, cyber fraud and cyber hacking for performing various experiments. In the proposed model we have achieved an accuracy of 97.6% while predicting cyber-criminal activities.

Keywords: Twitter web parsing · Text classification · Social network analysis · Crime prediction · Cyber crime

1 Introduction

In the past five years it has been identified that there is a steep increase in the count of data, cloud and devices which are the reasons for the security threats. Threat predictions have been acknowledged as authentic threats and unfortunately there is an approximation of further such occurrences in greater magnitude [11]. As per the cyber security ventures report in the year 2020, cybercrime is estimated to cost nearly $6 trillion per annum by the year 2021 [12]. Our paper puts forth a Machine learning and Neural Network based model for recognition of cyber-crimes, using social network analysis. We have collected the dataset from twitter, the social media network. We have studied the patterns of the tweets and using that data, constructed a framework to forecast major forms of cybercrimes (cyberbullying, cyber harassment, cyber fraud and cyber hacking). To create a prediction model and to categorize dataset as per the different kinds of crime we have used Logistic Regression, Nearest Neighbors (KNN), Random Forest Classifier and Neural Networks in our approach. Furthermore, along with these machine learning methods, we have generated results using word cloud library to find the most prevalent word frequency in tweets of each section of cybercrime. Figuring out which are the jargons used while identifying the kind of cybercrimes that can be ensued will be crucial in mitigating its occurrence. The results reveal that all four algorithms attain precision,

M. Singh et al. (Eds.): ICACDS 2022, CCIS 1614, pp. 160–172, 2022.
https://doi.org/10.1007/978-3-031-12641-3_14

recall and F-measure values greater than 0.9, with the Logistic Regression outperforming the others. The proposed model achieves higher accuracy than the existing systems.

2 Literature Survey

This study talks about how crime analysis has evolved into a fascinating profession which interacts with major things related to civic safety that are recognized worldwide. Analyzing tweets and investigating it is still a hot topic in the industry nowadays. The technique of extracting, analyzing, and classifying information is used for polarizing and classifying the twitter text sentiment in order to forecast cybercrime using aspect-based SA [1]. According to this research, a growing flow of network dataset is arriving from social platforms, which are utilized to develop a range of similarization of data for many types of investigations, including socially seen civic behaviors, crimes, and so on. Using data from social media websites, a framework is constructed to anticipate primary categories of social cybercrime. Working on EDA of data, classification, building of model, and working with the model to gauge future crimes are the three components that make up the proposed system [2]. It is no secret that monitoring criminal activities around the world is in the interest of every country so they can work on mitigating the issues that make the crime happen. Although the general public cannot get access to the data that can generate this insight easily which is why this paper has put forward how theft of information derived from approx. 400 articles of news are being used to prevent theft [3] Massive amounts of data are generated by social networks. Twitter, a microblogging platform, has over 230M everyday users who post millions of tweets. To estimate crime rates, this paper proposes analyzing public data from Twitter. In recent years, the rate of crime has risen. Although various different methods are utilized to minimize cybercrime, not any of the previous approaches have worked on utilizing the language used in Tweets (hateful or non-hateful) which acts like a source of information to predict the rate of crimes. In this work, we propose that evaluating the language used in tweets is a reliable indicator of city crime rates [4]. Cyberbullying is one of the law's violations where the offence is perpetrated on social media platforms like Twitter. If no one reports the tweet, this activity is impossible to notice. The goal of cyberbullying tweet detection is to categorize tweets that contain bullying. The SVM approach is used for classifying the dataset, with the goal of determining the division of the hyperplane in both the positive classes as well as the negative ones. This is a classification of the text study, and because a large number of data is utilized, higher columns of feature are created, this study also employs Information Gain method for selecting the feature method to exclude the ones which are irrelevant for classifying the data [5]. The goal of this particular study is to develop a method for figuring out the criminal Hubs in India, focusing more on the location of Jamtara. The map depicts the area with crime alarms depending on the nature of the reported incident. Classifying the support vector process is used in the proposed method. The finding of this paper generates a higher accuracy with respect to the statistical data showcase of criminal activity [6].

This particular study looks into text classification approaches for detecting tweets about alcohol consumption. The timeline of 24-h on New Year in 2012, they classified about 35k tweets based in NYU. The tweets were pre-processed into stemmed/unstemmed and unigram/bigram representations. The classification challenge was subsequently tackled using supervised algorithms. The methods work with areas which are under the operating curve of the receiver of values 0.6, 0.9, 0.93, and 0.9, respectively, using 10-fold cross-validation. For those particular text tweets, they compare the classifying of tweet dataset approach to a human-made Boolean search, and the text classification method outperforms the hand-crafted search [7]. When we look at the cyber-bullying on social networking mediums, for example, it is a sort of cyber victimization that have resulted in considerable threat to people, including suicide. Detecting the themes that trigger the experience of being a victim is a solid step toward protecting the internet society from victimization [8]. This study focuses on two issues. First, they look into predicting the spatial trajectories of people who use social networking sites. Current work on the problem concentrated on the utilization in networks with traces of cellular based social web services, which all output organized geographic data. Rich textual content, which users frequently post alongside structured data, has received less attention. They look into how to incorporate content with text into models which predict and already are built, and show that it improves next-place prediction significantly when compared to numerous published research baselines [9]. This research proposes a machine learning method for detecting crimes and their locations. As a first phase, they retrieved tweets relating to the crimes using predefined keywords. Then, to classify the tweets, an Artificial Neural Network (ANN) model was built. The geolocation and crime type are then obtained in the last stage. The proposed criminal detection approach has been proven effective in an empirical examination of the prototyping system [10]. This particular research does an in-depth analysis of what cybercrime really is and goes beyond the normal conceptions. It not only considers the effects of digital technologies that are in place today, but also discusses the limitations and varied nature of cybercrimes that come under its umbrella [13].

3 Model Description

The baseline of describing the model provides insight into a dataset's potential predictive power. To train and predict, the models frequently need significant analysis and computational power, making them a useful cross-check on the validity of an answer. The proposed system consists of three modules. Phase 1 focuses on processing the tweet dataset, and is taken forward as the solution and input for the Phase 2 module. Phase 2 Module generates a model which has been trained and the Phase 3 module, focuses on the prediction of output, as illustrated in Fig. 1. Phase 1 works as per the data mining process, which includes building a dataset, cleaning it, normalising it, pre-processing it before use. The second phase involves building a model suitable for our data prediction and checking its accuracy, precision, and working, and optimizing the features until we get the best result. In Phase 3, all the algorithms are compared and then based on the metrics, the best one is utilized for prediction of the variable as given in the Fig. 1 below.

In our research, we have built a model using four different algorithms. The novelty feature is an element that showcases an observation or result that leads to innovation or realization of a discovery. In this paper, the novel feature is an improvement on the already existing proposed methods present to identify cybercrimes of different kinds.

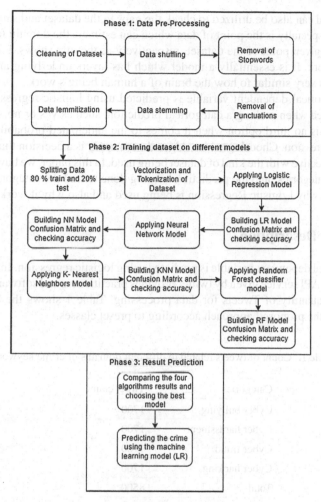

Fig. 1. Model procedure

Based on the kind of data used, and the efficiency of the model proposed, this is an augmentation in the prediction of cybercrime using Machine learning technology. The first one is Logistic Regression, a supervised algorithm. To define this algorithm, we need to know what a discrete result is, therefore when we use probability to describe that particular discrete result and model it using ML, it is called LR. The second one is Random Forest. The building of numerous basic bunches of decision trees within a stage where data is trained democratically of a major portion which measures using metric of mode through using the different classes divided into parts, these steps are the essential principles underpinning the random forest approach. This voting approach has the effect of adjusting for the undesired feature of decision trees to overfit data for training, among other things. Third one is K-Nearest Neighbors. When we talk of KNN, it is known for the usual process of regressing the data and classifying it. It is not a fixed method but can be flexible and can also be utilized to check the gaps in the dataset and sample it again. KNN's main specialty is the point of data which can estimate the domain for continuity results of the given point of the dataset, respectively as the title shows. Last one is the Neural Network. It is essentially a model which has layers underlying and is used in batches and is very similar to how the brain of a human being's work.

The categorical dependent variable is predicted using logistic regression. In other words, it's used when making a categorical prediction, such as yes or no, true or false, 0 or 1. There is no third option when it comes to the anticipated probability or output of logistic regression. Choosing which Algorithm to apply is a decision that needs to be taken in perspective with the kind of dataset being used. In this paper, we have divided the dataset into categories, and the prediction is being done to find the categorical variable. Hence, this is why Logistic Regression is being used and also why it works well.

4   Dataset Description

The categorical dependent variable is predicted using logistic regression. In other words, the streaming API service given by twitter provides practitioners and software developers with a large quantity of Tweets for data processing. Table 1 shows the 8,500 tweets collected for the proposed approach according to preset classes.

Table 1. Count of tweets related to different domains of crime keywords

Categories	Tweet count
Cyber bullying	1700
Cyber harassment	1700
Cyber fraud	1700
Cyber hacking	1700
Total	8500

The given number of tweets were deemed appropriate enough for the processing using the proposed approach. Using the web parsing extraction method, the numerous tweets were collected and then organized into a spreadsheet file, which was then taken through the process of data preprocessing which is discussed in detail in the below section. To extract each of the categories of cybercrime, a hashtag feature was used and the data for the prior few years was extracted.

5   Dataset Preprocessing

Following are the steps to analyze the crime prediction twitter dataset, starting with loading the dataset.

1. Removal of the noise and extra attributes from the dataset.

2. Removal of any set patterns which may lead to skewed results was done by data shuffling.
3. Removal of the stop words from the processed dataset. Stop words are the most frequently used terms in the text. The value of terms that appear frequently in the text is quite low. Some examples may include in, the, on.
4. Removal of punctuations, conversion of all the text to lowercase for ease of use, filtering out the alphanumeric and numeric characters to make the dataset cleaner.
5. Breaking the Text into root words by data stemming.
6. Lemmatization of words.
7. Splitting of Labels and text
8. Training and testing split of 80:20 ratios and creation of machine learning models for application of the respective algorithms.

Table 2 specifies the processed dataset which has been divided according to the four different categories of crime. It will be shuffled in the next stage.

Table 2. Processed dataset

	Description	Category
0	The management bullied, harassed me & lied…	CB
1	@EnclaveEmily Looks like you might need…..	CB
2	Proud dad moment today. The bully at the sch..	CB
3	Instead of this local fb group @Trimdontime…	CB
4	Empower your child to stand up to bullying…..	CB

Table 3, shows the process of data mining. The description or the text, through which the text classification is being done, is cleaned. The noise and alphanumeric characters are removed using python libraries.

Table 3. Dataset being cleaned

	Description	Category	Description_clean
2999	@flipkartsupport @_kaly…..	3	flipkartsupport kalyan……
650	Enjoy today @marshawri…..	2	Enjoy today marshaurla….
1112	Worth noting that detailed….	2	Worth note retail custo…..
2159	Its been fun to share some….	3	Its been fun online fra……
2600	Why you should report cry…	3	Why report crypto sca……

In this procedure, we have used text classification to work on the dataset because of the structure of the data that we have acquired. When we parse the social network site, the most important column which gives us and has the potential to provide the highest insight is the text description, which is why using the text classification-based approach is in our favor and will produce better results.

6 Implementation

Analyzing, comprehending, organizing, and going across the textual set of data is hard and takes a lot of time because it contains a lot of noise as it is raw in format. Most of the time, it is not being used up to its highest potential. Numerous insights can be drawn if only the dataset is carefully chosen and mined. In Machine Learning, this is the use case where the process of classifying the texts is being utilized. Researchers and businesses can take this method of text classification and then without any major cost or wastage

a. Prevalent words in tweets for all dataset

b. Cyber bullying

c. Cyber Harassment

d. Cyber Fraud

e. Cyber Hacking

Fig. 2. a. Prevalent words in tweets for all dataset, b. Cyber bullying, c. Cyber harassment, d. Cyber fraud, e. Cyber hacking

of time, apply it and distribute it across all relevant content, like emails, official papers, social networks, bots, etc. It majorly improves time cost benefit analysis of data tasks, which enhances the solutions designed by using data science analysis selection of this technology. In Fig. 2a, you can see the highest count words that were used in the dataset overall. The corresponding Fig. 2b and the consecutive Fig. 2c, describe the words from cyber fraud and hacking. In Fig. 2d and Fig. 2e, various different crime section dataset is showcased according to the highest count of words used respectively. Wordcloud library was used to figure out the frequency of text in the tweets divided by categories.

Hence, Exploratory Data analysis was done on the dataset. An EDA is essentially a detailed study that is used to identify the underlying structure of a data set. It is beneficial for our research as it conceptually showcases the variety of trends, designs, along with linkages which aren't easily visible. We are not able to conclude reliable results based on a lot of data by easily going across the data, instead, we have to judge the data methodologically using a structured analytic angle. Developing through practice of this method, information can aid in the detection of errors, the debunking of assumptions and the understanding of the correlations between various crucial elements. Such information finally led to the selection of a suitable predictive model. Then we analyzed and compared to find the best suitable model, which has the highest accuracy and which gives an accurate and realistic resultant insight.

7 Result

The pre-processed data is split into two groups in the ratio of 80: 20 where 80% is training data and 20% is testing data. Standard assessment criteria such as measuring the performance through parameters is done to determine the efficiency of the proposed model. Equation 1 showcases the ratio of TP versus Total positives, Eq. 2 showcases ratio of TP versus sum of FN and TP, in Eq. 3 ratio of twice multiple of Precision and Recall versus sum of the two values given below are used.

$$\text{Precision} = \text{true positives}/(\text{true positives} + \text{false positives}) \tag{1}$$

$$\text{Recall} = \text{true positives}/(\text{true positives} + \text{false negatives}) \tag{2}$$

$$\text{F} - \text{Measure} = (2 * \text{Precision} * \text{Recall})/(\text{Precision} + \text{Recall}) \tag{3}$$

The following tables give the performance measures for the four algorithms used in our approach:

1. KNN Model:
Accuracy: 95.84850691915513
Precision: 95.85
Recall: 95.85
F Score: 95.85

The details of the KNN Matrix model's confusion matrix can be seen in Table 4.

Table 4. KNN matrix

	Precision	Recall	F1-score	support
0	0.96	0.83	0.89	143
1	1.00	0.96	0.98	506
2	0.91	0.96	0.94	226
3	0.94	0.99	0.97	498
Accuracy	–	–	0.96	1373
Macro avg	0.95	0.93	0.94	1373
Weighted avg	0.96	0.96	0.96	1373

2. Random Forest:
Accuracy: 97.6693372177713
Precision: 97.67
Recall: 97.67
F Score: 97.67

The details of the Random Forest Matrix model's confusion matrix can be seen in Table 5.

Table 5. Random Forest classifier matrix

	Precision	Recall	F1-score	Support
0	0.95	0.85	0.90	143
1	1.00	1.00	1.00	506
2	0.93	0.97	0.95	226
3	0.99	1.00	0.99	498
Accuracy	–	–	0.98	1373
Macro avg	0.97	0.95	0.96	1373
Weighted avg	0.98	0.98	0.98	1373

3. Neural Network:
Accuracy: 96.50400582665696
Precision: 96.5
Recall: 96.5
F Score: 96.5

Table 6. Neural Network matrix

	Precision	Recall	F1-score	support
0	0.90	0.86	0.88	143
1	0.99	0.99	0.99	506
2	0.92	0.95	0.93	226
3	0.99	0.99	0.99	498
Accuracy	–	–	0.97	1373
Macro avg	0.95	0.95	0.95	1373
Weighted avg	0.97	0.97	0.97	1373

The details of the KNN Matrix model's confusion matrix can be seen in Table 6

4. Logistic Regression:
Accuracy: 97.66933372177713
Precision: 97.67
Recall: 97.67
F Score: 97.67

The details of the Logistic Regression's confusion matrix can be seen in Table 7

Table 7. Logistic Regression matrix

	Precision	Recall	F1-score	support
0	0.92	0.91	0.92	144
1	1.00	0.99	1.00	480
2	0.96	0.94	0.95	236
3	0.98	0.99	0.99	513
Accuracy	–	–	0.98	1373
Macro avg	0.97	0.96	0.96	1373
Weighted avg	0.98	0.98	0.98	1373

The visually descriptive Confusion matrix of Logistic regression, K-Nearest Neighbors, Random forest classifier and Neural Network can be seen consequently in Fig. 3a, Fig. 3b, Fig. 3c and Fig. 3d.

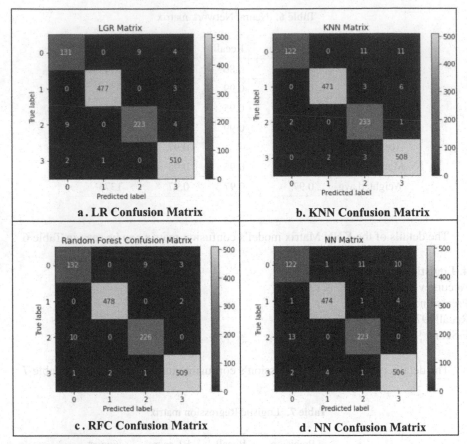

Fig. 3. a. LR confusion matrix, b. KNN confusion matrix, c. RFC confusion matrix, d. NN confusion matrix

8 Limitations

There are various limitations and situations where the proposed technology may fall short. One prime example can be, while choosing the dataset, if we focus on the wrong set of categorical variables, then the outcome and prediction may be vastly different than what is appropriate. Another possible limitation is that there is a certain limited scope up to which the proposed approach can be utilized. While fighting cybercrime realistically, there are a number of factors that we need to consider like the time, the cost and the impact the crime may have on the people involved in it.

9 Conclusion and Future Work

The results of the comparison test between four algorithms namely Logistic Regression, K Nearest Neighbors, Random Forest and Neural Network are shown in Table 8.

Table 8. Comparative analysis

Model/Algorithm	Precision	Recall	F-Score	Accuracy
Logistic Regression	0.9767	0.9767	0.9767	97.67%
KNN	0.958	0.958	0.958	95.8%
Random Forest	0.976	0.976	0.976	97.60%
Neural Network	0.965	0.965	0.965	96.50%

In the above table we can see that Logistic Regression and Random forest classifiers have the highest accuracy. Based on detailed comparison between the two, Logistic Regression has proved to be more effective for the dataset used. Table 9 shows the Comparative Analysis of the Proposed and Competitor Approach.

Table 9. Proposed and competitor approach comparison

Approach	Algorithm	Metric: accuracy
Proposed approach	Logistic Regression	97.6%
Competitor approach	Support Vector Machine	92%

The goal of this study was to use Twitter data to forecast cybercrimes in social media. We employed the text classification feature of machine learning with the logistic regression model. The research shows that the current state of the art yields better results. The suggested algorithm is now offline, but in the future, it might be modified to predict future crimes using real-time Twitter data streaming. To render the system more efficient and resilient, additional incident categories can be added. Concludingly, this paper showcases that with the dawn of the new age of technology, there are two sides which are prominent. Although there has been a plethora of opportunities that have become available, the misuse of this technology has become equally easier, given the medium of social media. Cybercrime is a new variant of crime, but nonetheless, it is no less lethal. Using the improved method of the proposed approach in this paper, we hope to mitigate and lessen the risks of various cybercrime happening by trying to predict its occurrence beforehand with the use of technology.

References

1. El Hannach, H., Benkhalifa, M.: WordNet based implicit aspect sentiment analysis for crime identification from Twitter. Int. J. Adv. Comput. Sci. Appl. **9**(12), 150–159 (2018)
2. Abbass, Z., Ali, Z., Ali, M., Akbar, B., Saleem, A.: A framework to predict social crime through twitter tweets by using machine learning. In 2020 IEEE 14th International Conference on Semantic Computing (ICSC), pp. 363–368. IEEE, February 2020

3. Arulanandam, R., Savarimuthu, B.T.R., Purvis, M.A.: Extracting crime information from online Newspaper articles. In: Proceeding of AWC 2014, Proceedings of the Second Australasian Web Conference, Auckland, New Zealand, vol. 155, pp. 31–38, 20–23 January 2014
4. Almehmadi, A., Joudaki, Z., Jalali, R.: Language usage on Twitter predicts crime rates. In: Proceedings of the 10th International Conference on Security of Information and Networks (2017), pp. 307–310, October 2017
5. Purnamasari, N.M.G.D., Fauzi, M.A., Indriati, L.S.D.: Cyberbullying identification in twitter using support vector machine and information gain based feature selection. Indones. J. Electr. Eng. Comput. Sci. **18**(3), 1494–1500 (2020)
6. Mahor, V., Rawat, R., Telang, S., Garg, B., Mukhopadhyay, D., Palimkar, P.: Machine Learning based detection of cyber crime hub analysis using Twitter data. In: 2021 IEEE 4th International Conference on Computing, Power and Communication Technologies (GUCON), pp. 1–5. IEEE, September 2021
7. Aphinyanaphongs, Y., Ray, B., Statnikov, A., Krebs, P.: Text classification for automatic detection of alcohol use-related tweets: a feasibility study. In: Proceedings of the 2014 IEEE 15th International Conference on Information Reuse and Integration (IEEE IRI 2014), pp. 93–97. IEEE, August 2014
8. Shoeibi, N., Shoeibi, N., Julian, V., Ossowski, S., Arrieta, A.G., Chamoso, P.: Smart cyber victimization discovery on twitter. In: Corchado, J.M., Trabelsi, S. (eds.) SSCTIC 2021. LNNS, vol. 253, pp. 289–299. Springer, Cham (2022). https://doi.org/10.1007/978-3-030-78901-5_25
9. Wang, M., Gerber, M.S.: Using twitter for next-place prediction, with an application to crime prediction. In: 2015 IEEE Symposium Series on Computational Intelligence, pp. 941–948. IEEE, December 2015
10. Sandagiri, S.P.C.W., Kumara, B.T.G.S., Kuhaneswaran, B.: Detecting crime related twitter posts using artificial neural networks based approach. In: 2020 20th International Conference on Advances in ICT for Emerging Regions (ICTer), pp. 5–10. IEEE, November 2020
11. Brar, H.S., Kumar, G.: Cybercrimes: a proposed taxonomy and challenges. J. Comput. Netw. Commun. (2018)
12. Ch, R., Gadekallu, T.R., Abidi, M.H., Al-Ahmari, A.: Computational system to classify cybercrime offences using machine learning. Sustainability **12**(10), 4087 (2020)
13. Gordon, F., McGovern, A., Thompson, C., Wood, M.A.: Beyond cybercrime: new perspectives on crime, harm and digital technologies. Int. J. Crime Justice Soc. Democr. **11**(1) (2022)

Real-Time Image Based Weapon Detection Using YOLO Algorithms

Manoj Gali[1], Sunita Dhavale[1(✉)], and Suresh Kumar[2]

[1] Defence Institute of Advanced Technology (DIAT), Pune, India
sunitadhavale@gmail.com
[2] Defence Institute of Psychological Research (DIPR), Delhi, India

Abstract. From last few years country faces major challenge in maintaining security standards particularly in public and highly sensitive places such as airports, movie theatres, stadiums, and national parks, etc. The offsite and onsite planers use many tactics to resist authority and disrupt and to add turmoil in order to achieve their goals and objectives. These tactics can be planned or unplanned. Many crowd management experts suggest that the availability of suspicious objects such as Camera, Handgun, Rifles, Dagger, Sword and Sticks at sight before or during the event can be an indication of upcoming threat or any unlawful activities and their identification may help security forces in their proactive management and control of any destructive activities. In this research work, we generated a novel dataset "DIAT-Weapon" Dataset for weapon object detection using web scraping techniques. DIAT-Weapon Dataset consists of 2712 images divided into six categories mainly: Camera, Handgun, Rifles, Dagger, Sword and Sticks. We customized and fine-tuned YOLOv4 models to classify and position the six types of harmful objects i.e. Camera, Handgun, Rifles, Dagger, Sword, and Sticks in real time. To achieve real-time faster performance and better detection accuracy, YOLOv4 is fine-tuned, and the preset anchors trained on DIAT-Weapon annotated dataset. Using a series of YOLOv4 object detection algorithms, we demonstrated experiments on our dataset, achieving 0.63 mAP. To our best knowledge, this is the first work that utilizes customized YOLOv4 model for real-time localization and classification of weapon objects into six different categories. To our best knowledge, this is the first work that utilizes customized YOLOv4 model for real time localization and classification of weapon objects into six different categories.

Keywords: Weapon detection · YOLOv4 · Deep learning · Real time object detection

1 Introduction

A minor security breach in identification of suspicious elements who carrying such harmful or restricted weapons/objects at such sensitive locations might have serious consequences in terms of national security. This becomes more challengeable particularly in public spaces such as airports, protests, movie theatres, stadiums, and national parks where maintaining security standards is necessary to ensure safety of infostructure and

M. Singh et al. (Eds.): ICACDS 2022, CCIS 1614, pp. 173–185, 2022.
https://doi.org/10.1007/978-3-031-12641-3_15

national themes including human life. DIPR studies [31, 32] suggest that the availability of violent material like riffle, churra, sword, sticks, handgun, grandees, bomb, etc. at location or even showing the violent material can escalate aggression and violent behavior in the people. The early detection of these material or the person who carrying these vulnerable materials may help security forces to apply their best crowd management strategies. Although the human visual framework has performed admirably in terms of monitoring, humans can be slow, expensive, and corruptible in the long run, can expose people on the ground to danger. With the escalation of technologies in hardware has given an opportunity to study and monitor the situations physically using videos and CCTV live footages but the amount of data generated was enormous, and studying each footage was a daunting process for humans. With the advancement of computational powers and availability of intensive dataset, training large deep learning networks for computer vision applications is possible. In the domain of object detection, CNN-based architecture algorithms like R-CNN (Regions with CNN features), Fast R-CNN, Faster R-CNN, SSD (Single Shot Multi-Box Detector), YOLO (You Only Look Once), and its versions, perform exceptionally well on real-time object detection, which in turn paved the way for machine-based surveillance systems (Fig. 1).

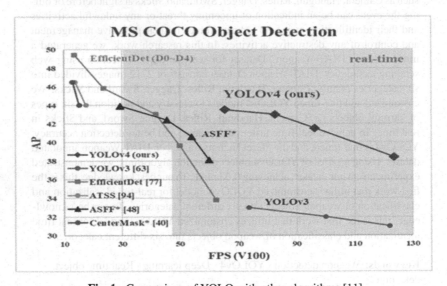

Fig. 1. Comparison of YOLO with other algorithms [11]

Building a customized weapon object detection algorithm poses major problems like, 1) lack of freely available weapon datasets, 2) necessity for a domain expert to decide harmful objects classes, and 3) real-time accurate classification and localization these objects from given video surveillance footage. In this work, we recognised various markers/harmful objects, such as cameras that can hamper privacy of sensitive locations like army base, and some hazardous weapons, such as Sticks, Daggers, Swords, Handguns, and Riffles, that can cause significant physical human injury. Further, we generated a novel dataset "DIAT-Weapon" Dataset along with annotations for threat object detection. The dataset consists of 2712 images divided into six categories. We customized one stage model – YOLO for the first time to detect weapons as it is outperforming other algorithms in terms of FPS, AP score and being size invariant, for getting real-time performance as any event like triggering gun in public places can happen in any moment. We fine-tuned YOLO algorithm to get better trade-off between accurate detection/localization and real-time performance.

2 Literature Review

Many object detection techniques based on deep learning have been proposed since 2018. These techniques can be based on 1) two-stage models, such as R-CNN [28], Fast R-CNN [25], Faster R-CNN [26], mask-RCNN [27], etc. mostly consisting, the region proposal network (RPN) network to select the approximate region of possible objects; followed by the object detection network to classify the candidate regions with accurate bounding; or 2) one-stage model, such as YOLO series [7, 8, 11] SSD [13], etc. where object detection is framed as a regression problem offering faster speed, with slightly lower accuracy.

Murugan et al. [1], has presented different object identification, object classification, and object tracking algorithms in the literature and has given methods for video summarization. Hu et al. [2], proposed a novel unified method for recognizing vehicle number plates and automobiles, gathering high energy frequency portions of images from digital camera imaging sensors with their proposed algorithm. For video surveillance Raghunandan et al. [3], has enhanced algorithms for various object detection techniques such as face detection, skin detection, color detection, shape detection, and target detection.

Elhoseny et al. [4], proposed a machine learning model for multi-object recognition and tracking that use an optimal Kalman filter [5] to track objects. Ahmad et al. [6], developed a framework for monitoring students during virtual tests, which employs YOLOv2 [7] and YOLOv3 [8] to detect things such as cell phones, laptops, iPads, and notebooks. Thoudoju et al. [9], uses YOLOv3 to detect objects in aerial images and satellite images. Kumar et al. [10], Detects vehicle classes such as automobile, truck, two-wheeler, and people using YOLOv3 and YOLOv4 [11]. Jose et al. [12], Detects things such as firearms and knives in suspicious regions using YOLO architecture to determine the likelihood of domestic violence. Though there are many variants of convolutional neural networks like SSD [13], R-FCN [14] performs well in terms of accuracy compared to YOLO, but their FPS (frame per second) is major drawback.

Though there are many object detection algorithms none has used to identify threat objects to ensure security standards. This study aims to build a real-time surveillance system that can detect dangerous objects in live CCTV feeds and, in the future, support these systems in surveillance robots. The paper is organized as follows: Sect. 3 describes about the dataset; Sect. 4 describes about the YOLO algorithms used and Sect. 5 consists of results and experiments.

3   DIAT Weapon Dataset

We chose stick, riffle, sword, handgun, camera, and dagger as our markers based on the severity of the damage that may be caused by utilizing these objects in prohibited locations, and it will offer the institution a perspective of what countermeasures they should do. Many weapons were brought under the umbrella of these classes, and subclass specifications were given in Table 1.

Table 1. Subclasses

Class	Sub classes
Sticks	Banton, lathi, hockey sticks, baseball bat
Riffle	Rifles, shotguns, muzzle loading firearms
Handgun	Pistols and revolvers
Camera	Surveillance cameras, digital cinema cameras, point and shoot cameras, DSLR e.tc
Daggers	Dagger, kitchen knives
Swords	Swords

The data was gathered from a variety of open sources, including the OIDV4 toolkit [15] and web scraping the images available in Internet ensuring diversity of collected data with respect to various conditions including different color, different shapes, different backgrounds, different time periods, variety of weather conditions, different occlusions, multiple perspective etc. Roboflow software is used to construct bounding boxes for each image to support for training the models in Darknet, TensorFlow and PyTorch. These classes are difficult to collect as there are few unique images for class in open domain, yet we managed 2,712 photos in total, divided into six classifications, with an average of 1.6 annotations per image. The sample images with annotation from generated DIAT-Weapon dataset are shown in Fig. 2. Those who are interested to get educational access to the DIAT-Weapon dataset, please send an e-mail request to "sunitadhavale@diat.ac.in" mentioning the subject: "DIAT-Weapon Image Dataset Educational Access Request" from their institutional e-mail id. This dataset will also be made publicly available at https://www.diat.ac.in/view-profile/?id=98.

Histogram plot of number objects for each class and number of annotations per image has given in Fig. 3 and 4. Data augmentation techniques are used to handle data imbalance problems. The Purpose of the data augmentation is to make model much robust towards the data. We Primarily focused on Photometric distortions such as random noise, Hue and Exposure. The image quality statistics were collected using Roboflow software, the average size of collected images were 0.7 mp, ranging from 0.01 to 20.90 mp and the median ratio of images is 1024 × 685. The aspect ratio histogram plot was given in Fig. 5, where majority of the images fell into the category of wider images. We conducted the experiments on images and resized to the size 416 × 416 and we added random noise of 5%, Hue and Exposure are between −25° to +25° as an augmentation technique.

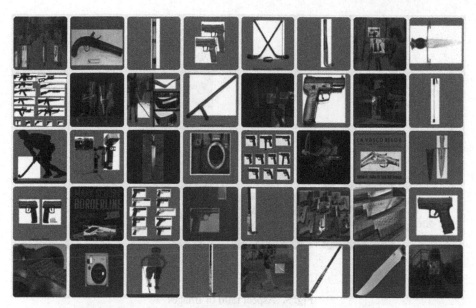

Fig. 2. Sample images

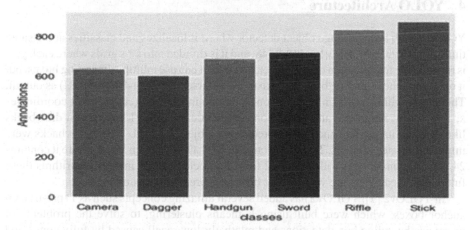

Fig. 3. Histogram of class balance

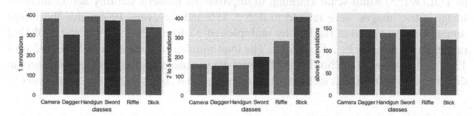

Fig. 4. Annotations distribution of classes

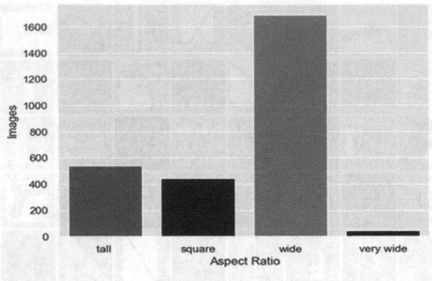

Fig. 5. Aspect ratio of images

4 YOLO Architecture

YOLOv1 [29] a single stage object detector where it localizes and classifies at the same time. Input size of YOLO v1 is 448 * 448, and it is divided into s * s grids where each grid is responsible for detection of one object, however it can use multiple bounding boxes, but it only gives the bounding box with maximum Intersection-over-Union (IOU) as output. The output will be position information of this bounding box (centre point coordinates x, y, width w, height h), and prediction confidence. YOLO v1 has certain drawbacks like constant image size and able to predict one object per grid. These drawbacks were improved in later versions. The architecture of YOLO v1 is given Fig. 6 where it contains 24 convolutional layers followed by 2 fully connected layers, in later algorithms these fully connected layers are with anchor boxes for predicting bounding boxes.

In YOLOv2 [7], YOLOv2 has added several amazing concepts such as 1) pre-trained anchor boxes, which were built using K-means clustering, to solve the problems of imprecise bounding box detections and relatively low recall caused by fully connected layers in YOLOv1. 2) Batch Normalization has improved 2% mAP score compared to YOLOv1. 3) Multi scale Training, to improve the model's stability across multiple image sizes, images were resized to 416 * 416 and for every 10 epochs, the image dimension was varied at random by multiples of 32 from 320 to 608 as the YOLOv2 model down samples by a factor of 2. The total number of detections in YOLOv2 is of 13 * 13 * number of anchor boxes. Darknet 19 was used as backbone architecture, to detect things faster, and it includes of 19 convolutional layers and 5 max pool layers.

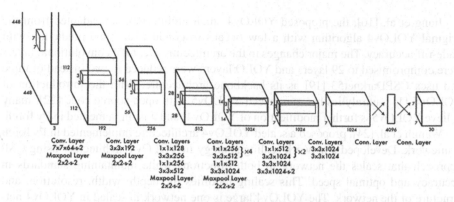

Fig. 6. YOLOv1 architecture [29]

In YOLOv3 [8], the Independent SoftMax layers was replaced with Independent logistic classifier for multi label classification to address the overlapping labels like Women and Person. YOLO v3 predicts in 3 feature scales like feature pyramid networks [30] as to improve prediction levels at large, medium, and small targets. At each scale it uses 3 boxes, and the shape of tensor is $N * N * (3 * (4 + 1 + C))$, where C is number of classes, 4 bounding box offsets and 1 objectiveness score. These feature maps are up sampled to concatenate with previous layer outputs. The backbone architecture Darknet-19 was modified with darknet-53 architecture because darknet-53 is size invariant. The convolutional layer of stride 2 is use instead of max pooling operation. YOLO v3 tiny is variation is YOLO v3 architecture where its backbone network consists of 7 convolutional layers and 6 max pooling layers, and it predicts in 2 scales. YOLO v3 tiny has compromised accuracy but it has a faster detection time.

Bochkovskiy et al. [11], proposed YOLOv4 architecture for object detection shown in Fig. 7, this architecture was implemented in Darknet framework. The YOLOv4 architecture was divided into 4 categories. 1) Input, it contains images, patches and video stream etc. 2) Back Bone, these convolutional Neural Network architectures were trained on Imagenet [18] Dataset and the author considers these networks CSPDarknet53 [19], CSPResNext50 [19] and EfficientNet-B3 [17] and finalizes CSPDarknet53 [19] as backbone network. 3) Neck, at neck combinations of various levels of backbone features are mixed and the author uses SPP [20] and PAN [21] as neck.4) Head, as a head the architecture uses YOLOv3 [8] for detection of objects.

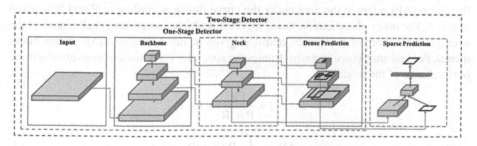

Fig. 7. Object detection algorithm architecture [11]

Jiang et al. [16], the proposed YOLOv4 small architecture was adapted from the original YOLOv4 algorithm with a few tweaks to conduct real-time prediction with trade-off accuracy. The major changes in the architecture are, total convolutional layers were compromised to 29 layers and YOLO layers were reduced to two instead of three and uses CSPDarknet53 [19] as its backbone architecture. With the introduction of YOLOv4 [11], ultralytics has announced YOLOv5 with open-source code [22], many believe YOLOv5 is further modification of YOLOv4, and it is implemented in PyTorch.

Wangh et al. [24], proposed a scaled YOLOv4 architecture implemented in PyTorch framework. Developed a network scaling strategy in YOLOv4 architecture using CSP approach that scales the network in both directions while maintaining standards in accuracy and optimal speed. This scaling modifies the depth, width, resolution, and structure of the network. The YOLOv4 large is one network in scaled in YOLOv4 networks, designed for cloud-based GPU to achieve high accuracy and it has variations like YOLOv4-P5, YOLOv4-P6 and YOLOv4-P7 where it detects objects in the scale of 3, 4, 5.

5 Experimental Results

In this section, the outcomes of the models were explained, as well as the metrics used to evaluate the models and the models' outputs. The DIAT Weapon image dataset is divided into train set, and test set in the ratio of 80:20. All these images are manually labelled using Roboflow software. All images are resized to 416×416 size. Due to small sized dataset, we used data augmentation like random horizontal translation, image flipping, and image distortion. The same transformation is performed on corresponding bounding boxes. All experiments are carried out on NVIDIA RTX-6000 GPU powered high end Tyrone workstation with 2 Intel Xeon processors, 256 GB RAM, 4 TB HDD configuration. For software stack, Python 3.7.2, CUDA 10.0, cuDNN 7.6.5, PyTorch 3.7.2, Darknet used.

In object detection models, we calculate precision based on the Intersection over Union metric [23]. Mean average precision (mAP) is the standard evaluation metric used for any object detection algorithms. Here we used mAP along with precision (P) and recall (R). Precision (P) ranged between [0, 1] and refers to the proportion of the correctly predicted 'True' labels in all the predicted 'True' labels. Recall (R) ranged between [0, 1] and represents the proportion of correctly predicted 'True' labels in the total number of actual 'True' labels. F1 Score is used to comprehensively measure the quality of algorithm in terms of both P and R as given in Eq. 1. For a category, average Precision (AP) refers to the area under the curve drawn according to P and R is given in Eq. 2. For multi-classification tasks, mAP of multiple categories is calculated as the average mAP score of all classes, and it is given in Eq. 3. Higher mAP means better model. For real-time classifications, frames per second (FPS) is used to measure real-time performance of the model.

$$F_1 = 2 * \frac{PR}{P + R} \in [0, 1] \tag{1}$$

$$AP_i = \int_0^1 P_i(R_i)dR_i \tag{2}$$

$$mAP = \frac{\sum_1^n AP_i}{n} \in [0, 1] \tag{3}$$

The metrics of the model are given in Table 2. All YOLO models pertained with MS-COCO benchmark dataset are fine-tuned to our weapon dataset. Whereas YOLOv4 outperformed other models in terms of mAP, Precision, and F1-score as 0.63, 0.77, 0.65 YOLOv4-csp outperformed other models in terms of recall with 0.67. The mAP value plots per batch for YOLOv4 and YOLOv4 tiny, trained in Darknet framework given Fig. 8 where x-axis represents number of batches and y-axis represent loss. For YOLOv5 and scaled YOLOv4 CSP, trained using PyTorch were given in Fig. 9 where x-axis represents epochs and y-axis for mAP score. Figures 10 and 11 show the results of YOLOv4 detecting single objects per image and multiple objects per image of six classes, where each object is identified by a bounding box and tagged with its class. In real time frames are extracted from videos using OpenCV and detection has done per image basis.

Table 2. Model results

Metric	YOLOv4-tiny	**YOLOv4**	YOLOv4-csp	YOLOv5
mAP@0.50	0.42	**0.63**	0.61	0.61
Precision	0.59	**0.77**	0.47	0.69
Recall	0.4	**0.56**	0.67	0.58
F1-Score	0.48	**0.65**	0.55	0.63

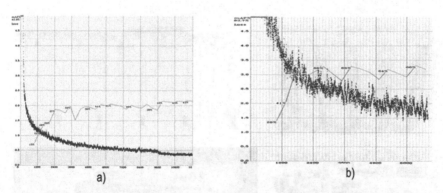

a)   b)

Fig. 8. Plot of mAP a) YOLOv4tiny b) YOLOv4

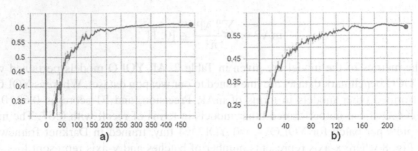

Fig. 9. Plot of mAP a) YOLOv5 b) scaled YOLOv4 CSP

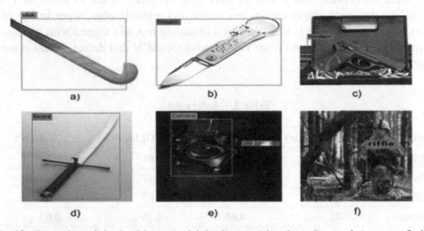

Fig. 10. Detection of single objects a) stick b) dagger c) handgun d) sword e) camera f) rifle

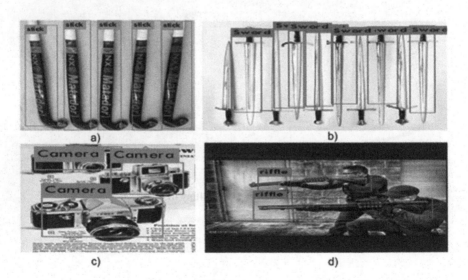

Fig. 11. Detection of multiple objects a) stick b) sword c) camera d) riffle

6   Conclusion and Future Scope

In this paper, we customized YOLO object detector for real time weapon detection. We introduced DIAT Weapon Dataset having 6 different classes of weapons i.e. Handgun, Sword, Camera, Riffle, Stick and Dagger. This dataset and learnt YOLO Model will be useful for harmful object detection to address the concerns of national security. During experimental analysis, it is found that YOLOv4 has achieved significant results in real time demonstration using videos as test dataset with more than 30 fps using OpenCV DNN module. Although YOLO v5 is lightweight than YOLO v4, accuracy of YOLO v4 was found good in real time performance. In future, we have two objectives, 1) we will add some more images and categories of weapon in our proposed dataset and will try to enhance prediction accuracy, generalization capability along with faster detection. To expand the data set, we will use Generative Adversarial Networks (GAN) network. 2) To integrate the trained models to work on real time CCTV feeds along with robots.

Acknowledgements. This research is supported by the Life Sciences Research Board (LSRB) in association with Defence Institute of Psychological Research (DIPR) sanction letter no. LSRB/o1/15001/M/LSRB-381/PEE& BS/2020, dated 15.03.2021. The authors want to thank NVIDIA for the academic GPU research grant. The authors wish to thank all the students who supported us during the data collection.

References

1. Senthil Murugan, A., Suganya Devi, K., Sivaranjani, A., Srinivasan, P.: A study on various methods used for video summarization and moving object detection for video surveillance applications. Multimedia Tools Appl. **77**(18), 23273–23290 (2018). https://doi.org/10.1007/s11042-018-5671-8
2. Hu, L., Ni, Q.: IoT-driven automated object detection algorithm for urban surveillance systems in smart cities. IEEE Internet Things J. **5**(2), 747–754 (2018). https://doi.org/10.1109/JIOT.2017.2705560
3. Raghunandan, A., Raghav, P., Ravish Aradhya, H.V.: Object detection algorithms for video surveillance applications. In: 2018 International Conference on Communication and Signal Processing (ICCSP). IEEE (2018)
4. Elhoseny, M.: Multi-object detection and tracking (MODT) machine learning model for real-time video surveillance systems. Circuits Syst. Signal Process. **39**(2), 611–630 (2020)
5. Welch, G., Bishop, G.: An introduction to the Kalman filter, pp. 127–132 (1995)
6. Ahmad, I.: A novel deep learning-based online proctoring system using face recognition, eye blinking, and object detection techniques. System **12**(10) (2021)
7. Redmon, J., Farhadi, A.: YOLO9000: better, faster, stronger. In: 2017 IEEE Conference on Computer Vision and Pattern Recognition (CVPR), pp. 6517–6525 (2017). https://doi.org/10.1109/CVPR.2017.690
8. Redmon, J., Farhadi, A.: YOLOv3: an incremental improvement. arXiv preprint arXiv:1804.02767 (2018)
9. Thoudoju, A.K.: Detection of aircraft, vehicles and ships in aerial and satellite imagery using evolutionary deep learning. Dissertation (2021). http://urn.kb.se/resolve?urn=urn:nbn:se:bth-22310

10. Kumar, B.C., Punitha, R., Mohana, M.: YOLOv3 and YOLOv4: multiple object detection for surveillance applications. In: 2020 Third International Conference on Smart Systems and Inventive Technology (ICSSIT), pp. 1316–1321 (2020). https://doi.org/10.1109/ICSSIT 48917.2020.9214094

11. Bochkovskiy, A., Wang, C.-Y., Liao, H.-Y.M.: YOLOv4: optimal speed and accuracy of object detection. arXiv preprint arXiv:2004.10934 (2020)

12. Jose, D.: Deep learning based gender responsive smart device to combat domestic violence. SPAST Abstr. **1**(01) (2021). https://spast.org/techrep/article/view/2933

13. Liu, W., et al.: SSD: single shot multibox detector. In: Leibe, B., Matas, J., Sebe, N., Welling, M. (eds.) ECCV 2016. LNCS, vol. 9905, pp. 21–37. Springer, Cham (2016). https://doi.org/10.1007/978-3-319-46448-0_2

14. Dai, J., Li, Y., He, K., Sun, J.: R-FCN: object detection via region-based fully convolutional networks. In: Advances in Neural Information Processing Systems, vol. 29. Curran Associates, Inc. (2016). https://proceedings.neurips.cc/paper/2016/file/577ef1154f3240ad5b 9b413aa7346a1e-Paper.pdf

15. Vittorio, A.: OIDv4_ToolKit: toolkit to download and visualize single or multiple classes from the huge Open Images V4 dataset. GitHub repository (2018). https://github.com/EscVM/OIDv4_ToolKit. Accessed 04 Apr 2022

16. Jiang, Z., et al.: Real-time object detection method based on improved YOLOv4-tiny. arXiv preprint arXiv:2011.04244 (2020)

17. Tan, M., Le, Q.: EfficientNet: rethinking model scaling for convolutional neural networks. In: Proceedings of the 36th International Conference on Machine Learning. PMLR, vol. 97, pp. 6105–6114 (2019)

18. Krizhevsky, A., Sutskever, I., Hinton, G.E.: ImageNet classification with deep convolutional neural networks. In: Advances in Neural Information Processing Systems 25, pp. 1097–1105 (2012). https://doi.org/10.1145/3065386

19. Wang, C.-Y., Mark Liao, H.-Y., Wu, Y.-H., Chen, P.Y., Hsieh, J.W., Yeh, I.H.: CSPNet: a new backbone that can enhance learning capability of CNN. In: 2020 IEEE/CVF Conference on Computer Vision and Pattern Recognition Workshops (CVPRW), pp. 1571–1580 (2020). https://doi.org/10.1109/CVPRW50498.2020.00203

20. He, K., Zhang, X., Ren, S., Sun, J.: Spatial pyramid pooling in deep convolutional networks for visual recognition. IEEE Trans. Pattern Anal. Mach. Intell. **37**(9), 1904–1916 (2015). https://doi.org/10.1109/TPAMI.2015.2389824

21. Liu, S., Qi, L., Qin, H., Shi, J., Jia, J.: Path aggregation network for instance segmentation. In: 2018 IEEE/CVF Conference on Computer Vision and Pattern Recognition, pp. 8759–8768 (2018) https://doi.org/10.1109/CVPR.2018.00913

22. YOLOv5: Ultralytics open-source research into future vision AI methods. https://github.com/ultralytics/yolov5. Accessed 04 Apr 2022

23. Rezatofighi, H., Tsoi, N., Gwak, J., Sadeghian, A., Reid, I., Savarese, S.: Generalized intersection over union: a metric and a loss for bounding box regression. In: 2019 IEEE/CVF Conference on Computer Vision and Pattern Recognition (CVPR), pp. 658–666 (2019). https://doi.org/10.1109/CVPR.2019.00075

24. Wang, C., Bochkovskiy, A., Liao, H.: Scaled-YOLOv4: scaling cross stage partial network. In: 2021 IEEE/CVF Conference on Computer Vision and Pattern Recognition (CVPR), Nashville, TN, USA, pp. 13024–13033 (2021). https://doi.org/10.1109/CVPR46437.2021.01283

25. Girshick, R.: Fast R-CNN. In: 2015 IEEE International Conference on Computer Vision (ICCV), pp. 1440–1448 (2015). https://doi.org/10.1109/ICCV.2015.169

26. Ren, S., He, K., Girshick, R., Sun, J.: Faster R-CNN: towards real-time object detection with region proposal networks. IEEE Trans. Pattern Anal. Mach. Intell. **39**(06), 1137–1149 (2017). https://doi.org/10.1109/TPAMI.2016.2577031

27. He, K., Gkioxari, G., Dollár, P., Girshick, R.: Mask R-CNN. In: 2017 IEEE International Conference on Computer Vision (ICCV), pp. 2980–2988 (2017). https://doi.org/10.1109/ICCV.2017.322
28. Girshick, R., Donahue, J., Darrell, T., Malik, J.: Rich feature hierarchies for accurate object detection and semantic segmentation. In: 2014 IEEE Conference on Computer Vision and Pattern Recognition, pp. 580–587 (2014). https://doi.org/10.1109/CVPR.2014.81
29. Redmon, J., Divvala, S., Girshick, R., Farhadi, A.: You only look once: unified, real-time object detection. In: 2016 IEEE Conference on Computer Vision and Pattern Recognition (CVPR), pp. 779–788 (2016). https://doi.org/10.1109/CVPR.2016.91
30. Lin, T.-Y., Dollár, P., Girshick, R., He, K., Hariharan, B., Belongie, S.: Feature pyramid networks for object detection. In: 2017 IEEE Conference on Computer Vision and Pattern Recognition (CVPR), pp. 936–944 (2017). https://doi.org/10.1109/CVPR.2017.106
31. Suresh, K.: Detection, analysis and management of atypical behaviour of crowd and mob in LIC environment. ST/14/DIP-732, DIPR/Note/No./714 (2017)
32. Suresh, K.: Predicting the probability of stone pelting in crowd of J&K. ST/14/DIP-732, DIPR/Note/No./719 (2018)

Audio Recognition Using Deep Learning for Edge Devices

Aditya Kulkarni[1]([✉]), Vaishali Jabade[1], and Aniket Patil[1,2]

[1] Electronics and Telecommunication, Vishwakarma Institute of Technology, Pune, India
{aditya.kulkarni18,vaishali.jabade}@vit.edu, Aniket.Patil@ifm.com
[2] Artificial Intelligence Team, ifm Engineering Private Limited, Pune, India

Abstract. This paper has proposed a methodology that creates an automatic speech recognition system, the task for which would be to recognize keywords. Deep learning was deployed for classifying the spoken words. The created audio data set consisted of short audio clips which were then converted to a Spectrogram by computing Short Time Fourier Transform (STFT) of each audio sample from the data. Spectrogram is a picture of spectrum of frequencies of a signal. Convolutional neural network is a deep learning algorithm, prominently used for classifying image data. In our case Spectrogram, which is the audio representation, was used to train the CNN model and the model which achieved a higher recognition rate was deployed on the hardware. The proposed research has been motivated from the requirement of an audio classification system that can be deployed on hardware and further based on the classification the hardware has been assigned certain task to be completed.

Keywords: Convolutional Neural Networks · Deep learning · Short Time Fourier Transform · Spectrogram

1 Introduction

Sound is the result of a phenomena that can be explained as, mechanical radiant energy carried through a material medium by longitudinal pressure wave. It is known that sound can be stored digitally on computers or machines that have the capability of recording and processing it. For a computer to store and process raw sound, it needs to be preprocessed and converted from analog to digital format. Audio Recognition or Speech Recognition [1] creates strategies and technology that makes computers capable to recognize and translate digitally stored audio or spoken language into text. Speech technology is already being used in a variety of industries, helping organizations and individuals improve their productivity and their lives are being efficient due to this change. There are numerous algorithms that have been developed for recognizing speech, some of the prominent ones are,

---

V. Jabade and A. Patil—These authors contributed equally to this work.

© The Author(s), under exclusive license to Springer Nature Switzerland AG 2022
M. Singh et al. (Eds.): ICACDS 2022, CCIS 1614, pp. 186–198, 2022.
https://doi.org/10.1007/978-3-031-12641-3_16

A Natural Language Processing (NLP): Natural language processing is a hybrid
   of artificial intelligence and semantics. NLP allows humans to interact with
   computers; it is most commonly employed with machine learning algorithms
   to identify voice and convert it to text.
B Hidden Markov Models (HMM): The main principle that is behind the Hidden Markov Models is the Markov chain. The Markov chain is based on the
   Markov property in probability theory and statistics.
C N-grams: These are the most fundamental language models, in which sentences or phrases are given a probability. An N-gram is a concatenation of
   N-words. For example, "start local setup" is a trigram or 3-gram. Grammar and the probability of specific word sequences are employed to improve
   recognition and accuracy.
D Artificial Neural Networks: Speech recognition systems use deep learning for
   recognizing audio [2]. Deep learning is based on artificial neural networks,
   it is a form of machine learning that uses numerous layers of processing to
   extract progressively higher-level properties from data.

We have attempted to create an ASR [3] system using [4–6] deep learning
algorithm convolutional neural network (CNN). CNNs take an image as input
and uses filters to learn the image's many features. This allows them to notice
the important points in the image and differentiate one from the other. The
ability of Convolutional Neural Network to pre-process information on its own
distinguishes it from other machine learning techniques [7]. The input to the
proposed CNN is a time-frequency representation of the audio signal known as
spectrogram as an image.The audio signal is converted to a spectrogram [8] using
the Short Time Fourier Transform (STFT). The frequency of a tonal sound is
interpreted as high or low pitch in any audio. Because the signals on the spectrum
may be visually inspected, the Fourier spectrum of a signal gives such frequency
content. In actuality, we work with discrete-time signals, hence the Discrete
Fourier Transform (DFT) is the time-frequency transform to use. Positive and
negative frequency components are complex conjugates of each other for real-
valued inputs. Non-stationary signals, on the other hand, are speech signals.
When we translate a spoken sentence to the frequency domain, we receive a

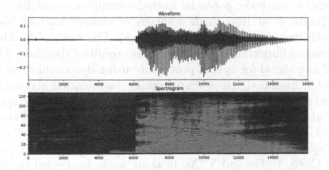

Fig. 1. This figure presents an example of a signal's waveform and its corresponding
spectrogram

spectrum that is an average of all phonemes in the sentence, despite the fact that we usually want to see the spectrum of each individual phoneme separately.

By separating the signal into shorter sections, we can focus on signal quality at a specific point in time, a technique known as windowing. By windowing and taking the discrete Fourier transform (DFT) of each window, we obtain the signal's short-time Fourier transform (STFT). The STFT [9] is one of the most often used tools in speech analysis and processing. It describes the evolution of frequency components over time. STFTs, like the spectrum of the audio itself, have a physical and intuitive explanation for their parameters. Another similarity is that the STFT output is complex-valued; however, where the spectrum output is a vector, the STFT output is a matrix. As a result, we can't see the complex-valued output right away. The log- spectra of STFTs, on the other hand, is widely employed to visualize them [10]. A spectrogram, can be viewed in Fig. 1, is a type of heat map that can be used to understand such two- dimensional log-spectra.

2   Related Work

In their study, J. Meng, J. Zhang, and H. Zhao introduce voice recognition and discuss its basic concepts and applications. The study analyses the classification of various voice recognition systems. Audio classification techniques using deep learning have been analyzed by M. S. Imran et al. The paper presents knowledge required to select the classification model for audio classification and detection. D. O'Shaughnessy in their paper explain the importance of creation of an Automatic Speech Recognition (ASR) system.

Deng et al. have summarized the invited and contributed papers. For review, the material has been divided into five pieces which mention importance of neural network hyperparameters. The paper by Malik et al. provides a thorough comparison of cutting-edge approaches currently being employed in the field of automatic voice recognition.

Hamid et al. present a detailed study on Convolutional Neural Networks used for creating an automatic speech recognition system. The research conducted by L. Deng and X. Li is organized around distinct ML paradigms that are either already popular or have the potential to make significant contributions to ASR technology. Tandel et al. have made a survey of various machine learning techniques in which deep learning is also mentioned. The work of N. Zheng and X. -L. Zhang presents importance about the phase result of the Short Time Fourier Transform of any signal for speech processing using deep neural networks.

G. Yu and J. Slotine derive a simple audio classification algorithm based on treating sound spectrogram as texture images. The importance to speech feature extraction is given in the paper by U. Khan, M. Sarim et al. It reviews different feature extraction techniques such as Principal Component Analysis (PCA), Mel Frequency Cepstral Coefficients (MFCC), Cepstral Analysis (CA), etc.

P. Li, M. Chen, F. Hu and Y. Xu in their work, have used spectrograms to develop a speech recognition system, the methodology consists of use of Local

Binary Patterns (LBP) operator for obtaining the features from the spectrogram used to train the deep neural network. A. Graves et al. mention the work of LSTM RNN for speech recognition with evaluation of the deep neural network on the TIMIT phoneme recognition benchmark dataset. K. J. Piczak has compared their proposed technique to a baseline system based on MFCCs using the largest public dataset of urban sound sources accessible for research.

Sumon et al. have mentioned work on three different deep neural networks for recognizing Bangla short speech commands. The authors have mentioned in detail the results of design of deep neural network. In Paper by Nanni et al. present ensembles of classifiers, using various data augmentations and tested the classifiers over three freely available environmental audio benchmark datasets. The paper suggests four audio image representations of which we are using the Mel Spectrogram. Andreadis et al. mention a convolutional neural network based solution for audio classification on resource constrained wireless edge devices.

N. Hailu, I. Siegert and A. Nürnberger, in their paper, have given importance to augmentation of audio data. The augmentation method consists of using different audio codecs with changed bit rate, sampling rate, and bit depth. The proposed study by A. Elbir et al. investigated the music genre categorization problem utilizing the Short Time Fourier Transform (STFT), which is one of the most useful time-frequency analysis tools. The chapter from "Speech Enhancement Techniques for Digital Hearing aids" by K. R. Borisagar, R. M. Thanki and B.S. Sedani discusses the Fourier transform (FT), the short- time Fourier transform (STFT), and the wavelet transform.

Kong et al. have experimented and evaluated various automatic speech recognition systems and compared the results with human perception results. One of the systems mentioned consists of usage of deep neural networks for which the input features extracted were Mel-scale log filter banks. G. R. Doddington discuss the intrinsic performance constraints of speaker and audio recognition, as well as an assessment of the performance attained by listening, visual spectrogram analysis, and automatic computer procedures.

The authors K. Minami et al. follow approach for classifying respiratory sounds which is divided into two parts. First, one-dimensional signals are transformed into two-dimensional time-frequency representation images using the short-time Fourier transform and the continuous wavelet transform. Second, convolutional neural networks are used to classify transferred images. In their study, J. Salmon and J. P. Bello addressed an investigation that yields a distinction between the aspects of the processing pipeline that influence performance and the particular characteristics of the urban auditory domain.

The approach used in this research aims to develop an automatic speech recognition system utilizing Artificial Neural Networks. To train the neural network, each sample is transformed to a spectrogram using the Short Time Fourier Transform. Various audio data augmentation techniques are used to improve the neural network's identification rate and to introduce variance between each sample. After that, the neural network was put to use on hardware.

3   Proposed Methodology

The proposed methodology is mentioned in the steps as follows, firstly we will take a look at the custom data that was created followed by that we will observe augmentation and pre-processing techniques for the collected data. The third step is about different architectures that were created and the final step consists of information about the deployment of the neural network on the hardware.

The spectrogram audio representation is used and is passed as feature vector [11] to the neural network, there are various other feature vectors and techniques one of the technique is mentioned in [12]. The audio representation that was chosen for training the neural network was STFT, it was chosen due to its simple and readily available implementation using TensorFlow for Convolutional Neural Networks. The hardware deployment was motivated from the idea of task based audio recognition system, each keyword spoken would be recognised by the hardware and the hardware would then perform a task which would be dependent on the keyword.

3.1   Dataset

The audio data that we have collected is stored digitally in the WAV format. The data contains short audio clips of certain speech commands. The recording was done on a Laptop, by changing different microphones and surrounding to ensure a perfect variation between the samples in the dataset. Firstly, the data set consisted of three keywords which is also the first data set, "get", "stop" and "wait" for which there were 30 samples collected per keyword, in total 180 files. The sole purpose of this data containing 30 WAV files per keyword was to get us acquainted with the basics of the Audio Recognition process using Deep Learning technology. It was created to understand the process of conversion of audio signal to spectrogram, training the network and deployment on hardware for the first time.

Further, the data was changed to consist only two keywords as "start local setup" and "stop local setup", this data set which can be considered as second data set consisted 330 samples per keyword and the network architecture was kept similar for this data. After training we observed the model to overfit and to prevent overfitting, we introduced data augmentation. The audio data augmentation technique that was utilized to add augmented samples was noise addition. After the augmentation of data, the final number of samples per keyword doubled to 660, in which 330 were normal recorded samples and other 330 were augmented samples, in total 1,320 files. The model architecture was changed again based on the training results from this data. The final model architecture which achieved a higher recognition rate was trained on the final data (Third data set) which consisted of 14 different keywords, this was considered to be the final data set. Each data contained 700 samples per keyword, in total 9800 files.

Other implementations such as [13–16], have used state of the art data sets which have a huge amount of data, which is desired for any machine learning use case. The purpose of choosing a small data set was to replicate and try to solve the problem that edge computing [17] and real world non consumer applications face, which do not have access to huge amounts of data. A neural network for such application where the data is not in abundance and the architecture must be as small as possible was the goal for the deployment on hardware.

3.2 Data Preprocessing

Data augmentation aids in the generalization of synthetic data from an existing data set, allowing the model's generalization potential to be increased. For audio data [18], the augmentation techniques such as noise addition, time shifting, changing pitch and speed can be done. Before going for pre-processing the audio dataset collected is passed through certain augmentations. The augmentation done is adding random noise to the audio files, with the help of Python's NumPy library the audio files are introduced with random noise.

After completion of augmentation the next step is to convert the audio to a Spectrogram. To create a Spectrogram, TensorFlow's Signal module [19] is used. The module consists of a function that computes the Short Time Fourier Transform of a signal. The Short Time Fourier Transform (STFT) [20] is used to perform time-frequency analysis, it is utilized to make representations that capture the signal's local time and frequency information. The STFT, like the Fourier transform [21], employs fixed basis functions to transform the signal; however, instead of using fixed-size time-shifted window functions, it uses fixed-size time-shifted window functions. In simple terms the process of STFT can be reviewed, firstly a chunk is taken out of the signal, the size of the chunk can be decided, in our case we took the chunk size to be same as the window size, then the chunk is multiplied by the window function, we used the Hann window function to generate a window. The next step is to multiply the window to the chunk, and then compute the Discrete Fourier Transform (DFT) of the multiplication result.

This small procedure of multiplying and computing DFT is repeated all over the signal. The overall result is a Spectrogram [22–24] which can also be said as a pictorial representation of the signal's strength.

3.3 Proposed Neural Network

Various machine learning techniques for audio classification include Logistic Regression, K-nearest neighbor, K-means [25], Random Forest and Support Vector Machine. Deep learning's Deep Neural Network is one of the many techniques for audio classification. There are various neural network architectures under deep learning that have been deployed. Certain state of the art fully connected deep neural networks such as AlexNet, VGG, Inception and ResNet have tremendously large architectures.

Our goal was to select a smaller architecture for our Convolutional Neural Network (CNN), there are various reasons for the same, the most important reason for us was deploying the model on the hardware. For cross platform deployment of neural networks we have used TensorFlow Lite. It is an open source framework which is used to inference machine learning models on mobile and edge devices. The proposed model architecture was different for all the three data sets and the architecture for the third dataset was considered to be the final architecture. For the second dataset the model architecture was kept similar as shown in Fig. 2.

After training the model on the first data, we changed the structure and keywords of the second data set. There were only two keywords, this was done to observe the performance of the model when there are less number of classes in the data set. The model architecture for first and second data set are the same.

For the final dataset the model architecture was changed due to the size of the data and the number of classes was 14. There were two different architectures that were observed first one consisted input resizing value as (32, 32) the architecture for the same is given in Fig. 3. The highest recognition rate was observed with the architecture that consisted input resizing value as (64, 64) which is shown in Fig. 4.

Layer (type)	Output Shape	Param #
resizing (Resizing)	(None, 32, 32, 1)	0
normalization (Normalizatio n)	(None, 32, 32, 1)	3
conv2d (Conv2D)	(None, 30, 30, 32)	320
conv2d_1 (Conv2D)	(None, 28, 28, 64)	18496
max_pooling2d (MaxPooling2D)	(None, 14, 14, 64)	0
dropout (Dropout)	(None, 14, 14, 64)	0
flatten (Flatten)	(None, 12544)	0
dense (Dense)	(None, 128)	1605760
dropout_1 (Dropout)	(None, 128)	0
dense_1 (Dense)	(None, 8)	1032

Fig. 2. The architecture created for the first dataset to understand the process of audio classification using Spectrograms

Layer (type)	Output Shape	Param #
resizing (Resizing)	(None, 32, 32, 1)	0
normalization (Normalization	(None, 32, 32, 1)	3
conv2d (Conv2D)	(None, 30, 30, 32)	320
conv2d_1 (Conv2D)	(None, 28, 28, 32)	9248
conv2d_2 (Conv2D)	(None, 26, 26, 64)	18496
conv2d_3 (Conv2D)	(None, 24, 24, 64)	36928
conv2d_4 (Conv2D)	(None, 22, 22, 128)	73856
max_pooling2d (MaxPooling2D)	(None, 11, 11, 128)	0
dropout (Dropout)	(None, 11, 11, 128)	0
flatten (Flatten)	(None, 15488)	0
dense (Dense)	(None, 128)	1982592
dropout_1 (Dropout)	(None, 128)	0
dense_1 (Dense)	(None, 14)	1806
activation (Activation)	(None, 14)	0

Total params: 2,123,249
Trainable params: 2,123,246
Non-trainable params: 3

Fig. 3. Final data set model architecture, input to this architecture is resized to 32 × 32.

3.4   Deploying Neural Network on Hardware

TensorFlow Lite is a cross-platform, open-source deep learning framework that transforms a TensorFlow pre-trained model to a unique format that can be adjusted for speed or storage. To make the inference at the edge, the transformed format model can be installed on edge devices such as mobile phones running Android or iOS, or Linux-based embedded devices such as Raspberry Pi or Micro-controllers. Before converting the model to tflite file, it needs to be stored locally where it was trained. There are two formats for storing a TensorFlow model, HDF5 and SavedModel, we used the SavedModel format to store the trained model. After saving the model it is converted to TFLite format using post-training qunatization which is provided by TensorFlow Lite, it creates a tflite model file which then is used to inference on a programmable graphic display which had an ARM quad core 64-bit processor with 1GByte of RAM as working memory and 8GByte Flash mass storage, with embedded Linux 4.14 as the operating system. It consists of an LED display with display resolution of 1280×480.

```
Input shape: (374, 129, 1)
Model: "sequential_1"

Layer (type)                    Output Shape              Param #
=================================================================
resizing_1 (Resizing)           (None, 64, 64, 1)         0

normalization_1 (Normalizati    (None, 64, 64, 1)         3

conv2d_5 (Conv2D)               (None, 62, 62, 32)        320

conv2d_6 (Conv2D)               (None, 60, 60, 32)        9248

conv2d_7 (Conv2D)               (None, 58, 58, 64)        18496

conv2d_8 (Conv2D)               (None, 56, 56, 64)        36928

conv2d_9 (Conv2D)               (None, 54, 54, 128)       73856

max_pooling2d_1 (MaxPooling2    (None, 27, 27, 128)       0

dropout_2 (Dropout)             (None, 27, 27, 128)       0

flatten_1 (Flatten)             (None, 93312)             0

dense_2 (Dense)                 (None, 128)               11944064

dropout_3 (Dropout)             (None, 128)               0

dense_3 (Dense)                 (None, 14)                1806

activation_1 (Activation)       (None, 14)                0
=================================================================
Total params: 12,084,721
Trainable params: 12,084,718
Non-trainable params: 3
```

Fig. 4. This figure represents the model architecture for final data set with input resizing layer getting a size as 64×64.

For inference, the hardware was programmed to carry out a task based on the classification. The process for the same can be described in steps, firstly audio is recorded on the hardware, then according to the data processing required for the model, in the similar manner the recorded audio is processed and the tflite model file is loaded into the hardware. In the next step after running inference on the processed recorded audio the predictions are obtained. Finally, based on these predictions the hardware has been programmed to complete the assigned task. The task is assigned on the basis of predictions for the classes, for example a particular task will perform if certain the prediction of certain class is observed higher than others. The result from the hardware is presented in Fig. 5

Fig. 5. Inference results on programmable graphic display show the confidence score of the prediction done on certain test samples.

4 Results

The neural network for the first data set which consisted only 3 keywords and for the second data set which consisted only 2 keywords was kept similar, only change that was done was in the second data which was due to the model overfitting. For the third data set the model was trained for more epochs with a lower learning rate. Throughout all the architectures the training set consisted 80% of data, for validation set we used 10% and same for the testing set.

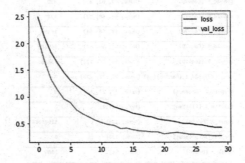

Fig. 6. After training the model using the architecture for final data set which consisted resizing layer value as 32×32, the loss curve for training is represented in this figure.

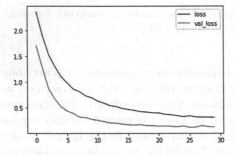

Fig. 7. This figure shows the loss curve which is plotted from the training of the model architecture that gets input resized to 64×64 through the resizing layer.

Figures 6 and   7 represent the loss curves for third data set for both the resized inputs 32×32 and 64×64 and the final model architecture that gave the higher recognition rate.

Table 1. Summary of all architectures

Neural network architecture for data set 1	
Number of hidden layers	2
Accuracy	90%
Total number of samples for training	180
Number of keywords	3
Number of epochs for training	10
Learning rate	0.01
testing accuracy	91%
Neural network architecture for data set 2 (Original files)	
Number of hidden layers	2
Accuracy	92%
Total number of samples for training	660
Number of keywords	2
Number of epochs for training	20
Learning rate	0.0005
testing accuracy	90%
Neural network architecture for data set 2 (Original | augmented files)	
Number of hidden layers	2
Accuracy	96%
Total number of samples for training	1320
Number of keywords	2
Number of epochs for training	20
Learning rate	0.0005
testing accuracy	94%
Neural network architecture for data set 3 (Original files)	
Number of hidden layers	5
Accuracy	93%
Total number of samples for training	1960
Number of keywords	14
Number of epochs for training	30
Learning rate	0.0003
testing accuracy	92%
Neural network architecture for data set 3 (Original + augmented files)	
Number of hidden layers	5
Accuracy	96
Total number of samples for training	9800
Number of keywords	14
Number of epochs for training	50
Learning rate	0.0005
testing accuracy	96%

5 Conclusion

From the results it can be observed that the model designed after numerous training sessions using the final dataset which consisted of large number of classes, the model achieved a final recognition rate of 96% with the architecture being simple, the model is portable on edge devices and is cross compiled on the mentioned hardware using TensorFlow Lite. Our goal, thus, was met which was to create a simple automatic speech recognition using a smaller model architecture which can recognize various keywords with a greater accuracy and should be compatible with the memory specifications of the hardware, so as to run inference on it. The presented research can be explored in various ways. Changes can be done to the data and the audio representations which were used to train the neural network. Using some more experimentation the model architecture could be kept smaller so as to do cross compilation without worrying for the memory restriction in hardware. Furthermore, the presented research allowed us to understand the importance of time frequency analysis of a non stationary signal for real world applications. The summary for all architectures is presented in Table 1.

Acknowledgments. I would like to thank Aniket Patil, and the AI Team of ifm Eng. Pvt. Ltd. for providing me with guidance and support at every phase of this research, as well as for imparting vital knowledge and teaching me the etiquettes of a professional employee. I would also want to convey my gratefulness to Vaishali Jabade, for her steady, valuable instruction, patience, constant care, and kind encouragement throughout the research work, which enabled me to deliver this paper in an effective manner.

References

1. Meng, J., Zhang, J., Zhao, H.: Overview of the speech recognition technology. In: 2012 Fourth International Conference on Computational and Information Sciences, 2012, pp. 199–202 (2012). https://doi.org/10.1109/ICCIS.2012.202
2. Imran, M.S., Rahman, A.F., Tanvir, S., Kadir, H.H., Iqbal, J., Mostakim, M.: An analysis of audio classification techniques using deep learning architectures. In: 2021 6th International Conference on Inventive Computation Technologies (ICICT), 2021, pp. 805–812. https://doi.org/10.1109/ICICT50816.2021.9358774
3. O'Shaughnessy, D.: Automatic speech recognition. In: 2015 CHILEAN Conference on Electrical, Electronics Engineering, Information and Communication Technologies (CHILECON), 2015, pp. 417–424. https://doi.org/10.1109/Chilecon.2015.7400411
4. Deng, L., Hinton, G., Kingsbury, B.: New types of deep neural network learning for speech recognition and related applications: an overview. In: 2013 IEEE International Conference on Acoustics, Speech and Signal Processing, 2013, pp. 8599–8603. https://doi.org/10.1109/ICASSP.2013.6639344
5. Malik, M., Malik, M.K., Mehmood, K., Makhdoom, I.: Automatic speech recognition: a survey. Multimed. Tools Appl. **80**(6), 9411–9457 (2020). https://doi.org/10.1007/s11042-020-10073-7

6. Abdel-Hamid, O., Mohamed, A., Jiang, H., Deng, L., Penn, G., Yu, D.: Convolutional neural networks for speech recognition. IEEE/ACM Trans. Audio Speech Lang. Process. **22**(10), 1533–1545 (2014). https://doi.org/10.1109/TASLP.2014.2339736

7. Deng, L., Li, X.: Machine learning paradigms for speech recognition: an overview. IEEE Trans. Audio Speech Lang. Process. **21**(5), 1060–1089 (2013). https://doi.org/10.1109/TASL.2013.2244083

8. Tandel, N.H., Prajapati, H.B., Dabhi, V.K.: Voice recognition and voice comparison using machine learning techniques: a survey. In: 2020 6th International Conference on Advanced Computing and Communication Systems (ICACCS), pp. 459–465 (2020). https://doi.org/10.1109/ICACCS48705.2020.9074184

9. Zheng, N., Zhang, X.-L.: Phase-aware speech enhancement based on deep neural networks. IEEE/ACM Trans. Audio Speech Lang. Process. **27**(1), 63–76 (2019). https://doi.org/10.1109/TASLP.2018.2870742

10. Yu, G., Slotine, J.: Audio classification from time-frequency texture. In: 2009 IEEE International Conference on Acoustics, Speech and Signal Processing, 2009, pp. 1677–1680. https://doi.org/10.1109/ICASSP.2009.4959924

11. Khan, U., Sarim, M., Bin Ahmad, M., Shafiq, F.: Feature extraction and modeling techniques in speech recognition: a review. In: 2019 4th International Conference on Information Systems Engineering (ICISE), pp. 63–67 (2019). https://doi.org/10.1109/ICISE.2019.00020

12. Li, P., Chen, M., Hu, F., Xu, Y.: A spectrogram-based voice print recognition using deep neural network. In: The 27th Chinese Control and Decision Conference (2015 CCDC), pp. 2923–2927 (2015). https://doi.org/10.1109/CCDC.2015.7162425

13. Graves, A., Mohamed, A., Hinton, G.: Speech recognition with deep recurrent neural networks. In: 2013 IEEE International Conference on Acoustics, Speech and Signal Processing, pp. 6645–6649 (2013). https://doi.org/10.1109/ICASSP.2013.6638947

14. Piczak, K.J.: Environmental sound classification with convolutional neural networks. In: 2015 IEEE 25th International Workshop on Machine Learning for Signal Processing (MLSP), pp. 1–6 (2015). https://doi.org/10.1109/MLSP.2015.7324337

15. Ahmed Sumon, S., Chowdhury, J., Debnath, S., Mohammed, N., Momen, S.: Bangla short speech commands recognition using convolutional neural networks. In: 2018 International Conference on Bangla Speech and Language Processing (ICBSLP), pp. 1–6 (2018). https://doi.org/10.1109/ICBSLP.2018.8554395

16. Nanni, L., Maguolo, G., Brahnam, S., Paci, M.: An ensemble of convolutional neural networks for audio classification. Appl. Sci. **11**, 5796 (2021). https://doi.org/10.3390/app11135796

17. Andreadis, A., Giambene, G., Zambon, R.: Convolutional Neural Networks for audio classification on ultra low power IoT devices. In: 2021 IEEE International Black Sea Conference on Communications and Networking (BlackSeaCom), 2021, pp. 1–6.https://doi.org/10.1109/BlackSeaCom52164.2021.9527865

18. Hailu, N., Siegert, I., Nürnberger, A.: Improving automatic speech recognition utilizing audio-codecs for data augmentation. In: 2020 IEEE 22nd International Workshop on Multimedia Signal Processing (MMSP), 2020, pp. 1–5. https://doi.org/10.1109/MMSP48831.2020.9287127

19. Abadi, M., et al.: TensorFlow: Large-Scale Machine Learning on Heterogeneous Distributed Systems (2016)

20. Elbir, A., İlhan, H.O., Serbes, G., Aydın, N.: Short time fourier transform based music genre classification. In: 2018 Electric Electronics, Computer Science, Biomedical Engineerings' Meeting (EBBT), 2018, pp. 1-4. https://doi.org/10.1109/EBBT.2018.8391437

21. Borisagar, K.R., Thanki, R.M., Sedani, B.S.: Speech Enhancement Techniques for Digital Hearing Aids. Springer, Cham (2019). https://doi.org/10.1007/978-3-319-96821-6

22. Kong, X., Choi, J., Shattuck-Hufnagel, S.: Evaluating automatic speech recognition systems in comparison with human perception results using distinctive feature measures. In: 2017 IEEE International Conference on Acoustics, Speech and Signal Processing (ICASSP), 2017, pp. 5810-5814. https://doi.org/10.1109/ICASSP.2017.7953270

23. Doddington, G.R.: Speaker recognition-Identifying people by their voices. Proc. IEEE **73**(11), 1651-1664 (1985). https://doi.org/10.1109/PROC.1985.13345

24. Minami, K., Lu, H., Kim, H., Mabu, S., Hirano, Y., Kido, S.: Automatic classification of large-scale respiratory sound dataset based on convolutional neural network. In: 2019 19th International Conference on Control, Automation and Systems (ICCAS), 2019, pp. 804-807 (2019). https://doi.org/10.23919/ICCAS47443.2019.8971689

25. Salamon, J., Bello, J.P.: Unsupervised feature learning for urban sound classification. In: 2015 IEEE International Conference on Acoustics, Speech and Signal Processing (ICASSP), 2015, pp. 171-175. https://doi.org/10.1109/ICASSP.2015.7177954

Comparative Analysis on Joint Modeling of Emotion and Abuse Detection in Bangla Language

Afridi Ibn Rahman, Farhan, Zebel-E-Noor Akhand(✉),
Md Asad Uzzaman Noor, Jubayer Islam, Md. Motahar Mahtab,
Md Humaion Kabir Mehedi, and Annajiat Alim Rasel

Department of Computer Science and Engineering, Brac University, 66 Mohakhali,
Dhaka 1212, Bangladesh
{afridi.ibn.rahman,farhan,zebel.e.noor.akhand,md.asad.uzzaman.noor,
md.jubayer.islam,md.motahar.mahtab,humaion.kabir.mehedi}@g.bracu.ac.bd,
annajiat@bracu.ac.bd

Abstract. Emotions are not linguistic entities, although they are easily articulated through language. Emotions influence our actions, ideas, and, of course, how we communicate. On the other hand, abusive text, such as undiscriminating slang, offensive language, and vulgarity, is more than just a message; it is a tool for very serious and brutal cyber violence. Hence, detection of such language has become very important in any language now-a-days. Therefore, many works and researches have been done on detecting emotional language, abusive language or both in many dialects including Bangla. This paper proposes to present a comparative analysis of different researches made on detecting emotional and abusive Bangla language. It further aims to present the best approach that tailors certain attributes of emotional and abusive language detection with respect to their prognosis performance and their implementation toughness in Bangla lingo. Potential enhancements for future study are presented in the paper, while the limitations of current researches are addressed and discussed. This work seeks to bring a fresh viewpoint to the joint modeling of emotional and abusive language detection in Bangla by examining and criticizing flaws and in order to offer future changes, poor design choices must be examined.

Keywords: NLP · CNN · RNN · MTL · BiGRU · LSTM · ICT · AI

1 Introduction

NLP is a branch of Machine Learning that concentrates on understanding and analyzing text or audio input that is analogous to human language [1]. While NLP is not an independent subject, it is a category of various fields like information engineering, computer science, linguistics and artificial intelligence (AI) [2]. As a concoction of artificial intelligence and linguistics in the 1950s, the journey

M. Singh et al. (Eds.): ICACDS 2022, CCIS 1614, pp. 199–209, 2022.
https://doi.org/10.1007/978-3-031-12641-3_17

of NLP began. In order to rapidly index and search large volumes of text, NLP was primarily separated from text information retrieval (IR), which makes use of scalable statistics-based methods [3]. NLP study has advanced from punch cards and batch processing (when processing a sentence took seven minutes) to the digital era of Google, where a huge number of web pages are prepared in less than a second [4]. This paper compares the performance of different Neural Network models on a specific NLP task: joint modelling of emotion and abuse detection in Bangla. This comparison includes the performance, accuracy, ease of implementation and complexity of each model that detects emotional and abusive Bangla language or both.

Aggressive and abusive online behavior can have substantial psychological consequences for victims [5]. 76.9% of victims endured psychological issues like anger, anxiety and agitation; 13.6% signified communal impacts; 4.1% showed corporeal effects while 2.0% claimed financial mislaying. Blocking the attacker was the most common victim response [6]. This emanates the necessity of automated systems in order to detect abusive language, an issue that, in recent times, aroused the natural language processing community's interest. Forms of abuse include racism, sexism, abusive remark, cyber bullying, harassment and other expressions that demonize or affront an individual or a group [7]. This is an evident fact that Facebook currently has over 2 billion monthly active users. Reddit, another social curation site, has over 330 million active members [8]. With such a high level of user interaction, internet conversation is prone to abuse and antisocial behavior. According to a recent poll, 41% of Americans have directly encountered internet harassment, with 18% having experienced severe kinds of harassment, such as violent threats and sexual harassment [9].

Emotional language is the usage of expressive words, frequently adjectives, to portray how an author perceives or feels about something, elicit an emotional reciprocation from the reader and persuade them. Emotions are not verbal beings, although they are easily articulated through words. Feelings impact our behaviors, ideas, and, of course, our communication style. It is imperative to understand that emotional language is biased, which signifies that even though the language is clear and concise, making it easier to grasp, the goal of the dialects is to subtly or explicitly alter our perspective about something [10]. As a result, the detection of emotions and evaluation is one of the most taxing and emanating problems in NLP. An active field of NLP studies include the detection of an individual's emotional condition [11].

Today, the majority of computer-based materials and technical magazines and researches are done on English. Because of the language barrier, the general public has significant challenges in reaping the full benefits of contemporary communication and information technology (ICT), as well as a vastly expanded English knowledge database throughout the world. The only technology that can be utilized to overcome this barrier is language processing in the native tongue [12]. Emotion identification and text abuse detection are two of the most prominent applications of Natural Language Processing (NLP). This is an important field of research to improve interactions between people and machine. Despite the fact that these studies/topics have gotten a considerable amount of attention in the English language. In the Bangla language, it is yet a relatively untapped region. Bangla language is rated seventh in the world. Around 210 million people in the world speak in Bangla, the majority dwelling in Bangladesh and in two states of India [13]. As a result, recognizing both emotional and abusive Bangla language has proven critical.

2   Goals

This paper seeks to accomplish the following objectives in order to conduct a proper comparative analysis:

- **1. Evaluate and Describe the chosen Neural Network (NN) models:**
  To achieve our goal, the most advantageous Neural Network models must be identified and made suitable for deployment with thorough research. The models' performance on typical NLP tasks, the consistency of the proposed model architecture, and the accessibility of open source code with an acceptable license for use in the research business settings are all significant factors to consider in this study.
- **2. Evaluate and compare all the models' performance results:**
  The models need to be instructed and tested in an equivalent and citable way using appropriate criteria. As a result, consistent scripts must be built for each model, along with comparison methodologies. If the models perform differently, the causes for the contrasts need to be studied properly with the knowledge that lies within the extent of this research.
- **3. Present a final verdict for the models that will help us reach our objective:**
  The models must be evaluated in terms of their performance and usefulness in the context of this paper using the intuition acquired from the execution and evaluation. As this is a scientific study, the emphasis should be on generalized qualities that may be used for future, independent research initiatives rather than on specific business aspects.

3   Selected Researches for Comparison

This section discusses the selected researches that have been taken for comparative analysis for this study. At present, the detection of abuse and emotional language in Bangla has been a prevalent and extensive experimental domain. In 2019, for recognizing several sorts of abusive writing in Bengali, Emon et al. suggested a variety of approaches based on Machine Learning (ML) and Deep Learning (DL) [14]. They imposed stemming rules for Bangla language by measuring the effectiveness of algorithms using specific Bangla grammatical rules.

As a result, better accuracy is gained when these stemming rules are applied to a small dataset. Secondly, again in 2019, Chakraborty et al. presented a study on recognizing abusive language in Bengali on the social network. They employed a considerably higher amount of data in their proposed identification algorithm, and they accepted both emoticons as well as the Bengali unicode as genuine inputs. Furthermore, the use of successive exclamation and question marks is taken into account [15]. Next in 2017, Rabeya et al. has given a methodology for extracting emotion from Bengali literature at the sentence level. They evaluated two basic emotions, "happiness" and "sadness", in order to discern emotion from Bengali text. Their proposed approach identifies emotion by analyzing the sentiment of each line with which it is associated [16]. Then in 2020, Rayhan et al. has used BiGRU and CNN-BiLSTM for detecting multiple emotion in Bangla text [17]. Finally, Rajamanickam et al. in 2020 has made a presentation on joint modelling of abuse and emotion detection. They evaluate different MTL architectures for this purpose and achieve a good outcome through the shared parameters [7]. Table 1 shows an overview of the objective, method and accuracy review results obtained by the chosen papers.

Table 1. Overview of the selected researches

Paper	Accuracy	Objective
Emon et al. [14]	82.20%	Bangla Abuse Detection
Chakraborty et al.[15]	78%	Bangla Abuse Detection
Rabeya et al. [16]	77.16%	Bangla Emotion Detection
Rayhan er al. [17]	77.78%	Bangla Emotion Detection
Rajamanickam et al. [7]	78.40%	Modeling Abusive and Emotional Language Together

4   Analyse and Describe Selected Neural Network (NN) Models

Among the selected researches, Emon et al. has achieved the highest accuracy using a relatively small dataset in detecting abusive Bangla text. They have applied two neural networks in their research. One is a Recurrent Neural Network with Long Short Term Memory, and the other is an Artificial Neural Network. It is to be noted that they have used Stemming approach for data preprocessing to get root form of Bengali word, and then they trained the Neural Networks with CountVectorize. The outcome that they have achieved is proven to be the best among the other approaches selected in this study.

On the other hand, Chakraborty et al. has also considered using unicode and emoticons. Moreover, they also gave extra attention to the exclamation and question marks for their detection system. To do so, they have made CNN-LSTM network to treat each phrase, emoticon, or series of exclamation or question marks as a symbol. The dataset is transformed to an integer value before being sent to the network. The model is then enhanced with four layers: an embedding layer, a convolution layer, an LSTM layer, and a fully connected layer. Finally, they employed different colors to differentiate each succeeding layer.

For Bangla emotional language detection, Rabeya et al. has focused more on lexicon backtracing than focusing more on any neural network. But the accuracy they have achieved is still as good as the other approaches.

Rayhan et al., on the other hand, proposed the CNN-BiLSTM model, which specifies the CNN model of two 1D CNN layers, trailed by a regularization dropout layer, and finally a max pooling layer. The output of the max pooling layer was then used as an input to the BiLSTM layer, which was then followed by a dropout layer. They also developed the Bidirectional Gated Recurrent Unit (BiGRU), a Recurrent Neural Network version (RNN) [17]. The feature extraction method was utilized as that of the BiGRU model's input. Following that, the embedding, vocabulary size, sequence length and maximum input embedding dimension, and maximum input are all set to zero. The BiGRU cell units were estimated based on the information supplied. A RELU activation function and also a dropout were used in the calculation.

After thoroughly analysing, describing and deeply comparing all the selected Neural Network, this paper is proposing the following Fig. 1. Methodology to achieve its aim detecting both abusive and emotional Bangla language detection in a joint model which has been created on the basis of the above comparison.

5   Analyse and Compare All the Models' Performance Results

This section depicts the performance comparison analysis of the researches that this study has selected. Table 2 represents, recall, F-1 Score and precision of RNN in abuse detection done by Emon et al.

Table 3 shows that the accuracy of Chakraborty et al.'s LSTM network increased with the size of the dataset, with a 77.5% accuracy for 5644 data points. We achieved 65.5% accuracy with 3604 data and 73% accuracy with 4933 data. It should be observed that this CNN-LSTM model performs better as the dataset size increases, which bodes well for constructing the joint model that this study is looking for.

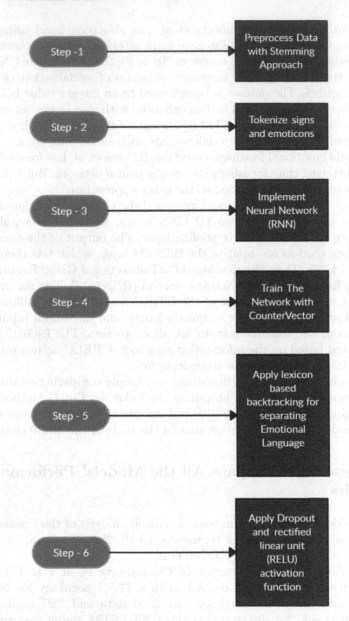

Fig. 1. Proposed methodology based on the comparative analysis.

Table 2. The overall class chart RNN recall, F-1 Score and precision [14]

Class name	Precision	Recall	F1 score	Support
Slang	0.89	0.92	0.91	89
Religious Hatred	0.83	0.84	0.84	70
Personal Attack	0.80	0.80	0.80	56
Politically Violated	0.78	0.89	0.83	65
Anti Feminism	0.91	0.67	0.77	30
Positive	0.88	0.81	0.84	112
Neutral	0.62	0..66	0.64	50

Table 3. Accuracy of CNN-LSTM [15]

Data	Accuracy %
3604	65.5
4933	73.0
5644	77.5

Table 4 shows the performance of the CNN-BiLSTM for detecting emotional Bangla language from Rayhan et al. study which has done better than the approach they have followed with Bigru.

Table 4. CNN-BiLSTM evaluation [17]

Emotion		Precision	Recall	F1-Score
Bangla	English			
আনন্দ	Happy	52.35	71.20	60.34
বিষন্নতা	Sad	62.60	56.62	59.46
ভয়	Fear	72.50	72.59	72.50
রাগ	Angry	71.43	63.43	67.19
ভালোবাসা	Love	77.78	68.75	72.99
আশ্চর্য	Surprise	72.55	67.27	69.81

Table 5 displays the results of the BiGRU model. According to the statistics, the maximum precision is 72.38%, which is greater than the other emotion classifications. The lowest accuracy, on the other side, is 53.42%.

Table 5. BiGRU evaluation [17]

| Emotion | | Precision | Recall | F1-Score |
Bangla	English			
আনন্দ	Happy	62.50	64.00	63.24
বিষন্নতা	Sad	53.42	57.35	55.32
ভয়	Fear	69.01	73.75	71.30
রাগ	Angry	72.38	56.72	63.60
ভালোবাসা	Love	70.00	68.75	69.37
আশ্চর্য	Surprise	64.52	72.73	68.38

6   Proposing a Final Verdict for Joint Modelling

After deeply analyzing the performances and usability of all models, we can conclude that a hybrid approach will give a better result for our study. Since each of the selected researches have their own unique approach to meet their goals, we are combining those unique approaches to achieve our desired output.

Firstly, from Emon et al.'s approach, we can apply stemming preprocessing which adds more to the accuracy for both small dataset and large dataset which will help us greatly in whichever size of Bangla dataset we use for our study.

Usage of exclamatory signs, question marks and the newly rising phenomenon among the millennials and the gen z of using 'emoji' which is essentially a graphical symbol in computer mediated communication hold a great significance in both abusive and emotional texts. These signs and emojis have unique semantic and emotional features. As a result, they provide contextualization cues, such as markers of positive or negative attitudes in the texts. Hence it is imperative to take these signs and emojis into account while detecting both abuse and emotions in texts. Therefore, employing Chakraborty et al.'s approach of considering emoticons, Bangla unicodes, and punctuations such as exclamation and question marks as genuine inputs will increase the possibility of achieving great results in detecting abusive and emotional Bangla language.

Again, for emotion detection, the approach from Rayhan et al. can be followed. To distinguish emotion from Bengali text, two primary emotions, "happiness" and "sadness", were used. The suggested method determines emotion by examining the sentiment of each line to which it is linked.

Finally, Rajamanickam et al.'s approach should be used for combined modeling of both abusive and emotional language. Multi-Task Learning (MTL) is the name of the approach, and it contains two optimization goals: one for identifying abuse and one for recognizing emotion. The two objectives for abuse detection and emotion recognition are weighted by a hyperparameter (Beta), which regulates the relevance allocated to each task.

The Fig. 2 depicts a visual representation of the final verdict for the join modelling.

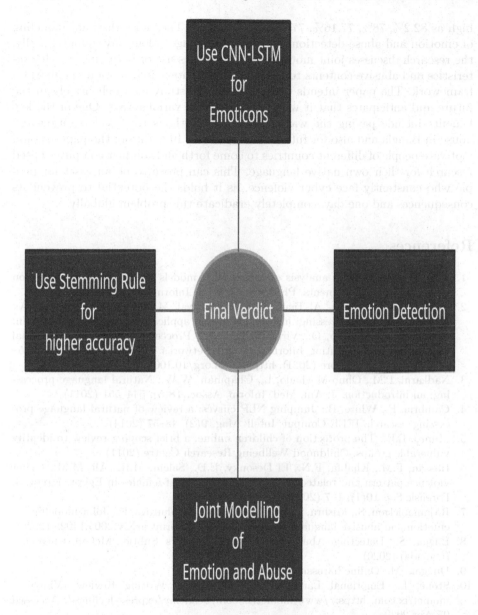

Fig. 2. Different approaches for higher accuracy

7 Conclusion

Through a comparative analysis in Bangla Language from the studies of other researchers, a new approach for both abuse identification and emotion detection has been provided in this paper. The researches analyzed, yielded accuracies as high as 82.2%, 78%, 77.16%, 77.78% and 78.40%. Therefore, the joint modeling of emotion and abuse detection in Bangla Language is logically viable. Finally, the research discusses joint modeling, which makes use of both emotive characteristics and abusive contents to gather supplementary information via an MTL framework. The paper intends to implement the study comprehensively in the future and anticipates that it will be beneficial in various ways. One of the key benefits include paving the way out for new methods to detect emotion and abuse in Bangla and also for future NLP research. In addition, the paper should motivate people of different countries to come forth and conduct computer based research for their own native language. This can prove to be an asset for people who constantly face cyber violence, as it holds the potential to prevent its consequences and one day, completely eradicate this problem globally.

References

1. Buz, T.: Comparative analysis of neural NLP models for information extraction from accounting documents. Ph.D. thesis, ETSI_Informatica (2018)
2. Johri, P., Khatri, S.K., Al-Taani, A.T., Sabharwal, M., Suvanov, S., Kumar, A.: Natural language processing: history, evolution, application, and future work. In: Abraham, A., Castillo, O., Virmani, D. (eds.) Proceedings of 3rd International Conference on Computing Informatics and Networks. LNNS, vol. 167, pp. 365–375. Springer, Singapore (2021). https://doi.org/10.1007/978-981-15-9712-1_31
3. Nadkarni, P.M., Ohno-Machado, L., Chapman, W.W.: Natural language processing: an introduction. J. Am. Med. Inform. Assoc. 18(5), 544–551 (2011)
4. Cambria, E., White, B.: Jumping NLP curves: a review of natural language processing research. IEEE Comput. Intell. Mag. 9(2), 48–57 (2014)
5. Munro, E.R.: The protection of children online: a brief scoping review to identify vulnerable groups. Childhood Wellbeing Research Centre (2011)
6. Hassan, F.M., Khalifa, F.N., El Desouky, E.D., Salem, M.R., Ali, M.M.: Cyber violence pattern and related factors: online survey of females in Egypt. Egypt. J. Forensic Sci. 10(1), 1–7 (2020)
7. Rajamanickam, S., Mishra, P., Yannakoudakis, H., Shutova, E.: Joint modelling of emotion and abusive language detection. arXiv preprint arXiv:2005.14028 (2020)
8. Bagga, S.: Detecting Abuse on the Internet: It's Subtle. McGill University (Canada) (2020)
9. Duggan, M.: Online harassment 2017 (2017)
10. Stone, L.: Emotional Language in Literature - Writing Review (Video) – mometrix.com. https://www.mometrix.com/academy/express-feelings/. Accessed 05 Apr 2022
11. Murthy, A.R., Anil Kumar, K.M.: A review of different approaches for detecting emotion from text. In: IOP Conference Series: Materials Science and Engineering, vol. 1110, p. 012009. IOP Publishing (2021)
12. Islam, M.S.: Research on Bangla language processing in Bangladesh: progress and challenges. In: 8th International Language and Development Conference, pp. 23–25 (2009)
13. Mehedy, L., Arifin, N., Kaykobad, M.: Bangla syntax analysis: a comprehensive approach. In: Proceedings of International Conference on Computer and Information Technology (ICCIT), Dhaka, Bangladesh, pp. 287–293 (2003)

14. Emon, E.A., Rahman, S., Banarjee, J., Das, A.K., Mittra, T.: A deep learning approach to detect abusive Bengali text. In: 2019 7th International Conference on Smart Computing and Communications (ICSCC), pp. 1–5. IEEE (2019)
15. Chakraborty, P., Seddiqui, M.H.: Threat and abusive language detection on social media in Bengali language. In: 2019 1st International Conference on Advances in Science, Engineering and Robotics Technology (ICASERT), pp. 1–6. IEEE (2019)
16. Rabeya, T., Ferdous, S., Ali, H.S., Chakraborty, N.R.: A survey on emotion detection: a lexicon based backtracking approach for detecting emotion from Bengali text. In: 2017 20th International Conference of Computer and Information Technology (ICCIT), pp. 1–7. IEEE (2017)
17. Rayhan, M.M., Al Musabe, T., Islam, M.A.: Multilabel emotion detection from Bangla text using BiGRU and CNN-BiLSTM. In: 2020 23rd International Conference on Computer and Information Technology (ICCIT), pp. 1–6. IEEE (2020)

Determining Dengue Outbreak Using Predictive Models

Darshan V. Medhane[1($\boxtimes$)] and Varun Agarwal[2($\boxtimes$)]

[1] MVPS's KBT College of Engineering, Nashik, India
darshan.medhane@gmail.com
[2] MIT Academy of Engineering, Pune, India
varunagarwal020@gmail.com

Abstract. Early prediction of contagious and infectious diseases can help health organizations in planning strategies to prevent disease transmission and thus can forestall the outbreak. Several works are there to predict the disease outbreak using climate data but our approach provides better results using univariate model. Our approach is to split the data in terms of variability and volume to find out the best forecasting model for predicting dengue cases with low variability and high-volume data points. For this we have also analyzed the correlation of climate factors with the number of the dengue cases and after comparing the competencies of different forecasting models, we found out that ARIMA is the best suitable model for low variability and high-volume data points with 5.4 RMSE and 3.6 MAE value. The predictive power of these models will be useful to authorities in taking preventive steps.

Keywords: Outbreak · Correlation matrix · Forecasting model · ARIMA

1 Introduction

Coronavirus comes out of nowhere and prompts such countless passing's, assuming we had earlier predictions about its outbreak, such a lot of misfortune doesn't have happened, we could have come with proper preventive measures and saved the lives of many. Coronavirus is one of the recent virus outbreaks but we have been suffering from disease outbreaks from an earlier stage like Spanish influenza killed 40–50 million in 1918, Asian influenza killed 2 million people in 1957, Hong Kong influenza killed 1 million people in 1968, chikungunya, malaria, zika virus and many more. These viruses wherever spread causes obliteration in terms of the lives and economy of the country. These outbreaks have one common thread that is factors contributing to the outbreak. Weather change or small climatic variations comes out to be the key factor for any outbreak of infectious disease transmission [1]. Thus, if prior information for an outbreak is available then it will be easier for doctors to treat patients and the government to make earlier moves.

Dengue fever transmitted by the Aedes mosquitos is greatly influenced by climatic variations throughout the region [2]. Most cases of dengue are not registered because

© The Author(s), under exclusive license to Springer Nature Switzerland AG 2022
M. Singh et al. (Eds.): ICACDS 2022, CCIS 1614, pp. 210–221, 2022.
https://doi.org/10.1007/978-3-031-12641-3_18

of asymptotic symptoms revolving around the closely related serotypes namely DEN-1, DEN-2, DEN-3, DEN-4 [3]. This vector-borne disease spreading across the globe and taking life is predominantly dependent on temperature change, as stated by NOAA - 2020 was globally the earth's warmest year and this disease is very sensitive to temperature change. To overcome this problem there is a need for dengue outbreak prediction model which helps in preventing epidemics.

In this paper, we have classified the data in terms of variability and volume and compared the competencies of different forecasting models to find out the best suitable model. For this we have used simple moving average, exponential smoothing, ARIMA, Fbprophet and XGBoost. Rest of the paper is structured as follows: Sect. 2 consists of the previous work in forecasting dengue outbreaks. In Sect. 3, methodology is discussed, Sect. 4 and Sect. 5 discusses algorithms applied and the results for the forecasting model. And in Sect. 6 we have concluded our work with its future aspects.

2 Literature Review

An increase in the number of cases of dengue in several tropical and non-tropical regions gained the interest of many researchers in analyzing the past data and making use of machine learning to forecast the cases. There are several methodologies proposed for the early prediction of dengue outbreaks. Nan et al. [2] proposed a methodology in which they had found five climate conditions responsible for dengue transmission and made predictions using various machine learning models, and found the best accuracy with the XGBoost model. In [4] Mishra et al. proposed a technique in which they used several machine learning models such as neural network, XGBoost, Linear Regression and they founded the best accuracy with Twofolds linear regression. In the above methods, there was no use of time series forecasting models and analyzing patterns in time series is of utmost importance for forecasting. Anggraeni et al. [5] used google trends for increasing the accuracy of prediction by 3% using the ARIMAX model by trying different combinations of p, d, q values. Sillabutra et al. [6] proposed a technique for forecasting dengue morbidity rates using the ARIMA model on finding optimal parameters $p = 3$, $q = 0$, $d = 1$. Anitha et al. in [7] used a decision tree for classification on an individualistic basis whether a person has dengue fever or not, no climate factors were considered in this, only individualistic features were considered. Nakvisut et al. [8] collected open-source data from governmental organizations of Thailand. They had included various climate factors such as average wind speed, maximum temperature, minimum humidity for building a two-step prediction model. For the first step time series forecasting models were used and for the next step used supervised learning, models were there such as linear regression, support vector machine, and neural networks and achieved a root mean squared error of 14.8. Anggraeni et al. [9] used weather data for predicting the number of dengue cases using an artificial neural network (ANN), trying with different combinations of parameters such as learning rate, units of hidden layers, training cycle, and then used Google API for visualization. Manivannan et al. [10] in their paper used a K-means clustering algorithm to make clusters of dengue using dengue serotypes based on the age group. Dinayadura et al. [11] finds the influence of climatic factors on dengue cases and developed a support vector regression model for risk area identification. Jain et al. [12] proposed scenarios where they mentioned the involvement of public traveling in the transmission of dengue cases and mentioned to need to add parameters involving mobility and their transmission. In [13] Makkar et al. developed an early warning system for predicting the number of dengue cases and prominent factors responsible for dengue transmission were pressure and rainfall. Rahmawati et al. [14] uses linear optimization and C-Support vector optimization to predict the dengue fever cases taking different climatic features in account and they had used grid search method for optimizing the parameters and model for better accuracy. Kristianto et al. [15] proposed a technique of combining genetic algorithm and triple exponential smoothing to predict the dengue cases and concluded how GA-TES helped them in increasing accuracy to 8% compared to simple triple exponential smoothing model. Anggraeni et al. [16] used the data of Malang regency and divided it into 3 parts of lowlands, middle and highlands, and found that cases in lowlands and middle lands are higher than highlands cases. They also found that rainfall is also affecting the number of dengue cases because in these villages' dengue cases are higher in the rainy season. Rachata et al. [17] converted the daily data into weekly data and applied entropy technique for feature extraction and to give input to the neural network and use neural network for prediction whether outbreaks occur or not. Chovatiya et al. [18] proposed a technique for the prediction of dengue cases using weather information and used a recurrent neural network that gives an accuracy of 94% as stated by them. Sasongko et al. [19] use various algorithms of backpropagation to predict the early detection of dengue. It included gradient descent, FGS Quasi-Newton, Conjugate Gradient Descent - Powel, Resilient Backpropagation (RB), and Levenberg Marquardt out of which Levenberg Marquardt was the most efficient. Rahim et al. [20] proposed a technology stating using a nonlinear autoregressive moving average with exogenous input and the selection criteria for the parameter of this time series model were AIC, FPE, and Lipschitz, in which they got the accuracy of 88.40%.

3   Methodology

Forecasting algorithm finds pattern in the historical data points and helps in predicting the number of dengue cases in the future time period. The flow of the framework in shown in Fig. 1. We have taken the historical data having weekly number of dengue cases along with the weather conditions in those subsequent areas. We have split the data into four quadrants based on its coefficient of variability vs. volume and targeted the months with more less variability and more volume for data prediction and then performed descriptive analysis on data to have a sanity check on the data and find out the stationarity of the data so that the forecasting algorithms can find patterns in the data while predicting the number of cases. Descriptive analysis of data involves checking for null values, finding modality, skewness and kurtosis of the data to find the outliers in the data and handle them. Also, stationarity of data is checked using the Dicky fuller test to find out if the data is stationary or not using calculation of p-value. Further which we have analyzed different forecasting algorithms underlying their advantages and disadvantages in order to use those models and compared the results of these models to find out the best fit model.

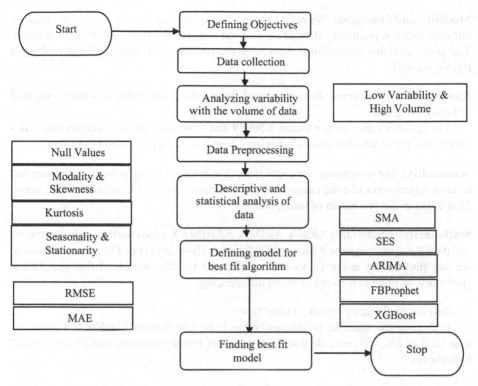

Fig. 1. Framework for prediction of number of dengue cases

3.1 Dataset

The dataset used for this research purpose consists of data of two cities one is the capital of Puerto Rico i.e. San Juan and another city is Iquitos, a city in Peru.

Data consists of many climate factors such as precipitation, humidity, temperature, relative humidity, average temperature, minimum, and maximum temperature, etc. Data is very little in terms of data points so we have to apply different algorithms and analyze patterns [21].

3.2 Descriptive Analysis and Preprocessing

After data collection process we have done the following descriptive analysis of data to understand the statistics of data and do the preprocessing so that algorithm can be defined based upon the statistical results:

Null Values: Data is checked for any type of null values and we have null values, as our data is less so we cannot simply drop that data. Therefore 3 methods are used in filling the null values that are replacing it with mean, median, or mode, and the corresponding accuracy is checked and we found that median is a better method because there are outliers in the data and mean is very much prone to outliers.

Modality and Skewness: We have unimodal data because we have only one peak in our data and it is positively skewed i.e. skewed towards the right and mean > median. The graph contains unimodality along with the profusion of outliers that is making it rightly skewed.

Kurtosis: It tells us about the outliers and checks the tail whether it is light weighted or heavy weighted.

The kurtosis value for our data is **0.30089** and from this, we can interpret that it is a leptokurtic curve which means it is having a heavy tail and profusion of outliers.

Seasonality: For seasonality, we check if the data is showing any pattern for the summer term or winter term like the cases are high in summer and less in winter or vice versa. That helps in the prediction of future cases.

Stationarity: Models like ARMA, ARIMA, SARIMAX needs the data to be stationery and data is considered to be stationary if it does not show any type of trend or seasonality, we can find this by using Dicky-Fuller Test, and by this, we found that our data is stationary, so there is no need of doing differencing.

Test for Stationarity – Dicky Fuller Test

In this, we calculate the p-value and it has to be less than 0.05 or 5%. If it is greater than 0.05 or 5%, we conclude that the time series has the unit root and accept the null hypothesis.

3.3 Exploratory Data Analysis

EDA for data is necessary as it tells us a story about the data which is useful in selecting what type of processing is required by the data and to understanding the data for choosing optimal algorithm. In our case we have data for two cities i.e. San Juan and Iquitos so

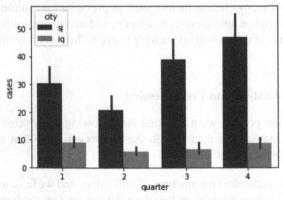

Fig. 2. Dataset consist of two cities so segmented the data and analyzed the number of dengue cases in quarterly manner to see in which quarter there are more number of dengue cases and what is the pattern in both the cities. This also gives us the significant months in which dengue cases were reported more

our data requires segmentation because weather conditions and the number of cases of both these cities are different, so to make sure the data of one city is not affecting the data of another city we will do segmentation (Fig. 1).

San Juan is affected most in quarter 4 i.e. for October, November, December, and Iquitos is most affected at the near end of quarter 4 and quarter 3 i.e. December, January, February, and March (Fig. 2).

We had plotted a scatter plot for finding the relation between temperature and cases in this we can see that as the number of cases increases the temperature also increases so this is showing a correlation and we can use this relation in our multivariate forecasting.

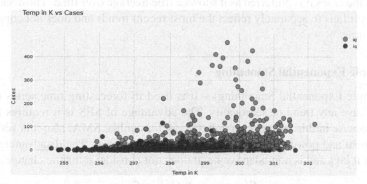

Fig. 3. Relation between temperature and the number of cases shows the positive correlation in data points i.e. in San Juan specifically when the temperature increases number of cases also increases but in a erratic nature

Likewise in the Fig. 4, we can visualize from this graph that cases decrease significantly as the precipitation increase so this is having a negative correlation with the cases. Humidity is having a positive relationship with the number of cases as its relation with the temperature and so these three are climatic factors that we have observed showing correlation with the number of cases.

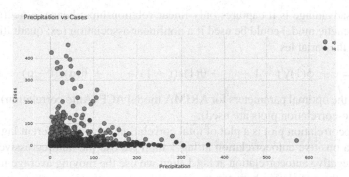

Fig. 4. Relation between Precipitation and the number of cases shows negative correlation

4 Algorithms

For predicting the dengue outbreaks, we have checked the accuracy for following forecasting models to find out the best suitable model discussing their underlying advantages and disadvantages in predicting on historical data.

4.1 Simple Moving Average

SMA (Simple Moving Average) - It is used in forecasting time series data by adding up earlier cases and dividing them by the total number of cases. If there is an unusual change in the cases it is preferred as it shows a true average over time. The disadvantage of SMA is it fails to accurately reflect the most recent trends and does not consider any other features.

4.2 Simple Exponential Smoothing

SES (Simple Exponential Smoothing) – It is used in forecasting time series data that does not have any trend or seasonality. The advantage of SES is it requires less data storage because in this we work on only two factors unlike SMA, also it is very simple to implement and powerful because of its weighting process. The disadvantage of this process is it lags and is non-adaptive i.e. it does not consider dynamic changes.

$$S_t = \alpha y_{t-1} + (1 - \alpha)S_{t-1} \tag{1}$$

4.3 Auto-regressive Integrated Moving Average

It is used in the prediction of time series data. It comprises 3 main terms i.e., Auto-Regressive which is lags of variables itself, integrated which is several steps required to make data stationery, and Moving Average Lags which are lags of previous information. The assumption is taken by ARIMA that the data is stationary.

The advantage of ARIMA is they are more accurate and reliable in case of erratic data.

The disadvantage is it captures only linear relationships; hence, a neural network model or genetic model could be used if a nonlinear association (ex: quadratic relation) is found in the variables.

$$y't = c + \phi(1)y't - 1 + \cdots + \theta(1)\varepsilon(t - 1) + \cdots + \theta(q)\varepsilon(t - q) + \varepsilon(t) \tag{2}$$

To find the optimal parameters for ARIMA model ACF (Auto correlation) and PACF (partial auto-correlation plots are used).

An autocorrelation plot is a plot of total correlation between different lag functions. If there is a positive autocorrelation at lag 1 then we use the autoregressive model. If there is a negative autocorrelation at lag 1 then we use the moving average model.

PACF is the correlation between two variables under the assumption that we know and take into account the values of some other sets of variables. If this model drops off at lag n, then use an AR(n) model and if the drop in PACF is more gradual then we use the moving average term.

4.4 FbProphet

Fbprophet is a time series forecasting algorithm developed by Facebook. It takes four factors into account i.e. seasonal, trends, holidays, and error or event effect. In this model non-linear trends are fit with yearly, weekly, seasonality, and holiday effects

$$Y(t) = g(t) + s(t) + h(t) + \varepsilon t \tag{3}$$

The Fbprophet model is effective in cases when data is having seasonality's, outliers in form of important events or Holiday which had led to a high number of cases, and historical trend changes and data must be at least of 1 year.

Advantages – It handles data with uneven time intervals, handles the null values, handles seasonality, and works well by default setting.

Disadvantages – Fails in forecasting erratic data, cannot handle exogenous variables, multiplicative models and data in prophet need to be fees in a pre-defined format.

5 Algorithms and Results

In the above graph, we had taken the training data frame and plotted the earlier cases, moving average on 20, 50, 100 rolling windows and we can see that 20-day rolling window is performing better in forecasting and it can catch patterns but for rolling window 50 and 100 we are seeing some erratic patterns that are worsening the forecast (Fig. 5).

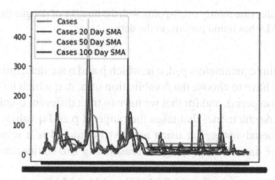

Fig. 5. Simple moving average model shows that there is significant error in the number of cases and the 20-, 50- and 100-days moving average

In exponential smoothing, we can see the 20 and 50-day exponential moving average and in this, we have tried with different values of the alpha parameter to reduce the error in forecasting and this also 20-day rolling window give us good results but not better (Fig. 6).

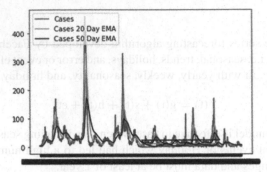

Fig. 6. Exponential smoothing model results shows how there is difference in the actual number of cases and the cases forecasted through exponential smoothing i.e. 20 days exponential moving average and 50 days moving average

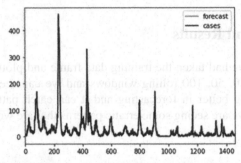

Fig. 7. Forecasted values are mostly overlapping with the number of dengue cases which is stating how accurately ARIMA has found patterns in the data

ARIMA takes three parameters p, d, q in which p and q are determined with ACF and PACF plots and we have to choose the combination of p, d, q which have minimum AIC (Acyle information criteria), and for that we have to try a different combination of p, d, q values or use auto Arima model that takes the range of p and q values and searches for best parameters, one advantage for using auto-Arima model is it removes seasonality, stationarity from the data after applying the seasonality and stationarity check (Table 1) (Fig. 7).

Simple moving average having 22.7 RMSE and 16.3 MAE and exponential smoothing having 14.01 RMSE and 7.02 MAE showing very high error because they are not able to analyse the dengue cases patterns in the past year in specific months. Fbprophet showing high inaccuracy because of the erratic nature of the data points and not able to predict the data points. But we can see that ARIMA is performing better than other time series forecasting models and Ensemble models i.e. XGBoost because in ARIMA it is taking the correlation of months of this year with months of the past year. The values of p, d, q are chosen by looking at the partial autocorrelation plot, autocorrelation plot, such that final model is having minimum AIC value and used in almost correct prediction of the number of dengue cases.

Table 1. Results from algorithms used for forecasting

S. no.	Algorithms used	RMSE	MAE
1	Simple Moving Average	22.68	16.29
2	Exponential Smoothing	14.01	7.024
3	ARIMA	5.38	3.59
4	FBProphet	18.58	13.1
5	XGboost	17.71	12.83

6 Conclusion

Analysis of the data showed how climate factors are affecting the number of cases and maintain a direct and indirect relationship. From our approach we found that data which lie in less variability and more volume quadrant is of the months for which dengue cases are highly reported and have a strong correlation with the climate factors but after applying the aforementioned algorithms, we found out that ARIMA is the best fit model for this type of data with 5.4 RMSE and 3.6 MAE. There is scope of reducing this error if the data points were more so that we can find out the missing patterns in the data. Currently, we had prepared the model that contains only two cities, so it requires segmentation at one level only and then runs the best fit model, which will give the algorithm the most optimal results. For future scope, we can expand our model to the data including multiple granularities, and segment the data area wise to be optimally utilized by the concerned authorities.

References

1. Andrick, B., Clark, B., Nygaard, K., Logar, A., Penaloza, M., Welch, R.: Infectious disease and climate change: detecting contributing factors and predicting future outbreaks. In: 1997 IEEE International Geoscience and Remote Sensing Symposium Proceedings, IGARSS 1997. Remote Sensing - A Scientific Vision for Sustainable Development, Singapore, vol. 4, pp. 1947–1949 (1997). https://doi.org/10.1109/IGARSS.1997.609159
2. Nan, J., et al.: Using climate factors to predict the outbreak of dengue fever. In: 2018 7th International Conference on Digital Home (ICDH), Guilin, China, pp. 213–218 (2018). https://doi.org/10.1109/ICDH.2018.00045
3. Ebi, K.L., Nealon, J.: Dengue in a changing climate. Environ. Res. 151, 115–123 (2016). ISSN 0013-9351
4. Mishra, V.K., Tiwari, N., Ajaymon, S.L.: Dengue disease spread prediction using twofold linear regression. In: 2019 IEEE 9th International Conference on Advanced Computing (IACC), Tiruchirappalli, India, pp. 182–187 (2019). https://doi.org/10.1109/IACC48062.2019.8971567
5. Anggraeni, W., Aristiani, L.: Using Google Trend data in forecasting number of dengue fever cases with ARIMAX method case study: Surabaya, Indonesia. In: 2016 International Conference on Information & Communication Technology and Systems (ICTS), Surabaya, pp. 114–118 (2016). https://doi.org/10.1109/ICTS.2016.7910283

6. Sillabutra, J., Soontornpipit, P., Viwatwongkasem, C., Satitvipawee, P., Phuthomdee, S.: Forecasting model for dengue morbidity rate in Thailand. In: 2018 International Electrical Engineering Congress (iEECON), Krabi, Thailand, pp. 1-4 (2018). https://doi.org/10.1109/IEECON.2018.8712202

7. Anitha, A., Wise, D.C.J.W.: Forecasting dengue fever using classification techniques in data mining. In: 2018 International Conference on Smart Systems and Inventive Technology (ICSSIT), Tirunelveli, India, pp. 398–401 (2018). https://doi.org/10.1109/ICSSIT.2018.8748864

8. Nakvisut, A., Phienthrakul, T.: Two-step prediction technique for dengue outbreak in Thailand. In: 2018 International Electrical Engineering Congress (iEECON), Krabi, Thailand, pp. 1–4 (2018). https://doi.org/10.1109/IEECON.2018.8712258

9. Anggraeni, W., et al.: Artificial neural network for health data forecasting, case study: number of dengue hemorrhagic fever cases in Malang regency, Indonesia. In: 2018 International Conference on Electrical Engineering and Computer Science (ICECOS), Pangkal Pinang, pp. 207–212 (2018). https://doi.org/10.1109/ICECOS.2018.8605254

10. Manivannan, P., Devi, P.I.: Dengue fever prediction using K-means clustering algorithm. In: 2017 IEEE International Conference on Intelligent Techniques in Control, Optimization and Signal Processing (INCOS), Srivilliputhur, pp. 1–5 (2017). https://doi.org/10.1109/ITCOSP.2017.8303126

11. Dinayadura, N.S., Mikler, A.R., Muthukudage, J.: An efficient approach of outbreak preparedness for dengue. In: 2017 IEEE International Conference on Healthcare Informatics (ICHI), Park City, UT, p. 327 (2017). https://doi.org/10.1109/ICHI.2017.16

12. Jain, R., Sontisirikit, S., Iamsirithaworn, S., et al.: Prediction of dengue outbreaks based on disease surveillance, meteorological and socio-economic data. BMC Infect. Dis. **19**, 272 (2019)

13. Makkar, G.: Real-time disease forecasting using climatic factors: supervised analytical methodology. In: 2018 IEEE PuneCon, Pune, India, pp. 1-5 (2018). https://doi.org/10.1109/PUNECON.2018.8745369

14. Rahmawati, D., Huang, Y.: Using C-support vector classification to forecast dengue fever epidemics in Taiwan. In: 2016 International Conference on System Science and Engineering (ICSSE), Puli, pp. 1–4 (2016). https://doi.org/10.1109/ICSSE.2016.7551552

15. Kristianto, R.P., Utami, E.: Optimization the parameter of forecasting algorithm by using the genetical algorithm toward the information systems of geography for predicting the patient of dengue fever in district of Sragen, Indonesia. In: 2017 2nd International conferences on Information Technology, Information Systems and Electrical Engineering (ICITISEE), Yogyakarta, pp. 45–50 (2017). https://doi.org/10.1109/ICITISEE.2017.8285548

16. Anggraeni, W., et al.: Modelling and forecasting the dengue hemorrhagic fever cases number using hybrid fuzzy-ARIMA. In: 2019 IEEE 7th International Conference on Serious Games and Applications for Health (SeGAH), Kyoto, Japan, pp. 1–8 (2019). https://doi.org/10.1109/SeGAH.2019.8882433

17. Rachata, N., Charoenkwan, P., Yooyativong, T., Chamnongthal, K., Lursinsap, C., Higuchi, K.: Automatic prediction system of dengue haemorrhagic-fever outbreak risk by using entropy and artificial neural network. In: 2008 International Symposium on Communications and Information Technologies, Lao, pp. 210–214 (2008). https://doi.org/10.1109/ISCIT.2008.4700184

18. Chovatiya, M., Dhameliya, A., Deokar, J., Gonsalves, J., Mathur, A.: Prediction of dengue using recurrent neural network. In: 2019 3rd International Conference on Trends in Electronics and Informatics (ICOEI), Tirunelveli, India, pp. 926–929 (2019). https://doi.org/10.1109/ICOEI.2019.8862581

19. Sasongko, P.S., Wibawa, H.A., Maulana, F., Bahtiar, N.: Performance comparison of Artificial Neural Network models for dengue fever disease detection. In: 2017 1st International Conference on Informatics and Computational Sciences (ICICoS), Semarang, pp. 183–188 (2017). https://doi.org/10.1109/ICICOS.2017.8276359
20. Abdul Rahim, H., Ibrahim, F., Taib, M.N.: A novel prediction system in dengue fever using NARMAX model. In: 2007 International Conference on Control, Automation and Systems, Seoul, pp. 305–309 (2007). https://doi.org/10.1109/ICCAS.2007.4406927
21. Dengue Forecasting. https://dengueforecasting.noaa.gov/

Subjective Examination Evaluation Based on Spelling Correction and Detection Using Hamming Distance Algorithm

Madhav A. Kankhar[(⊠)] and C. Namrata Mahender

Department of Computer Science and Information Technology, Dr. Babasaheb Ambedkar
Marathwada University, Aurangabad, Maharashtra, India
madhavkankhar.07@gmail.com, cnamrata.csit@bamu.ac.in

Abstract. The usage of Online examination systems in education is not a new
concept for the past several years, Objective assessments have been conducted
using examination systems. This research examines E-examinations that include
an E-assessment system that can be used for subjective questions. The present
work aims to investigate the spelling errors, for experiment 12th standard Business studies paper is collected from a CBSC school. The exam was conducted on
Microsoft teams. Hamming distance for word matching or spelling mistakes is
deployed on one word and one sentence. Types of Error considered while evaluating spell mistakes are Inserting, Missing, Replacement or Substituting and
Transposition error or Swap which resulted in a 46.62% correction on the overall
result of subjective inspection for spell mistake in answer assessment.

Keywords: Online subjective exam · Question answering system · Hamming
distance · Online examination · NLP

1 Introduction

The educational system has experienced various changes in the recent decade, the most
notable of which is the shift in learning and examination methodologies. Students' and
parents' conceptions of learning are changing as educational institutions steadily move
toward online instructional techniques and tests. The need for an automated online subjective testing and evaluation process has grown as a result of the scenario in Covid -19.
While designing and evaluating a completely automated Question answering system,
phrasing is a challenge [1]. Exams are an important part of a student's education since
they evaluate their knowledge and understanding of a subject. As a result, An examination system must include the preparation of a fresh paper for each student as well
as follow-up assessments. With the rapid growth of modern education, the notion of
an E-learning system was established to improve online course teaching by allowing
teachers to administer online assessments through virtual classrooms. Electronic learning addresses a number of problems that students face, including the expensive expense
of traditional academic courses [2]. Exam paper preparation and assessment take a lot

M. Singh et al. (Eds.): ICACDS 2022, CCIS 1614, pp. 222–234, 2022.
https://doi.org/10.1007/978-3-031-12641-3_19

of time and effort, use up a lot of resources, and place a lot of pressure on course instructors. As a result, E-examination systems are crucial in colleges and institutions since they allow all students in diverse locations to take electronic tests. Electronic assessments for massive open online courses (MOOCs) have been developed by universities such as MIT, Berkeley, and Stanford [3]. E-examination systems can electronically verify and set exam papers, assign scores, and grade answers swiftly and efficiently. These systems require fewer resources and less work on the part of the consumers. Traditional examination systems, on the other hand, necessitate the use of physical resources such as pens and paper, as well as more useful work and time. Exams containing objective questions are now evaluated only by existing electronic-examination systems. However, researchers have recently found the necessity to use this method to examine subjective questions [4]. Manually generated electronic text is full of errors, including spelling and typing errors. Internet search engines, for example, have been chastised for neglecting to spell check the user's query, which would have averted a plethora of pointless searches if the user had misspelled one or more query terms. An approximate word-matching algorithm is required to spot errors in queries with little or no contextual information and give words that are most similar to each misspelled word. [5].

Spelling error correction has been a long-standing Natural Language Processing (NLP) difficulty due to the vast amount of informal and unedited text generated online, such as web forums, tweets, blogs, and emails. It's become particularly significant recently as a result of the several potential applications for the vast volume of unstructured and unedited material generated online, such as web forums, tweets, blogs, and email. Misspellings in such material can cause increased sparsity and inaccuracy in several NLP applications, such as text summarization, sentiment analysis, and machine translation [6].

2 Subjective Examination

Depending on the length of the question, subjective questions can be answered in a few paragraphs or a few pages. As a result of the necessity to use an assessing technique while evaluating those questions, examining subjective replies will take time. Examiners may become frustrated as the action is repeated multiple times and students' comments become increasingly ludicrous. There were mistakes committed to evaluating the subjective questions' responses. As a result, a number of mechanisms are required to modify subjective responses. Human evaluation can be more effective when dealing with subjective difficulties such as sensitive judgments, sophisticated reasoning, and attitude expression. The human evaluator, on the other hand, invested a significant amount of time, sensitivity, and skill in evaluating the responses in order to receive feedback after a delay for scoring an inexperienced examiner would reduce the accuracy with which subjective answers were evaluated and the result in a plethora of redundant processes to correct the evaluation. The SQ&A System efficiently assesses subjective responses to save time. The system also assists examiners in enhancing the accuracy of subjective question judgment by running the algorithms. Aside from that, the strategy may assist professors in keeping track of their students' records, allowing them to create a more accurate graph of their academic status [7].

3 Related Work

Maram et al. [8] present an Arabic-language AEE stands for Automatic Evaluation of an Essay. The system makes use of a hybrid approach that combines the LSA and the RST (rhetorical structure theory) algorithms. The LSA method aids in the semantic analysis of the essay, whereas the RST method assesses the writing method as well as the essay's cohesion. Even if two texts do not include similar words, the LSA approach calculates their similarity ratio. A training phase and a testing phase are used by the system to process the input essay. The LSA method is used after calculating the average number of words in each essay, determining the top ten visible terms on a given topic, and determining the average number of words in each essay. The following steps are included in the testing phase: 1) Calculate the LSA distance 2) Counting how many words a vernacular has. 3) Counting the number of times a sentence is repeated. 4) Determining the length of the essay. 5) Counting how many spelling mistakes there are 6) Using the RST method 7) Examining the essay's overall coherence on the subject. The final score is then calculated using two phases and the LSA cosine distance between the input and training essays. The system assigns grades to schoolchildren's essays based on three criteria: 40% for writing approach, 50% for essay cohesion, and ten percent for spelling and grammar errors.

An automatic evaluation approach for descriptive English answers with several phrases is proposed by Anirudh et al. [9]. For questions in professional courses, the system evaluates the student's response using an answer key. Among the natural language processing methods used are Wu and Palmer's Longest Common Substring (LCS), LSA Cosine Similarity, and Pure PMI-IR. The similarity scores are then extracted from algorithms and blended using logistic regression to get a score that the instructor recommends. Each word in the student's answer is compared to each word in the answer key using the Wu-Palmer method. If both words are found in the English dictionary, the Wu-Palmer approach calculates a similarity score. Otherwise, if both terms aren't in the dictionary, the edit distance is used to compare them. LCS was used to compare both the student's answer and the answer-key phrases. Using the similarity matrix method, the LCS similarity score was combined with a Wu-Palmer approach similarity score. To determine the degree of similarity between the student's answer and an answer key, the algorithms compare them.

Ishioka and Kameda [10] propose the jess system, which is an automated Japanese essay rating method. In Japan, the system is used to grade essays for university entrance exams. The essay is graded on three criteria: eloquence, content, and organization Rhetoric is a syntactic variety that evaluates readability, lexical diversity, the number of large words, and the percentage of passive sentences. The process of presenting and integrating concepts in an essay is referred to as "organization." Jess evaluates the document's logical structure and looks for clear conjunctive sentences for the organization assessment. Content refers to material that is relevant to the issue, such as the precise information presented and the word used. Jess uses a technique called LSA to analyze content, which can be used to determine whether the contents of a written essay are appropriate for the essay topic. Jess uses learning models based on editorials and articles from the Mainichi Daily News newspaper.

In [11] To analyse online descriptive type students' replies, proposes utilising the Hyperspace Analog to Language (HAL) methodology and the Self-Organizing Map (SOM) method. To assess a learner's response, the student writes it down and sends it to HAL as input From an n-word vocabulary, HAL generates a high-dimensional semantic matrix. A method for building a matrix that involves the corpus motivating a window of length "1" by incrementing one word at a time. HAL disregards punctuation and sentence breaks, transforming each word into numeric vectors that express information about its meanings. Inside the window, "d" denotes the distance between two words, whereas "(1 − d + 1)" denotes the weight of a word association. The terms in this matrix are displayed in order of their lexical co-occurrence. Every word in the row vector appears based on the co-occurrence data for words that come before it, and every word in the column vector appears based on the co-occurrence data for words that come after it. The SVD function is used to turn the matrix into a singular value. The HAL-generated vector is fed into the Self-Organizing Map as an input (SOM). SOM stands for neural technique. SOM creates a document map using vectors. The document will then be shared with neighboring neurons. Other clustering algorithms such as Farthest First, Expectation-Maximization (EM), Fuzzy c-Means, k-Means, and Hierarchical were compared to SOM's results. They came to the conclusion that SOM rewards exceptional performance.

Raheel and Christopher [12] provide a one-of-a-kind solution for automatically marking short answer questions. The authors describe the system's architecture, which is comprised of three phases that address the student's response and compute the grade for the student's response. The first step is to use an Open Source spell checker like JOrtho to verify and correct your spelling. 2) Parsing the student's response with the Stanford Parser. This statistical parser can generate extremely precise parses. The parser outputs the following findings, which are part of the speech tagged text and design-dependent grammatical relationships between singular words. 3) The third part of the processing answer is a comparison of the tagged text with syntactical structures provided by writers in Question and Answer Language. This phase is managed by the syntax analyzer. The design also incorporates a grammatical relation analyzer, which examines the grammatical relations in the student's response to the examiner's grammatical relations. The final responsibility in the comparison phase is to pass the data aggregated from the syntax analyzer and the grammatical relation analyzer to the marker, who calculates the final grade of the answer.

For a student's answer test of a short essay, Mohd et al. [13] developed an automatic marking method. The system was used to process sentences written in the Malay language, which required the use of technology. Grammatical Relations (GR) from Malay sentences are represented using the syntactic annotation and dependency group approaches proposed in [12]. Tokenizing, recognizing, collocating, and extracting the GRs to process the sentences from the marking scheme and the students' answers are all entries to the Computational Linguistic System (CLS). The system incorporates a database with a table of Malay words and their Parts of Speech to assist the CLS (POS). To compute the grade for the student's answer, compare the GR received from the students' responses to the GR for the marking scheme. Compare the following sentence components: subject to subject, verb to verb, object to object, and phrase to phrase, to put it another way. The authors put the system through its paces to evaluate how it stacks up against human-awarded grades. To assign a grade to each question, they picked Malaysian teachers with prior experience in marking the scheme. The test conditions had been established. The test examines if the system can provide grades that are comparable to those provided by professors.

In their Indonesian essay evaluation, Lahitani et al. [14] used the TF-IDF method and the cosine similarity methodology to determine the degree of similarity. The test datasets consisted of ten student documents obtained from an e-learning source that were examined for similarity to the documents provided by their five experts. The best degree of cosine similarity was 0.39, and it was calculated using a ranked-based methodology.

4 Types of Spelling Error

Error patterns, or patterns of spelling errors, were used to develop techniques. As a result, various research on the types and trends of spelling errors have been done. Damerau's research is the most well-known of all of them. There are two types of spelling problems, according to his research. Typographical errors and cognitive errors [15].

4.1 Typographical Errors

This error occurs when the correct spelling of a word is known yet the term is mistyped by accident. These errors do not comply with any linguistic criterion because they are usually always related to the keyboard. Damerau's research found that 80% of typographic errors fit into one of four categories.

1) Inserting a single letter, such as "Obsolete for Obpsolete".
2) Deleting a single letter, such as "Obsolete for Osolete".
3) Substituting a single letter, such as "Obsolete for Obselete".
4) Transposition of two adjacent letters, such as typing "Obsolete for Oboslete".
Single errors can be generated by any of the editing operations listed above.

4.2 Cognitive Error

These are mistakes that arise when the correct spelling of a term is unknown. In the instance of cognitive errors, the misspelled word's pronunciation is identical or almost identical to the intended right word's pronunciation [16].

5   Method

Microsoft team was used to perform the test. The question paper for 12th Business studies was prepared which had a combination of objective (MCQ) and subjective questions, here subjective questions were limited to one word to one sentence responses. The response of 63 students was recorded for performing the experiment. Figure 1 shows the sample of the student answer sheet.

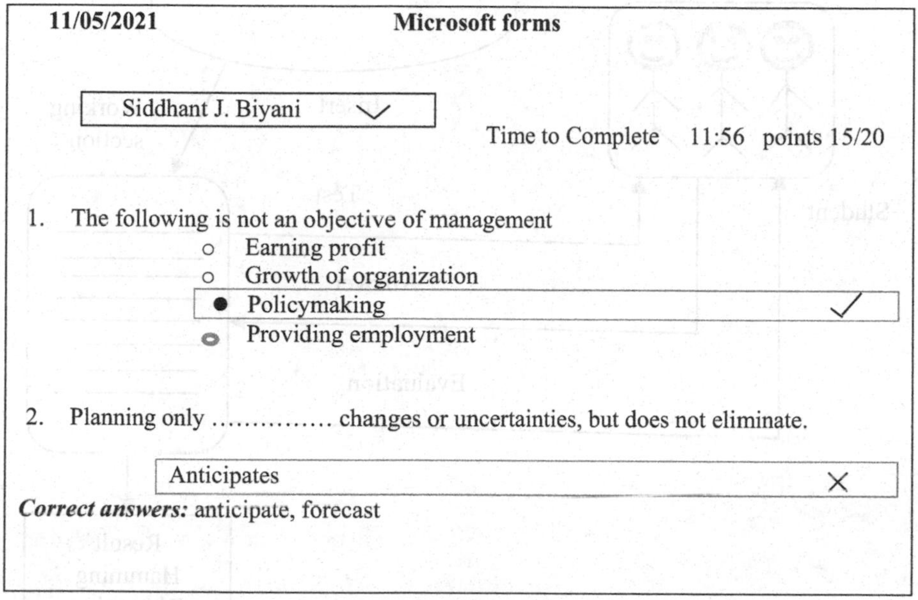

11/05/2021	**Microsoft forms**	

Siddhant J. Biyani   ⌄

Time to Complete   11:56   points 15/20

1.   The following is not an objective of management
      o   Earning profit
      o   Growth of organization
      ●   Policymaking   ✓
      ◔   Providing employment

2.   Planning only .... changes or uncertainties, but does not eliminate.

Anticipates   ✕

Correct answers: anticipate, forecast

Fig. 1.   Sample of student answer sheet

6   Proposed System

The question paper is provided, together with a model answer, and is used as a template for evaluating the student's answer sheet. The system's working model is depicted as a block diagram in Fig. 2.

Question file along with the module answer act as a template, which is stored separately, and the responses of students are collected at one place for evaluation of objective done, objective is done auto, while the one word and one sentence are evaluated using our proposed method as follows.

a)   For one word the word is checked for spelling mistakes, case sensitivity is removed before performing hamming distance.
b)   For one sentence, the sentence POS tagged & it's grammatically checked & then only it passed to our spell checker as discussed above in 'a'.

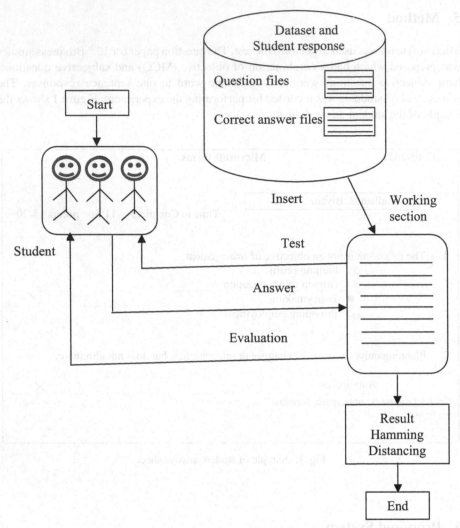

Fig. 2. Block diagram of the working model of the system

7 Hamming Distance

The Hamming Distance metric examines the similarity of any two texts of the same length, where the Hamming distance is the number of differences between the corresponding letters. To grasp the idea of hamming distance, consider any two texts.

The abbreviations "ABCDEF" and "ABCDDSQ" stand for "ABCDEF" and "ABCDDSQ," respectively. The distance between character A in the first position of the text "ABCDEF" and character A in the first position of the text "ABCDSQ" is zero. In the second, third, and fourth locations in both texts, the characters "BCD" are identical, and the Hamming distances are zero. The characters E and F of the first text differ from the characters S and Q of the second text at the same fifth location, hence the Hamming distances are 1. The Hamming distance is used to calculate the distance between binary vectors in binary texts. The Hamming distance is used for error repair and detection in intra-network data transmissions [17].

Using hamming distance we have calculated four types of error

1) Insert: - Means one extra character add here.
2) Missing: - It means some character missing there.
3) Replacement/Substituting: - It means any particular character replaced with another word.
4) Transposition error/Swap: - It means that the character position is changed from the actual alphabet position in a word.

The percentage error calculated for the responses collected is as follows, the percentage error rate is shown in Table 2.

Example:

See Table 1.

Table 1. Shows the example four types of error

Word	Insert	Missing	Replacement	Swap
Obsolete	Obsoleete	Obslete	Obxolete	Obsolete
Supervision	Supervviision	Supervison	Supercision	Supervisino

Table 2. The percentage error for subjective examination

Types of error	Percentage (%)
Inserting	27.72
Missing	11.97
Replacement/Substituting	12.6
Transposition error/Swap	6.3

While performing spell check we came across that position of the character wrongly inserted or swapped or substituted makes a difference while evaluating the answers for example color or colors may be taken as correct answers for the expected answers colors. Table 4 shows the error at the position in the word.

Due to misspelled alphabet change the meaning of actual word or meaning less word showing in Table 3.

Example:

Table 3. Show changes meaning of the word

Sr. No.	Word	Change meaning
1	Desert	Dessert
2	Heal	Heel
3	Mail	Male
4	Accept	Except

Table 4. Number of characters making error

Error position (in no. of character) in word	Percentage of words
1	40.95
2	8.19
3	3.15
4	1.89

8   Evaluation and Result

Table 5. Shows the overall evolution of spell mistake

Sr. no.	Question	Total question	Right answer	Wrong answer	Spell mistake	Blank answer
1	Anticipate, forecast	63	1	61	1	0
2	Market orientation	63	38	17	3	6
3	Responsibility	63	39	13	5	1
4	Supervision	63	27	35	0	1
5	Threat	63	8	14	37	5
6	Unity Of command	63	29	27	2	5
7	Obsolete	63	11	36	2	14
8	Motivation	63	57	3	3	0

(continued)

Table 5. (*continued*)

Sr. no.	Question	Total question	Right answer	Wrong answer	Spell mistake	Blank answer
9	Casual callers	63	38	9	15	1
10	Compensation	63	19	43	1	0
11	Workforce analysis	63	21	33	5	4
Overall Calculation		**756**	**288**	**291**	**74**	**37**

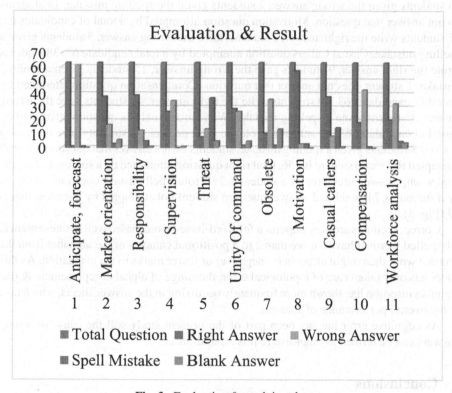

Fig. 3. Evaluation & result in column

The above Table 5 shows the result of subjective evaluation, in our system, we have constructed the exam like one sentence, one-word answer, multiple-choice question (MCQ), filling the blanks. We only use for evaluation in a survey that is an objective question all over 11 questions are there. Anticipate, forecast question attempted by total candidate 63, 1 student-written the right answer, 61 students given the wrong answer, 1 student given the spelling mistake. Market Orientation question attempted by a total of candidates 63, 38 students write the right answer, 17 students given the wrong answer, 3 students given the spelling mistake, 6 students do not answer that question. Responsibility question attempted by a total of candidates 63, 39 students written the right answer, 13 students gave the wrong answer, 5 students given the spelling mistake, 1 student does not answer that question. Supervision question attempted by a total of candidates 63, 27 students write the right answer, 35 students given the wrong answer, 1 student does not answer that question. Threat question attempted by a total of candidates 63, 8 students wrote the right answer, 14 students given the wrong answer, 37 students given the spelling mistake, 5 students do not answer that question. Unity of Command question attempted by total candidate 63, 29 students written the right answer, 27 students given the wrong answer, 2 students given the spelling mistake, 5 students do not answer that question. Obsolete question attempted by total candidate 63, 11 students write the right answer, 36 students given the wrong answer, 2 students given the spelling mistake, 14 students do not answer that question. Motivation question attempted by a total of candidates 63, 57 students write the right answer, 3 students gave the wrong answer, 3 students given a spelling mistake. Casual Callers question attempted by a total candidate 63, 38 students wrote the right answer, 9 students gave the wrong answer, 15 students given a spelling mistake, 1 student does not answer that question. Compensation question attempted by a total of candidates 63, 19 students write the right answer, 43 students gave the wrong answer, 1 student given a spelling mistake. Workforce Analysis question attempted by a total of candidates 63, 21 students wrote the right answer, 33 students gave the wrong answer, 5 students given a spelling mistake, students do not answer that question. Overall attempted this exam student is 756, total right question attempted by a student is 288, the total wrong answer attempted by a student is 291, total spell mistake question attempted by a student is 74, and blank answer question student not attempted by a student that is 37 (Fig. 3).

Correct spellings are very important for word-based word-based well as one sentence, misspelled words if have a more than 2 to 3 positional changes of the alphabet from the original word then might affect in (− negative) or fewer marks to the evaluation. As this paper has only taken care of topological error, the range of alphabet replacement & one alphabet insertion has shown more frequently occurring in the answer sheets, which cost to the overall performance of students.

As cognitive error has not been part of the present study still the cause of wrong answers can be termed as cognitive & it is around 38%.

9   Conclusions

The Question Answering System is the most effective technique for keeping track of valid and correct replies to candidate questions that are answered in natural language rather than via a query. The basic purpose of QAS, like the other queries submitted by users, is to receive accurate answers. Each individual's grasp of the subject is influenced by their expressive power, language used, and comprehension of the issue, and all of these courses have significant differences in subject and writing style, i.e. the reaction varies from person to person. The work focuses on the Question Answering System's spelling error. Error patterns, or patterns of spelling errors, were used to develop techniques. As a result, various research on the types and trends of spelling errors have been done. According to

this experiment, the overall mistake corrected is 46.62% when using hamming distance for word matching or spelling mistakes.

Acknowledgment. The authors would like to acknowledge thanks to Chattrapati Shahu Maharaj Research Training & Human Development Institute (SARTHI), Pune they awarded fellowship, CSRI DST Major Project sanctioned No. SR/CSRI/71/2015(G), and also thanks to Computational and Psycholinguistic Research Lab Facility supporting this work and Department of Computer Science and Information Technology, Dr. Babasaheb Ambedkar Marathwada University, Aurangabad, Maharashtra, India.

References

1. Kankhar, M.A., Mahender, C.N.: Word level similarity auto-evaluation for an online question answering system. J. Eng. Res. (Kuwait) (2021)
2. Alrehily, A.D., Siddiqui, M.A., Buhari, S.M.: Intelligent electronic assessment for subjective exams. ACSIT, ICITE, SIPM, pp. 47–63 (2018)
3. Dreier, J., Giustolisi, R., Kassem, A., Lafourcade, P., Lenzini, G., Ryan, P.Y.: Formal analysis of electronic exams. In: 2014 11th International Conference on Security and Cryptography (SECRYPT), pp. 1–12. IEEE, August 2014
4. Kudi, P., Manekar, A., Daware, K., Dhatrak, T.: Online examination with short text matching. In: 2014 IEEE Global Conference on Wireless Computing & Networking (GCWCN), pp. 56–60. IEEE, December 2014
5. Hodge, V.J., Austin, J.: A comparison of standard spell-checking algorithms and a novel binary neural approach. IEEE Trans. Knowl. Data Eng. **15**(5), 1073–1081 (2003)
6. Farra, N., Tomeh, N., Rozovskaya, A., Habash, N.: Generalized character-level spelling error correction. In: Proceedings of the 52nd Annual Meeting of the Association for Computational Linguistics, vol. 2, Short Papers, pp. 161–167, June 2014
7. Kankhar, M.A., Sayyed, S.N., Mahender, C.N.: Challenges in Online Subjective Examination Systems: An Overview (2021)
8. Al-Jouie, M.F., Azmi, A.M.: Automated evaluation of school children's essays in Arabic. Procedia Comput. Sci. **117**, 19–22 (2017)
9. Kashi, A., Shastri, S., Deshpande, A.R., Doreswamy, J., Srinivasa, G.: A score recommendation system towards automating assessment in professional courses. In: 2016 IEEE Eighth International Conference on Technology for Education (T4E), pp. 140–143. IEEE, December 2016
10. Ishioka, T., Kameda, M.: Automated Japanese essay scoring system: jess. In: Proceedings of 15th International Workshop on Database and Expert Systems Applications, 2004, pp. 4–8. IEEE (2004)
11. Meena, K., Raj, L.: Evaluation of the descriptive type answers using hyperspace analog to language and self-organizing map. In: 2014 IEEE International Conference on Computational Intelligence and Computing Research, pp. 1–5. IEEE, December 2014
12. Siddiqi, R., Harrison, C.: A systematic approach to the automated marking of short-answer questions. In: 2008 IEEE International Multitopic Conference, pp. 329–332. IEEE, December 2008
13. Ab Aziz, M.J., Dato'Ahmad, F., Ghani, A.A.A., Mahmod, R.: Automated marking system for short answer examination (AMS-SAE). In: 2009 IEEE Symposium on Industrial Electronics & Applications, vol. 1, pp. 47–51. IEEE, October 2009

14. Lahitani, A.R., Permanasari, A.E., Setiawan, N.A.: Cosine similarity to determine similarity measure: study case in online essay assessment. In: 2016 4th International Conference on Cyber and IT Service Management, pp. 1–6. IEEE, April 2016
15. Greene, D., Parnas, M., Yao, F.: Multi-index hashing for information retrieval. In: Proceedings 35th Annual Symposium on Foundations of Computer Science, pp. 722–731. IEEE, November 1994
16. Geisler, J.: Error Detection & Correction. Taylor University, Computer and System Sciences Department (2005)
17. Al-azani, S.A., Mahender, C.N.: Improve Hamming character difference based-on derivative lexical similarity and right space padding

Automated Classification of Sleep Stages Using Single-Channel EEG Signal: A Machine Learning-Based Method

Santosh Satapathy[1]([✉]), Shrinibas Pattnaik[2], Badal Acharya[2], and Rama Krushna Rath[3]

[1] Department of Information and Communication Technology, Pandit Deendayal Energy University (PDEU), Gandhinagar, India
Santosh.Satapathy@sot.pdpu.ac.in

[2] Department of Electrical and Electronics Engineering, Gandhi Institute of Science and Technology, Rayagada, Odisha, India

[3] Department of Computer Science and Engineering, IIIT, Chittoor, Andhra Pradesh, India

Abstract. One of the major contributors to improper sleep patterns is the rapidly occurring changes in today's lifestyle. We aimed to develop an automated algorithm based on to classify the sleep stages during sleep hours. Maintaining such unhealthy sleep patterns for a longer period may lead to different neurological disorders. Delay in diagnosis further worsens the condition and leads to other serious health issues. The first step in analyzing any sleep-based abnormalities is the proper classification of the sleep stages. The proposed study obtains, a single-modal channel of electroencephalogram (EEG) signals as input to the model. The main objective is to screen the pertinent features which can assist in identifying the irregularities that occurred during sleep hours. The entire experiment was carried out on two different subgroups of the ISRUC-Sleep dataset and finally, we considered the support vector machine (SVM) for the classification of sleep stages. The proposed model yielded the best classification accuracy of 97.73%, and 96.51% with subgroup-I, and subgroup-III subjects, respectively. The proposed model is effective for automated multi-class sleep state classification method is developed for different medical-conditioned subjects. Compared to gold standard polysomnography, our algorithm doesn't require any additional electrodes and which are especially valuable in improving the sleep staging classification performance.

Keywords: EEG · Sleep staging · Feature selection · Machine learning

1 Introduction

The human body contains different components, which function with several physiological processes. It has been seen that most of these physiological processes represent

M. Singh et al. (Eds.): ICACDS 2022, CCIS 1614, pp. 235–247, 2022.
https://doi.org/10.1007/978-3-031-12641-3_20

themselves in the form of signals and these signals are of different types, viz. biochemical, biomedical, biomechanical, etc. Generally, biomedical signals hold a lot of information about the changes in the physiological activities of our bodies [1–3]. Disease and ailments may create a disturbance in the biological system of humans which causes disturbance in the smooth functioning of our daily activities. It has been observed that these diseases and ailments in the human biological system are counter-productive leading to several pathological conditions which affect the health and well-being of the system [4–6]. Analyzing these biomedical signals was quite difficult a few years ago since the obtained signals were contaminated with different types of irrelevant noise and powerline interferences. Sometimes it is also required to obtain their spectral properties to understand the characteristics of the signals' behavior in terms of the different frequency ranges, and modeling of the different feature representation and parameterization. The other major challenge concerning biomedical signal processing is the experience and expertise of the clinicians in analyzing and interpreting the signals in a proper way [7].

However, this entire process is subjective. In recent research developments, digital signal processing and pattern recognition have been the backbone of biomedical research applications [8, 9]. Another crucial aid to biomedical research is a computer-aided diagnostic system which assists the domain experts to make important decisions during the diagnostic process plus it also offers the possibility of online monitoring for critically ill patients. Thus, there is a huge demand for more reliable, effective, and affordable clinical and health services. Hence there are so many novel medical equipment, techniques being developed to support such diagnosis, monitoring, and treatment of irregularities, abnormalities, and ailments in the human body. In the present world, one of such major health issues has been sleeping deprivation and associated illnesses which adversely affect the quality of life across the global population irrespective of age group. Hence applications based on sleep studies and patterns become crucial to diagnosing and proposing solutions to such ailments. EEG signals are most appropriate to identify irregularities in the regular sleep pattern more effectively. Additionally, it also has some more advantages that are it is high speed, time-efficient, and also non-invasive. Therefore, the diagnosis with EEG signals is inexpensive and widely accepted in the diagnosis of different sleep-related diseases [10]. It has been seen from the recent research developments in the signal processing techniques, an improved approach for classification on the basis of PSG signals has been reported. These enhanced techniques can help to recognize the abnormal sleep patterns and irregularities in sleep leading to sleep-related diseases. An automatic sleep staging system is obtained in the diagnosis of several sleep disorders, which is considered one of the best tools apart from the visual inspection of the EEG signals by physicians [11].

1.1   Related Work

Some of the latest works on automated sleep stage classification making use of single-channel and multiple channels are presented in Table 1.

Table 1. A brief analysis of state-of-the-art works for the sleep staging

Study and year	Classifier	Signal	Number of channels	Results (%)
Oboyya et al. [12]	Fuzzy c-means algorithm	EEG	One channel	85%
Güneş, K. Polat et al. [13]	K-means clustering	EEG	One channel	83%
Aboalayon et al. [14]	SVM	EEG	One channel	90%
Hassan, A. R. et al. [15]	Bootstrap aggregating	EEG	One channel	92.43%
Diykh, M. et al. [16]	SVM	EEG	One channel	95.93%
Kristin M. Gunnarsdottir et al. [17]	Decision tree	EEG, EOG, and EMG	One channel	80.70%
Sriraam, N. et al. [18]	Feedforward neural network	EEG	One channel	92.29%
Memar, P. et al. [19]	Random forest	EEG	One channel	86.64%
Da Silveira et al. [20]		EEG	One channel	90%
Xiaojin Li et al. [21]		EEG	One channel	94.4%
Zhu, G et al. [22]	SVM	EEG	Two channels	87.50%
Satapathy, S. K. et al. [23]	Ensemble learning model	EEG	Two channels	91.10% 90.11%
Satapathy, S. K. et al. [24]	Stacking ensemble learning model	EEG	One channel	90.8%

From the state-of-the-art works, it can be seen that most of the studies have focused on single-channel and multiple-channel. Some of the studies were well performed with the input of single-channel, additionally, it gives well comfortable for the subject during sleep recordings. This advantage gives an improvement in sleep staging accuracy. But it has been also observed that major of the studies highly suffered from data imbalance problems, misclassification of the N1 sleep stage, and improper selection of features for the classifier.

Therefore our proposed work has different from other similar works in this context; we have obtained two different session recordings for the subject who was affected with a sleep disorder. This work aims to contribute to the comparative analysis of sleep staging methods by assessing the classification accuracy of statistical moments derived from single-channel EEG signals. With regards to classification techniques, we used SVM for testing them between participating subjects in the research work. We report all Cohen's kappa values for various sleep stages for the comparison of the classification algorithm. The proposed algorithm can be observed to present high values of accuracy and Cohen's kappa of the data of healthy subjects.

Further, our proposed research work is organized as follows: Sect. 2 presents the data used in the experimental work. In Sect. 3 we briefly present the proposed methodology. Section 4 briefly about the outcome of the experimental analysis. Section 5 remarks the concluding remarks.

2 Experimental Data

In this study, for experimental analysis, we have acquired two different categories of the patient's records, one from the subjects who were already suffered from mild sleep-related problems and the other category of the subjects, who were completely healthy controlled. The whole dataset is an accumulation of three types of sleep recordings belonging to different patients. The entire recordings were collected by a group of expert clinicians [25]. The whole recordings were obtained from different patients, who were already faced various kinds of sleep-related problems, and the other part of the recordings were obtained from healthy subjects.

In this work, for our experimental study, we used the C3-A2 channel for computing sleep stage classification. Table 2 presents a detailed account of the percentage of sleep epoch distribution in different sleep stages.

Table 2. Description of the percentage of epoch per individual sleep stages

Subject number	W%	N1%	N2%	N3%	REM%
Subject-1 Subgroup-I	22.00%	8.40%	23.07%	30.80%	15.73%
Subject-2 Subgroup-I	30.80%	9.60%	30.13%	19.60%	9.87%
Subject-3 Subgroup-I	15.87%	18.93%	25.87%	16.80%	22.53%
Subject-4 Subgroup-I	2.53%	5.87%	43.60%	28.53%	19.47%
Subject-5 Subgroup-I	32.67%	13.87%	26.53%	21.87%	5.07%
Subject-9 Subgroup-I	9.6%	19.06%	42%	18.13%	11.2%
Subject-16 Subgroup-I	17.07%	16.67%	37.33%	16.00%	12.93%
Subject-23 Subgroup-I	28.27%	13.20%	36.00%	8.67%	13.87%
Subject-1 Subgroup-III	19.87	12.13	35.60	21.07	11.33
Subject-5 Subgroup-III	67.00%	8.67%	38.27%	33.47%	10.67%
Subject-6 Subgroup-III	7.20%	14.80%	34.80%	32.93%	10.27%
Subject-7 Subgroup-III	27.47%	7.07%	20.53%	34.67%	10.27%
Subject-8 Subgroup-III	45.73%	9.73%	15.20%	15.47%	13.87%

(*continued*)

Table 2. (*continued*)

Subject number	W%	N1%	N2%	N3%	REM%
Subject-9 Subgroup-III	13.47%	19.33%	31.47%	30.00%	5.60%
Subject-10 Subgroup-III	16.80%	29.07%	23.07%	13.07%	18.00%

This study includes three categories of subjects. One is affected by sleep problems with one session recording sleep data. The second category of data was extracted from subjects with sleep disorders in two sessions of recording. Finally, the third category includes healthy subjects. The authors have obtained the EEG channel of C3-A2 signals of 6 subjects. The stage annotations are also provided in the data repository according to AASM rules. The unscored epochs are not considered for further analysis in the experimental work. Each epoch is considered a 30 s time length in this present work. For analysis, of the sleep behavior, we have presented the samples of all sleep stages of EEG signals extracted from different categories of subjects. The sleep EEG signals behavior changes according to the sleep cycle covered by the subject. Every stage of sleep is characterized by different behavior from sleeping brains such as low amplitude, mixed frequency, sawtooth waveforms, low amplitude in muscle movements, and rapid eye movements. However, the EEG signal behaviors are complex since they are not periodic and also because of the continuous changes of amplitude, frequency, and phase range according to the sleep stage. In the proposed method, the authors have applied the Butterworth bandpass filter to reduce the undesired segments from the input signal. The filter is a second-order Butterworth bandpass filter, and the frequency ranges applied range from 0.5 Hz to 35 Hz.

3 Methodology

This study demonstrates a unique method of two-stage sleep scoring using a single EEG channel. Figure 1 represents the workflow of the proposed sleep stage classification system.

3.1 Proposed Automatic Sleep Stage Detection Method

The sleeping study can be divided into four layers. These are the subject layer, classification layer, section layer, and instance layer. The subject layer is divided into two sections. One section contains information about EEG, and another section represents sleep information. These two sections are the main domains of this study and compose the classification layer. The section layer represents the associated concepts related to the EEG signals and sleep sections. While the EEG section defines the used electrode name, obtained features, and participants enrolled for this experimental work, the sleep section addresses the obtained sleep stage rules for sleep quality analysis. Finally, the instance layer considered the specific settings associated used the experiments conducted in the research and is defined according to the other layers. In this proposed study, two different medical condition subjects' details from the ISRUC-Sleep dataset from the University Hospital of Coimbra, Portugal are evaluated. We have made use of the OSFS systems for feature selection [26, 27]. The final feature selection output is presented in Table 3.

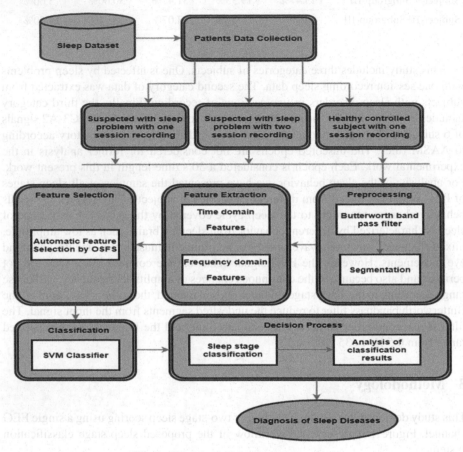

Fig. 1. Workflow of the proposed method

Algorithm 1
Automated SleepEEG Classification Test.
Procedure: *Sleepstage_Scoring = SleepStage_Classify (Feature List, Labels).*
Input: *Patient Records* [Extracted from Sleep_Dataset]

$$P_V = \{P_1, P_2, \cdots \cdots \cdots \cdots, P_N\}$$

$$P_{SI} = \{S_1, S_2, \cdots \cdots \cdots \cdots \cdots, S_M\}$$

$$P_{SE} = \{S_{I1}, S_{I2}, \cdots \cdots \cdots \cdots \cdots, S_{IZ}\}$$

%N = Number of Patients enrolled for SleepEEG test.

%M = *Number of Session recordings considered per individual enrolled subjects.*
%Z = *Number of Sleep stages epochs.*
Pre_Predict_Label = *pre-predicted annotations of sleep stages.*
Output: Sleepstage_Scoring = *Correct predicted sleep stage sequences.*
Step-1: Let P be the enrolled patient's records from different health conditions.
Step-2: Let K be the total sleep recordings of enrolled subjects from a single channel of EEG signal.
Step-3: For each EEG recording from individual enrolled patients in the SleepEEG study do.
Step-4: $P = P + 1$.
Step-5: Divide the whole K segments of sleep recordings into 30s epochs-wise.
Step-6: Extracting both time and frequency domain features from each epoch and stored into specific feature vectors per individual enrolled patients.

Feature Extraction (Extracted Feature Set)

$$FV_{P1} = \{P1_{F1}, P1_{F2}, \ldots \ldots \ldots, P1_{FE}\}$$
$$FV_{P2} = \{P2_{F1}, P2_{F2}, \ldots \ldots \ldots, P2_{FE}\}$$

$$\ldots \ldots \ldots \ldots \ldots \ldots \ldots \ldots \ldots \ldots \ldots \ldots \ldots \ldots \ldots$$

$$FV_{PN} = \{PN_{F1}, PN_{F2}, \ldots \ldots \ldots, PN_{FE}\}$$
$$F_j = \{y_1, y_2, y_3, \ldots \ldots \ldots \ldots, y_T\}$$

%N=Patients number, %E=number of feature types, %T=number of feature elements

Step-7: Forwarded the extracted features to the proposed feature selection techniques and select the suitable features and store them into the selected feature vector patients-wise.

Feature Selection (Selected Feature Set)

$$FS_{P1} = \{P1_{y1}, P1_{y2}, P1_{y3}, \ldots \ldots \ldots \ldots, P1_{yT}\}$$
$$FS_{P2} = \{P2_{y1}, P2_{y2}, P2_{y3}, \ldots \ldots \ldots \ldots, P2_{yT}\}$$

$$\ldots \ldots \ldots \ldots \ldots \ldots \ldots \ldots \ldots \ldots \ldots \ldots \ldots$$

$$FS_{PN} = \{PN_{y1}, PN_{y2}, PN_{y3}, \ldots \ldots \ldots \ldots, PN_{yT}\}$$

%Piyj: an element of feature selection set

Step-8: Forwarded selected feature vector list and predicted sleep stage segments into the proposed classification model.

Sleepstage_Scoring = SleepStage_Classify (Feature_Selection_PN, Pre_Predict_Label).

Step-9: if the Sleep stage classification accuracy is suitable then recommend the sleep stage scoring approach in the diagnosis of sleep-related problems, go to step 11.

Step-10: else go to step 6 [For new feature extraction from sleep recordings of the enrolled subjects].

Step-11: end if.

Output: *Automated Corrected Sleep Stage Sequences of the Subjects.*

Table 3. Final feature selection list

Subject	Selected features	Total
Subject1	$F1_1, F2_1, F3_1, F4_1, F5_1, F7_1, F9_1, F10_1, F11_1, F13_1, F14_1, F15_1, F22_1,$ $F25_1, F27_1$	15
Subject2	$F1_2, F5_2, F6_2, F7_2, F8_2, F9_2, F10_2, F11_2, F12_2, F13_2, F14_2, F16_2,$ $F18_2, F20_2, F27_2$	15
Subject3	$F1_3, F5_3, F6_3, F7_3, F8_3, F10_3, F12_3, F13_3, F17_3, F18_3$	10
Subject4	$F1_4, F2_4, F3_4, F4_4, F10_4, F12_4, F14_4, F15_4, F16_4, F17_4, F18_4, F22_4,$ $F24_4, F28_4$	14
Subject5	$F1_5, F5_5, F6_5, F7_5, F8_5, F9_5, F10_5, F12_5, F13_5, F14_5, F15_5, F17_5,$ $F18_5, F23_5, F28_5$	15
Subject9	$F1_9, F5_9, F6_9, F7_9, F8_9, F9_9, F10_9, F12_9, F13_9, F14_9, F18_9, F28_9$	12
Subject16	$F1_{16}, F2_{16}, F3_{16}, F4_{16}, F5_{16}, F7_{16}, F9_{16}, F10_{16}, F11_{16}, F13_{16}, F14_{16},$ $F15_{16}, F22_{16}, F25_{16}, F27_{16}$	15
Subject23	$F1_{23}, F2_{23}, F3_{23}, F4_{23}, F5_{23}, F7_{23}, F11_{23}, F12_{23}, F15_{23}, F16_{23},$ $F17_{23}, F20_{23}, F24_{23}, F26_{23}$	14
SubjectH1	$F1_{H\_01}, F5_{H\_01}, F6_{H\_01}, F8_{H\_01}, F10_{H\_01}, F16_{H\_01}, F17_{H\_01},$ $F18_{H\_01}, F22_{H\_01}$	9
SubjectH2	$F1_{02}, F2_{02}, F3_{02}, F4_{02}, F5_{02}, F7_{02}, F8_{02}, F9_{02}, F11_{02}, F12_{02}, F13_{02},$ $F14_{02}, F21_{02}, F22_{02}, F25_{02}$	15
SubjectH5	$F1_{05}, F2_{05}, F3_{05}, F4_{05}, F5_{05}, F7_{05}, F8_{05}, F9_{05}, F11_{05}, F12_{05}, F13_{05},$ $F14_{05}, F21_{05}, F22_{05}, F25_{05}$	15
SubjectH6	$F1_{H\_06}, F3_{H\_06}, F4_{H\_06}, F5_{H\_06}, F8_{H\_06}, F9_{H\_06}, F13_{H\_06}, F14_{H\_06},$ $F15_{H\_06}, F16_{H\_06}, F17_{H\_06}, F18_{H\_06}, F22_{H\_06}$	13
SubjectH7	$F1_{H\_07}, F5_{H\_07}, F6_{H\_07}, F7_{H\_07}, F8_{H\_07}, F10_{H\_07}, F16_{H\_07}$ $F18_{H\_07}, F28_{H\_07}$	9
SubjectH8	$F1_{H\_08}, F10_{H\_08}, F11_{H\_08}, F12_{H\_08}, F13_{H\_08}, F14_{H\_08}, F15_{H\_08},$ $F17_{H\_08}, F18_{H\_08}, F19_{H\_08}$	10
SubjectH9	$F1_{H\_09}, F4_{H\_09}, F5_{H\_09}, F6_{H\_09}, F7_{H\_09}, F8_{H\_09}, F10_{H\_09}, F12_{H\_09},$ $F13_{H\_09}, F16_{H\_09}, F17_{H\_09}$	11
SubjectH10	$F1_{H\_10}, F5_{H\_10}, F6_{H\_10}, F8_{H\_10}, F10_{H\_10}, F13_{H\_10}, F15_{H\_10}, F17_{H\_10}, F18_{H\_10}, F22_{H\_10}$	10

4 Experimental Results and Discussion

This complete methodology procedure of this proposed work is explained in pseudo-code format, which is described in Algorithm 1. The complete illustrations of the experimental steps of this proposed work are shown in Fig. 1. For each sampled patient, the experimental results were concluded separately. Multiple evaluation metrics such as accuracy (Acc) [28], recall (Rec) [29], specificity (Spe) [30], precision (Pre) [31], F1score (F1sc)

[32], and Cohen's Kappa Score [33, 34] to analyze the performance of the proposed sleep studies.

4.1 Classification Accuracy of Category-I Subjects ISRUC-Sleep Database

Table 4. Performance evaluation results for ISRUC-Sleep subgroup-I

C3-A2/SVM	Pre	Rec	Spe	F1Sc	Acc	Kappa score
Subject-1	92.3%	96.72%	71.51%	94.48%	91.2%	0.71
Subject-2	94.72%	96.71%	87.87%	97.7%	94%	0.84
Subject-3	94.4%	96.66%	69.74%	95.51%	92.4%	0.70
Subject-4	98.62%	99.04%	47.36%	98.81%	97.73%	0.5
Subject-5	96.16%	94.64%	92.24%	95.4%	93.85%	0.84
Subject-6	96.17%	96.60%	63.88%	96.37%	93.45%	0.60
Subject-16	95.1%	99.48%	81.60%	97.22%	95.60%	0.84
Subject-23	93.09%	95.1%	82.08%	94.10%	91.47%	0.77

The results for subject-1, subject-2, subject-3, subject-4, subject-5, subject-6, and subject-16 show an overall classification accuracy of 91.2%, 94%, 92.4%, 97.73%, 93.86%, 93.46%, 95.60% and 91.46% using SVM classifier. The higher precision value is presented for subject-4 using the SVM method of 98.63%. The highest specificity and F1-score performance results are 92.24% (Subject-5) and 98.83% (Subject-4) using the SVM method. The Kappa scores for subjects-2, 5, and 16 are found in excellent agreement with the subject to the best accuracy for investigation of sleep irregularities. The reported evaluation metrics performances for all subjects for both the session recordings are described in Table 4.

4.2 Classification Accuracy of Category-III Subjects ISRUC-Sleep Database

Finally, the final experimental results for the proposed SleepEEG method are done through the analysis of signals taken from healthy subjects, who haven't been on medication of any sort prior to the experiment. The one-session recording was taken into consideration for monitoring the sleep irregularities and the performances achieved for all the healthy controlled subjects are presented in Table 5.

It can thus be observed that the model achieves an overall accuracy of 92.5%, 92.26%, 91.73%, 91.6%, 96.53%, 66.13%, 95.06% and 93.06% through SVM classifier with ISRUC-Sleep subgroup-III data for subject-1, 2, 5, 6,7,8,9 and 10 respectively. The highest precision was reported from subject-07 (97.26%). The specificity performance was achieved as the highest form subject-07 (92.71%). Furthermore, the F1-Score performance reports the highest value of 97.75% from subject-02. The kappa score for subject-02 and subject-07 is well suitable with SVM classification models. A comparative system performance with respect to other sleep studies has also been performed by the authors and is presented in Table 6.

Table 5. Performance evaluation with ISRUC-Sleep subgroup-III data

C3-A2/SVM	Pre	Rec	Spe	F1sc	Acc	Kappa coefficient
SubjectH-1	93.32%	97.67%	71.81%	95.44%	92.53%	0.74
SubjectH-2	96.82%	98.70%	84.84%	97.75%	92.26%	0.86
SubjectH-5	93.61%	97.28%	49.42%	95.41%	91.73%	0.53
SubjectH-6	92.82%	98.56%	1%	95.60%	91.6%	0.007
SubjectH-7	97.26%	97.97%	92.71%	97.61%	96.53%	0.91
SubjectH-8	70.73%	64.12%	68.51%	67.26%	66.13%	0.32
SubjectH-9	96.08%	98.30%	74.25%	97.18%	95.06%	0.77
SubjectH-10	95.85%	96.47%	79.36%	96.16%	93.6%	0.76

Table 6. Result analysis in between the proposed method with previously contributed research works

Studies	Models		Accuracy
Ref. [35]	SVM		95%
Ref. [36]	Bayesian classifier		83%
Ref. [37]	SVM		93.97%
Ref. [38]			86.75%
Ref. [39]			81.74%
Ref. [40]	Random forest		75.29%
Ref. [41]	Stacked Sparse Auto-Encoders (SSAE)		82.03%
Ref. [42]	SVM		83.33%
Proposed	**SVM**	**ISRUC-Sleep (Subgroup-I)**	97.73%
		ISRUC-Sleep (Subgroup-III)	96.53%

5   Conclusion and Future Directions

Maintaining proper sleep quality is crucial for both physical and mental health. The best way of measuring sleep quality is to classify the sleep stages using the single-modal signal. However manual inspection of sleep stages classification is quite difficult, consumes more time, and is highly subject-oriented. To overcome these difficulties, an automated sleep staging system was proposed which considers the basic steps such as pre-processing of signals, feature extraction, feature selection, and classification. Proper feature extraction and selecting the most suitable features play an important role during the automated sleep staging process and it ultimately affects the classification process. Besides a good amount of work has already been contributed in this field, but still, some gaps need to be dealt with. In this work, an attempt was made to address a few of these issues by presenting appropriate solutions for feature combinations, feature screening, and generalized classification models. Additionally, we also analyzed the effectiveness of the proposed methodology via various evaluation metrics. The following points are summarized from the proposed sleep staging classification results. The most significant features are chosen to improve the sleep staging classification performance compared to existing contributions and other methods considered in this research study.

References

1. Panossian, L.A., Avidan, A.Y.: Review of sleep disorders. Med. Clin. N. Am. **93**, 407–425 (2009). https://doi.org/10.1016/j.mcna.2008.09.001
2. Smaldone, A., Honig, J.C., Byrne, M.W.: Sleepless in America: inadequate sleep and relationships to health and well-being of our nation's children. Pediatrics **119**, 29–37 (2007)
3. Hassan, A.R., Bhuiyan, M.I.H.: Automatic sleep scoring using statistical features in the EMD domain and ensemble methods. Biocybern. Biomed. Eng. (2016). https://doi.org/10.1016/j.bbe.2015.11.001
4. Aboalayon, K., Ocbagabir, H., Faezipour, T.: Efficient sleep stage classification based on EEG signals. In: Systems Applications and Technology Conference (LISAT), pp. 1–6 (2014)
5. Obayya, M., Abou Chadi, F.: Automatic classification of sleep stages using EEG records based on Fuzzy C-means (FCM) algorithm. In: Radio Science Conference (NRSC), pp. 265–272 (2014)
6. Alickovic, E., Subasi, A.: Ensemble SVM method for automatic sleep stage classification. IEEE Trans. Instrum. Measur. (2018). https://doi.org/10.1109/TIM.2018.2799059
7. Abeyratne, U.R., Swarnkar, V., Rathnayake, S.I., Hukins, C.: Sleep-stage and event dependency of brain asynchrony as manifested through surface EEG. In: Proceedings of the 29th IEEE Annual International Conference of the Engineering in Medicine and Biology Society, pp. 709–712 (2007)
8. Rechtschaffen, A., Kales A.: A Manual of Standardized Terminology, Techniques and Scoring Systems for Sleep Stages of Human Subjects. U.G.P. Office, Public Health Service; Washington, DC, USA (1968)
9. Iber, C., Ancoli-Israel, S., Chesson, A.L., Quan, S.F.: The AASM manual for the scoring of sleep and associated events: rules, terminology and technical specification. In: American Academy of Sleep Medicine (2007)
10. Satapathy, S.K., Loganathan, D.: Machine learning approaches with heterogeneous ensemble learning stacking model for automated sleep staging. Int. J. Comput. Digit. Syst. Univ. Bahrain J. https://doi.org/10.12785/ijcds/100109
11. Cogan, D., Birjandtalab, J., Nourani, M., Harvey, J., Nagaraddi, V.: Multi-biosignal analysis for epileptic seizure monitoring. Int. J. Neural Syst. (2017). https://doi.org/10.1142/S0129065716500313
12. Obayya, M., Abou-Chadi, F.: Automatic classification of sleep stages using EEG records based on Fuzzy C-means (FCM) algorithm. In: Radio Science Conference (NRSC), pp. 265–272 (2014)
13. Güneş, S., Polat, K., Yosunkaya, Ş: Efficient sleep stage recognition system based on EEG signal using k-means clustering based feature weighting. Expert Syst. Appl. **37**, 7922–7928 (2010)
14. Aboalayon, K., Ocbagabir, H.T., Faezipour, M.: Efficient sleep stage classification based on EEG signals. In: Systems, Applications and Technology Conference (LISAT), pp. 1–6 (2014)
15. Hassan, A.R., Subasi, A.: A decision support system for automated identification of sleep stages from single-channel EEG signals. Knowl.-Based Syst. **128**, 115–124 (2017)

16. Diykh, M., Li, Y., Wen, P.: EEG sleep stages classification based on time domain features and structural graph similarity. IEEE Trans. Neural Syst. Rehabil. Eng. **24**(11), 1159–1168 (2016)

17. Gunnarsdottir, K.M., Gamaldo, C.E., Salas, R.M.E., Ewen, J.B., Allen, R.P., Sarma, S.V.: A novel sleep stage scoring system: combining expert-based rules with a decision tree classifier. In: 40th Annual International Conference of the IEEE Engineering in Medicine and Biology Society (EMBC) (2018)

18. Sriraam, N., Padma Shri, T.K., Maheshwari, U.: Recognition of wake-sleep stage 1 multi-channel EEG patterns using spectral entropy features for drowsiness detection. Australas. Phys. Eng. Sci. Med. **39**(3), 797–806 (2018). https://doi.org/10.1007/s13246-016-0472-8

19. Memar, P., Faradji, F.: A novel multi-class EEG-based sleep stage classification system. IEEE Trans. Neural Syst. Rehabil. Eng. **26**(1), 84–95 (2018)

20. Da Silveira, T.L.T., Kozakevicius, A.J., Rodrigues, C.R.: Single-channel EEG sleep stage classification based on a streamlined set of statistical features in wavelet domain. Med. Biol. Eng. Comput. **55**(2), 343–352 (2016). https://doi.org/10.1007/s11517-016-1519-4

21. Wutzl, B., Leibnitz, K., Rattay, F., Kronbichler, M., Murata, M.: Genetic algorithms for feature selection when classifying severe chronic disorders of consciousness. PLoS ONE **14**(7), e0219683 (2019)

22. Zhu, G., Li, Y., Wen, P.P.: Analysis and classification of sleep stages based on difference visibility graphs from a single-channel EEG signal. IEEE J. Biomed. Health Inform. **18**(6), 1813–1821 (2014)

23. Satapathy, S.K., Bhoi, A.K., Loganathan, D., Khandelwal, B., Barsocchi, P.: Machine learning with ensemble stacking model for automated sleep staging using dual-channel EEG signal. Biomed. Signal Process. Control **69**, 102898 (2021). https://doi.org/10.1016/j.bspc.2021.102898

24. Satapathy, S.K., Loganathan, D.: Prognosis of automated sleep staging based on two-layer ensemble learning stacking model using single-channel EEG signal. Soft. Comput. **25**(24), 15445–15462 (2021). https://doi.org/10.1007/s00500-021-06218-x

25. Khalighi, S., Sousa, T., Santos, J.M., Nunes, U.: ISRUC-Sleep: a comprehensive public dataset for sleep researchers. Comput. Methods Programs Biomed. **124**, 180–192 (2016)

26. Eskandari, S., Javidi, M.M.: Online streaming feature selection using rough sets. Int. J. Approximate Reasoning **69**, 35–57 (2016)

27. İlhan, H.O., Bilgin, G.: Sleep stage classification via ensemble and conventional machine learning methods using single channel EEG signals. Int. J. Intell. Syst. Appl. Eng. **5**(4), 174–184 (2017)

28. Sanders, T.H., McCurry, M., Clements, M.A.: Sleep stage classification with cross frequency coupling. In: 36th Annual International Conference of the IEEE Engineering in Medicine and Biology Society (EMBC), pp. 4579–4582 (2014)

29. Bajaj, V., Pachori, R.: Automatic classification of sleep stages based on the time-frequency image of EEG signals. Comput. Methods Programs Biomed. **112**(3), 320–328 (2013)

30. Hsu, Y.-L., Yang, Y.-T., Wang, J.-S., Hsu, C.-Y.: Automatic sleep stage recurrent neural classifier using energy features of EEG signals. Neurocomputing **104**, 105–114 (2013)

31. Zibrandtsen, I., Kidmose, P., Otto, M., Ibsen, J., Kjaer, T.W.: Case comparison of sleep features from ear-EEG and scalp-EEG. Sleep Sci. **9**(2), 69–72 (2016)

32. Berry, R.B., et al.: The AASM manual for the scoring of sleep and associated events: rules, terminology and technical specifications. In: American Academy of Sleep Medicine (2014)

33. Sim, J., Wright, C.C.: The kappa statistic in reliability studies: use, interpretation, and sample size requirements. Phys. Ther. **85**(3), 257–268 (2005)

34. Liang, S.-F., Kuo, C.-E., Kuo, Y., Cheng, Y.-S.: A rule-based automatic sleep staging method. J. Neurosci. Methods **205**(1), 169–176 (2012)

35. Khalighi, S., Sousa, T., Oliveira, D., Pires, G., Nunes, U.: Efficient feature selection for sleep staging based on maximal overlap discrete wavelet transform and SVM. In: Annual International Conference of the IEEE Engineering in Medicine and Biology Society (2011)
36. Simões, H., Pires G., Nunes U., Silva V.: Feature extraction and selection for automatic sleep staging using EEG. In: Proceedings of the 7th International Conference on Informatics in Control, Automation and Robotics, vol. 3, pp. 128–133 (2010)
37. Khalighi, S., Sousa, T., Santos, J.M., Nunes, U.: ISRUC-sleep: a comprehensive public dataset for sleep researchers. Comput. Methods Programs Biomed. **124**, 180–192 (2016)
38. Sousa, T., Cruz, A., Khalighi, S., Pires, G., Nunes, U.: A two-step automatic sleep stage classification method with dubious range detection. Comput. Biol. Med. **59**, 42–53 (2015)
39. Khalighi, S., Sousa, T., Pires, G., Nunes, U.: Automatic sleep staging: a computer assisted approach for optimal combination of features and polysomnographic channels. Expert Syst. Appl. **40**(17), 7046–7059 (2013)
40. Tzimourta, K.D., Tsilimbaris, A.K., Tzioukalia, A.T., Tzallas, M.G., Tsipouras, L.G.: EEG-based automatic sleep stage classification. Biomed. J. Sci. Tech. Res. **7**(4), 6032–6037 (2018)
41. Najdi, S., Gharbali, A.A., Fonseca, J.M.: Feature transformation based on stacked sparse autoencoders for sleep stage classification. In: Camarinha-Matos, L.M., Parreira-Rocha, M., Ramezani, J. (eds.) DoCEIS. IAICT, vol. 499, pp. 191–200. Springer, Cham (2017). https://doi.org/10.1007/978-3-319-56077-9_18
42. Kalbkhani, H., Ghasemzadeh, P.G., Shayesteh, M.: Sleep stages classification from EEG signal based on Stockwell transform. IET Signal Process. **13**(2), 242–252 (2018)

Tuning Proximal Policy Optimization Algorithm in Maze Solving with ML-Agents

Phan Thanh Hung, Mac Duy Dan Truong, and Phan Duy Hung[✉]

Computer Science Department, FPT University, Hanoi, Vietnam
{hungpthe140966,truongmddhe141711}@fpt.edu.vn, hungpd2@fe.edu.vn

Abstract. The proximal Policy Optimization algorithm is the ML-Agents toolkit's default reinforcement algorithm. This algorithm can alternate between sampling data via environmental interaction and applying stochastic gradient descent to optimize a "surrogate" cost function. Although when creating a new machine learning model, it is tough to know the optimal model architecture for a given project immediately. In most cases, we can use the default given values from the algorithm's creator or use the machine to carry out this evaluation and choose the best model architecture automatically. Hyperparameters define the model architecture; thus, searching for the best model is called hyperparameter tuning. We focus on comparing four hyperparameters: Beta, Epsilon, Lambd, Num_epoch of PPO algorithm in solving a maze. The results obtained in the training process show the difference in the selection of hyperparameters. The modification of hyperparameters will depend on the maze's complexity and the complexity of the Agent's actions. This paper will help to make appropriate choices at hyperparameters in concrete and practical projects.

Keywords: Reinforcement learning · Proximal policy optimization · ML-agents · Maze-solving

1 Introduction

Artificial Intelligence (AI) has moved one step closer to its aim of imitating the human brain, thanks to advances in computing technology and the development of new sophisticated algorithms. In that field, a branch that is becoming more and more important is reinforcement learning (RL) [1]. RL is a promising machine learning method rising in popularity in the video game industry in recent years with many breakthroughs, especially in games, making robots do tasks instead of humans or working in an unknown environment [2].

RL is applied very strongly to games, so the Machine Learning Agents Toolkit (ML-Agent) library was born as a bridge for using RL into Unity [3]. Unity ML-Agent is a "open-source" Unity plugin that allows intelligent Agents to be trained through games and simulations. It also provides PyTorch implementations of cutting-edge algorithms, allowing game creators and enthusiasts to easily train intelligent Agents for 2D, 3D, and virtual reality/augmented reality games [4]. RL has been used to train artificial intelligence for play games; such as Dota2 with OpenAI Five in 2017, Chess and Go in 2017, and StarCraft in 2019 [5].

However, improvements in RL algorithms as well as development platforms like Unity's ML-Agents allow for the simulation of real-world surroundings and Agents. As a result, an ecosystem has been created in which AI may be trained utilizing RL techniques like Proximal Policy Optimization (PPO) [6], which OpenAI developed, has made a new turning point in reinforcement learning.

Today, more projects use automated software as a substitution for humans. One of them is a perplexing maze made up of several branches of tunnels in which the solver must locate the most effective route to the destination in the lowest amount of time [7]. That's when artificial Intelligence is critical in determining the most effective strategy to solve any maze. That is when RL came in handy.

The sensitivity to hyperparameters is a critical aspect of an RL method's applicability [8]. Complex tasks might take hours or days to learn, and fine-tuning hyperparameters is time-consuming. Thus, this research focuses on changing the hyperparameters in configuration for a well-trained PPO algorithm in maze solving with ML-Agent library and Unity. This paper also provides a helpful guideline for tuning hyperparameters when redeployment algorithms on novel environments in the future.

2   Related Works

In recent years, several techniques to RL using neural network function approximators have been proposed. John Schulman's [6] research proposes that the current situation be improved by providing an algorithm that achieves data efficiency and reliable performance, called Proximal Policy Optimization. To optimize policies, the researcher cycles between collecting information from the policy and running multiple epochs of optimization on the sampled data. They came to the conclusion that these methods offer the same level of stability and reliability as trust-region solutions, but are significantly easier to apply. They simply take a few lines of code change to a standard policy gradient implementation, are more generalizable, and perform better overall [9]. However, the algorithm was successful on various problems without tuning hyperparameter values, meaning that the results still did not achieve the best possible outcome.

Despite the fact that earlier research have indicated that SAC [10] outperforms PPO, hyperparameter adjustment can have a considerable impact on the performance of these algorithms. In a search-and-retrieve assignment generated in the Unity 3D game engine, Usguta Vohra [11] investigated the impact of the number of layers and nodes in SAC as well as PPO algorithms. On four separate evaluative situations with various target and distractors ratios, the SAC and PPO models were compared. The results showed that PPO outperformed SAC in all test scenarios when the number of layers and units in the design was the fewest. It also implied that when comparing models created with DRL algorithms, identical hyperparameter settings might well be employed. PPO benefited from increasing the number of nodes while keeping the hidden layers the same in all model settings. They also provide new information on how to assess the model's performance in various testing scenarios, which can be utilized as a starting point for evaluating deep reinforcement learning Agents.

Several self-contained game-playing Agents have also been developed using a range of approaches, including rule-based systems and machine learning. Many efforts have also been made to create Agents that behave in a manner that is as near to that of a human player as possible. The research team created a collection of PPO-based reinforcement learning Agents to evaluate the player completion rate of many stages in Tactile Games' Lily's Garden [2]. It looked at how the Agent's number of steps for finishing the levels correlated with the behavior of ~900,000 players. The results, which are based on ~60% of the game mechanics, show that the two-step training scenario produces the highest skilled Agent. In contrast, the Agent attains the most significant correlation to real players' completion rates with the one-step curriculum. Because the research was limited to a small number of levels with default hyperparameter values, the results might not have been the best outcome [10, 11].

All the above studies showed that PPO Is an efficient technique in various problems, but the main focus is testing video games or mimicking how humans play. Although most research still uses the default or just a little tuning, proper hyperparameter initialization and search can improve results.

3 Design Maze and Agent in Unity

This paper aims to build the environment using the game engine Unity. The toolkit enables the creation of learning environments rich in physical and sensory complexity, delivers appealing cognitive challenges, and facilitates dynamic multi-agent interaction by using Unity as a simulation platform. Unity also helps in interface design quickly; instead of UI code, objects can be dragged, dropped, and arranged scientifically and quickly [12].

To create deep learning models, the ML-Agents toolkit uses the OpenAI Gym [13, 14] environments as a wrapper as well as interacts between both the Python API and the Unity C# Engine. Although there have been significant modifications to the way the toolkit works in the most recent release, 2.0.0, the ML toolkit's essential functionality has remained unchanged. At the time of writing this research, version 2.0.1 had several new features, including a redesigned interface, the removal of several previously identified as obsolete techniques, and the replacement of those methods with their supported counterparts.

The framework makes it simple for game creators to incorporate machine learning Agents into their creations, as well as providing AI researchers with a flexible platform to test their theories. The concept distinguishes three types of brains: player, heuristic (programmed action), and learning [15]. These brains are in charge of the environment's Agents. The ML-Agents toolbox selected to use a Proximal-Policy Optimization-based baseline reinforcement learning method.

Maze design for training Agent (Fig. 1):

- The maze map has 8 × 8 cells. A cell includes four walls and one floor.
- There are four distinct actions available to an Agent: up, down, left, and right. Each action requires it to enter a cell.
- There is a destination for the Agent to complete the maze.

- The Agent has four raycasts (to detect collisions with the maze walls) on four sides around the Agent. The length of the raycast is one cell.
- The Agent has four 3D Ray Perception Sensors [16] - the Agent's observations, arranged according to the other four raycasts.
- The total number of observations created is: (Observation Stacks) * (1 + 2 * Rays Per Direction) * (Num Detectable Tags + 2) = 1 * (1 + 2 * 2) * (1 + 2) = 15.

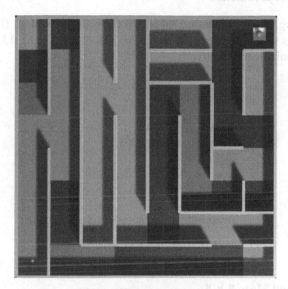

Fig. 1. Simple 8 × 8 maze.

The Agent moves in 4 directions (up and down, left, right). When the Agent moves in a specific direction and that side's raycast detects the wall, but the Agent still decides to go in that direction, deduct 1 point.

Agent when entering a cell will award or punished:

- Entering for the first time, Agent gets 3 points, and that background box turns yellow.
- Entering the second time, the Agent deducts 0.5, and the background cell turns orange.
- The Agent deducts 1 point the third time, and the background box turns purple.
- Entering from the fourth time onwards, the Agent has deducted 2 points, and the background is still purple. Purple is the final penalty level when entering.
- When colliding with the end of the maze, Agent will be awarded 100 points and finish solving the maze.

4   Experiments and Results

4.1   Implementation

Mazes and Agents use The software Unity Ver.2020 and ML-Agent library to build. All experiments were performed on NVIDIA 1080Ti GPU. The PPO algorithm with the hyperparameters is varied differently, with the max-step during training being 5000000.

There are two conditions to ending an episode:

- The Agent moves in the maze with enough steps of 3000, the episode will end, and the maze will reset.
- The Agent moves to the destination and ends the episode, and the maze will reset.

4.2 Config Hyperparameters

To perform tuning, the hyperparameters are taken to default values (Fig. 2) and then change each of its values (Tables 1, 2, 3, 4, 5, 6, 7, 8, 9, 10, 11, 12, 13, 14, 15 and 16) to observe and evaluate the results.

hyperparameters:
    batch_size: 128
    buffer_size: 2048
    learning_rate: 0.0003
    beta: 0.005
    epsilon: 0.2
    lambd: 0.95
    num_epoch: 3
    learning_rate_schedule: linear

Fig. 2. Default configuration hyperparameters.

Result of Fixed Maze 8 × 8

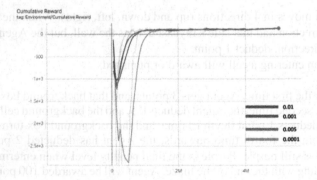

Table 1. Compare the results when changing the hyperparameter Beta.

Beta	Reward	Time cost
0.01	122.8	49 m 45 s
0.001	134.1	43 m 10 s
0.005	143.1	43 m 53 s
0.0001	143.2	43 m 30 s

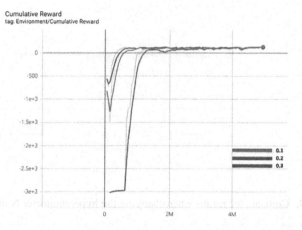

Table 2. Compare the results when changing the hyperparameter Epsilon.

Epsilon	Reward	Time cost
0.1	111	43 m 12 s
0.2	143.1	43 m 53 s
0.3	132.3	43 m 37 s

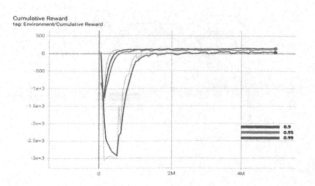

Table 3. Compare the results when changing the hyperparameter Lambd.

Lambd	Reward	Time cost
0.9	34.16	49 m 5 s
0.95	130.8	43 m 53 s
0.99	140.2	43 m 4 s

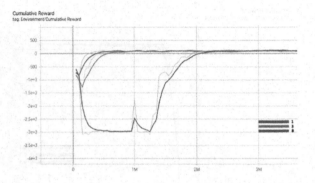

Table 4. Compare the results when changing the hyperparameter Num_epoch.

Num_epoch	Reward	Time cost
1	117.6	25 m 52 s
3	134.5	43 m 53 s
8	125.2	1 h 17 m 0 s

Results of Random Maze 4 × 4

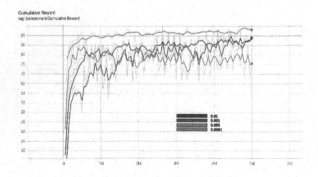

Table 5. Compare the results when changing the hyperparameter Beta.

Beta	Reward	Time cost
0.01	91.27	1 h 3 m 20 s
0.001	90.83	1 h 29 m 20 s
0.005	92.49	1 h 20 m 34 s
0.0001	76.26	2 h 6 m 18 s

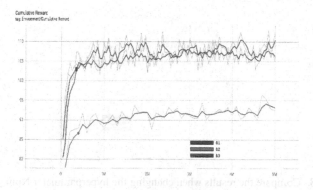

Table 6. Compare the results when changing the hyperparameter Epsilon.

Epsilon	Reward	Time cost
0.1	104.9	1 h 14 m 28 s
0.2	92.49	1 h 20 m 34 s
0.3	112.7	2 h 6 m 18 s

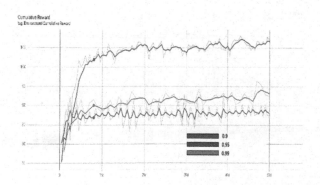

Table 7. Compare the results when changing the hyperparameter Lambd.

Lambd	Reward	Time cost
0.9	87.54	1 h 47 m 8 s
0.95	92.49	1 h 20 m 34 s
0.99	106.6	1 h 15 m 16 s

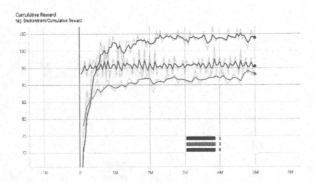

Table 8. Compare the results when changing the hyperparameter Num_epoch.

Num_epoch	Reward	Time cost
1	103.6	57 m 32 s
3	92.49	1 h 20 m 34 s
8	95.91	2 h 15 m 26 s

Results of Random Maze 6 × 6

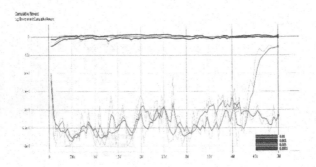

Table 9. Compare the results when changing the hyperparameter Beta.

Beta	Reward	Time cost
0.01	−1949	1 h 4 m 13 s
0.001	−17.02	1 h 5 m 26 s
0.005	−227.4	1 h 9 m 22 s
0.0001	43.44	1 h 12 m 37 s

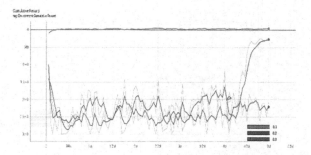

Table 10. Compare the results when changing the hyperparameter Epsilon.

Epsilon	Reward	Time cost
0.1	42.2	1 h 6 m 50 s
0.2	−268.7	1 h 9 m 22 s
0.3	−2056	1 h 26 m 5 s

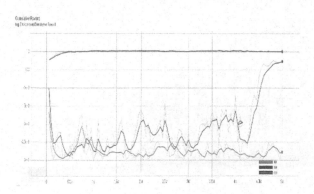

Table 11. Compare the results when changing the hyperparameter Lambd.

Lambd	Reward	Time cost
0.9	−2740	1 h 1 m 45 s
0.95	−227.4	1 h 9 m 22 s
0.99	4.037	1 h 36 m 0 s

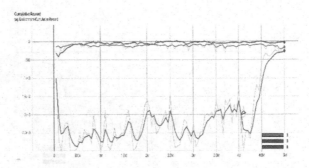

Table 12. Compare the results when changing the hyperparameter Num_epoch.

Num_epoch	Reward	Time cost
1	−121.3	56 m 52 s
3	−227.4	1 h 9 m 22 s
8	−64.08	1 h 48 m 21 s

Results of Random Maze 8 × 8

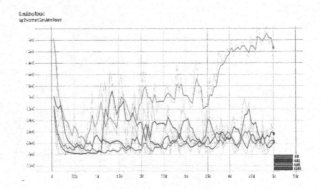

Table 13. Compare the results when changing the hyperparameter Beta.

Beta	Reward	Time cost
0.01	−2785	1 h 4 m 47 s
0.001	−2687	1 h 35 m 10 s
0.005	−1370	1 h 39 m 53 s
0.0001	−2690	1 h 13 m 11 s

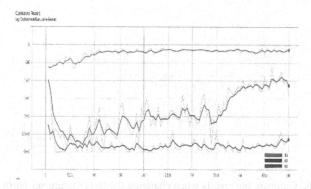

Table 14. Compare the results when changing the hyperparameter Epsilon.

Epsilon	Reward	Time cost
0.1	−114.1	1 h 20 m 35 s
0.2	−1370	1 h 39 m 53 s
0.3	−2540	1 h 32 m 50 s

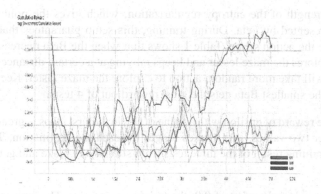

Table 15. Compare the results when changing the hyperparameter Lambd.

Lambd	Reward	Time cost
0.9	−2510	2 h 30 m 6 s
0.95	−1370	1 h 39 m 53 s
0.99	−2661	1 h 0 m 33 s

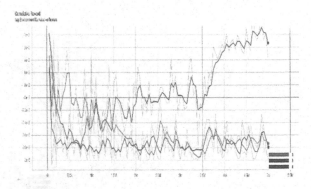

Table 16. Compare the results when changing the hyperparameter Num_epoch.

Num_epoch	Reward	Time cost
1	−2818	41 m 11 s
3	−1370	1 h 39 m 53 s
8	−2990	3 h 10 m 29 s

4.3   Analysis Results of Tuning Process

Beta: The strength of the entropy regularization, which gives the policy "more random," is represented by beta. During training, this setup guarantees that Agents correctly explore the action space. Table 1 shows that when the Beta decreases to 0.0001, the Agent explores the maze lesser and keeps moving at a certain distance. Increase the Beta; Agent will take more random action to explore the maze faster. Keep training for a long time; the smallest Beta gets the most reward out of 4 tests.

Epsilon: The reward of epsilon with a value of 0.3 in the first steps is greatly subtracted from the other two values, and it takes longer to reach the destination. The value 0.2 has the best training result of the three tests from solving the maze and getting the most points.

Lambd: With a Lamda value of 0.9, the training is inferior. The reward is much less than the other two values and the training time is also a bit more. Agent solves the maze about 1 million steps slower. Lambd values from 0.95 - 0.99 give good results, and Agent learns faster.

Num_epoch: Changing this value will make the model train fast or slow and significantly affect the model's performance quality. Num_epoch has a small value (equal to 1) that makes the training unstable, even taking 2 million steps to solve the maze, much worse than the other two values. Increasing this value makes the Agent learn faster and update more consistently. However, because of the number of passes performed through the buffer before the gradient decline step is implemented, the training time will increase.

5 Conclusion and Perspectives

This paper gives the tuning for PPO algorithm through hyperparameters Beta, Epsilon, Lambd, and Num_epoch. These values are changed and RL learning results are evaluated with the maze solving problem. The results show a clear difference between the training process and the hyperparameters. The change is based on different cases according to the complexity of the maze. Therefore, it is necessary to choose reasonable hyperparameters to set the best training results.

This paper also provides a helpful reference for tuning hyperparameters when redeployment PPO algorithm on novel environments in the future or applying it to machine learning problems with new methods [17, 18].

References

1. Elgeldawi, E., Sayed, A., Galal, A.R., Zaki, A.M.: Hyperparameter tuning for machine learning algorithms used for Arabic sentiment analysis. Informatics 8(4), 1–21 (2021)
2. Kristensen, J.T., Burelli, P.: Strategies for Using Proximal Policy Optimization in Mobile Puzzle Games. arXiv:2007.01542 (2020)
3. Juliani, A. et al.: Unity: A General Platform for Intelligent Agents. arXiv:1809.02627 (2020)
4. Jonsson, A.: Deep Reinforcement learning in medicine. Kidney Dis. (Basel) 5(1), 18–22 (2019). https://doi.org/10.1159/000492670. Epub 12 October 2018. PMID: 30815460; PMCID: PMC6388442
5. OpenAI et al.: Dota 2 with Large Scale Deep Reinforcement Learning, arXiv:1912.06680 (2019)
6. Schulman, J., Wolski, F., Dhariwal, P., Radford, A., Klimov, O.: Proximal Policy Optimization Algorithms. arXiv:1707.06347 (2017)
7. Sadik, A.M.J., Dhali, M.A., Farid, H.M.A.B., Rashid, T.U., Syeed, A.: A comprehensive and comparative study of maze-solving techniques by implementing graph theory. In: Proceedings of the International Conference on Artificial Intelligence and Computational Intelligence, vol. 1, pp. 52–56 (2010)
8. Hamalainen, P., Babadi, A., Ma, X., Lehtinen, J.: PPO-CMA: proximal policy optimization with covariance matrix adaptation. In: Proceedings of the IEEE International Workshop on Machine Learning for Signal Processing, MLSP (2020). https://doi.org/10.1109/MLSP49062.2020.9231618
9. Bellemare, M.G., Naddaf, Y., Veness, J., Bowling, M.: The arcade learning environment: an evaluation platform for general agents. J. Artif. Intell. Res. 47, 253–279 (2012). https://doi.org/10.1613/jair.3912

10. Kristensen, J.T., Valdivia, A., Burelli, P.: Estimating player completion rate in mobile puzzle games using reinforcement learning. In: Proceedings of the IEEE Conference Computational Intelligence and Games, pp. 636–639 (2020).https://doi.org/10.1109/CoG47356.2020. 9231581

11. Kim, T., and Lee, J.H.: Effects of hyper-parameters for deep reinforcement learning in robotic motion mimicry: a preliminary study. In: Proceedings of the 16th International Conference on Ubiquitous Robots, pp. 228–235 (2019).https://doi.org/10.1109/URAI.2019.8768564 (2019)

12. Unity - Manual: Creating user interfaces (UI). https://docs.unity3d.com/Manual/UIToolkits. html. Accessed 01 Feb 2022

13. Torrado, R.R., Bontrager, P., Togelius, J., Liu, J., Perez-Liebana, D.: Deep reinforcement learning for general video game AI. In: Proceedings of the IEEE Conference Computational Intelligence and Games (2018). https://doi.org/10.1109/CIG.2018.8490422

14. Johansen, M., Pichlmair, M., Risi, S.: Video game description language environment for unity machine learning agents. In: Proceedings of the IEEE Conference Computational Intelligence and Games (2019). https://doi.org/10.1109/CIG.2019.8848072

15. Jafri, R., Campos, R.L., Ali, S.A., Arabnia, H.R.: Visual and infrared sensor data-based obstacle detection for the visually impaired using the google project tango tablet development kit and the unity engine. IEEE Access **6**, 443–454 (2017). https://doi.org/10.1109/ACCESS. 2017.2766579

16. Zhu, W., Rosendo, A.: A functional clipping approach for policy optimization algorithms. IEEE Access **9**, 96056–96063 (2021). https://doi.org/10.1109/ACCESS.2021.3094566

17. Su, N.T., Hung, P.D., Vinh, B.T., Diep, V.T.: Rice leaf disease classification using deep learning and target for mobile devices. In: Al-Emran, M., Al-Sharafi, M.A., Al-Kabi, M.N., Shaalan, K. (eds.) ICETIS 2021. LNNS, vol. 299, pp. 136–148. Springer, Cham (2022). https://doi. org/10.1007/978-3-030-82616-1_13

18. Hung, P.D., Giang, D.T.: Traffic light control at isolated intersections in case of heterogeneous traffic. In: Kreinovich, V., Hoang Phuong, N. (eds.) Soft Computing for Biomedical Applications and Related Topics. SCI, vol. 899, pp. 269–280. Springer, Cham (2021). https:// doi.org/10.1007/978-3-030-49536-7_23

Efficient Approach to Employee Attrition Prediction by Handling Class Imbalance

M. Prathilothamai(✉), Sudarshana(✉), A. Sri Sakthi Maheswari(✉),
A. Chandravadhana(✉), and R. Goutham(✉)

Department of Computer Science, Amrita Vishwa Vidhyapeetham, Coimbatore, India
{m_prathilothamai,cb.en.u4cse18152,cb.en.u4cse18257,
cb.en.u4cse18310,cb.en.u4cse18345}@cb.amrita.edu

Abstract. The key success of any organization lies on its employees and thus ability to monitor employee attrition efficiently becomes important. Our work aims to use machine learning models to accurately predict on whether an employee would decide to leave a company or not. Main algorithms used for this purpose include Logistic Regression, KNN and Weighted Decision tree. Another aspect that needs to be looked into is the amount of classification imbalance present in this problem statement. Our work has applied multiple data based and algorithmic approaches to resolve this imbalance. The use of SMOTE, Near-Miss, ADASYNC and other techniques provide an efficient and balanced prediction of employee attrition. From all the experiments we have done, we arrive at a proposed hybrid model that combines data based and algorithmic method to resolve imbalance that accurately predicts attrition of up to 98.4%, proves recall rate of 98.43%, shows precision of 97.69% and an excellent F1 score of 99.17% . Our analysis would help companies understand the important areas that needs to be focused on to avoid attrition and hence lead to growth and higher achievements.

Keywords: Employee attrition · Feature selection · Class imbalance · ADASYN · SMOTE · KNN · Logistic regression · Weighted decision tree · Hybrid method

1 Introduction

In the emerging world of competition where workplaces expand and become complex, expectations from HR departments also rise. The need of workplaces to understand that its employees are their vital resource is the need of the hour and hence must give importance to HR analytics, especially to employee attrition. Employee Attrition refers to the number of employees that leave their organization, the cause of this can be voluntarily or involuntarily [8]. Various causes for employees to leave include resignation (due to personal and professional reasons), termination, retirement.

It is important for the HR department to understand and predict which of their employee is likely to leave and to try and reduce the attrition rate in order to have a better work graph. "It takes a lot of time and energy to build a great employee and only

M. Singh et al. (Eds.): ICACDS 2022, CCIS 1614, pp. 263–277, 2022.
https://doi.org/10.1007/978-3-031-12641-3_22

a second to lose one."[9]. Companies need to know the well-being of their employees and to know if they are overworked in order to retain employees.

The goal of this research is to predict employee attrition beforehand so as to help the HR understand their workforce and take necessary actions to hold on to their key employees. The chosen dataset comes with a challenge of handling class imbalance which has been resolved using sampling techniques. In the process of predicting employee attrition, data preprocessing and analysis, feature selection are done.

The Second section of the paper contains the details and overview of related works. The methodology used is discussed in section three. The fourth and fifth section describes Data Preprocessing techniques, Exploratory Data Analysis respectively. The sixth section deals with various feature selection algorithms. The next section deals with resolving data imbalance using Data Handling methods (undersampling and oversampling) and Algorithmic method. The Machine Learning models used are discussed in section eight, followed by Results and analysis in section nine, Conclusion ins section ten and References in the end.

2 Literature Survey

Employees can leave the organization for many personal as well as professional reasons [10]. Some of the reason of leaving the organization could be better-paying job outside, pursuing higher studies, workload etc. Various Feature Selection methodologies are used to identify common factors influencing employee attrition. Research [1] employs Recursive Feature Elimination (RFE) and SelectKBest approaches. Classification and Regression Tree (CART) analysis could also be considered to add to ranking variable importance based on how they contribute to building of the tree [2]. Selecting just the top variables based on their importance will also reduce train and test time. In [4], T-test method is implemented for feature ranking. Authors of [6], perform feature engineering to generate 3 New features - Tenure per job, Years without change and compa Ratio. After which correlation of attributes is found and top features were identified to be Age, Gender, and Tenure. Other research [6, 7] also identify Age, Gender, Monthly Income and distance from home as the top factors. [4] explains the significant difference a properly preprocessed data can have on a model.

The chosen dataset has major class imbalance [17–19], which affects model performance. The research also highlights the use of ADASYN and manual undersampling approaches to eliminate class imbalance, which improves the performance from a 0.50 F1 Score to a 0.93 F1 Score when imbalance was handled. Research [9], delves into methodologies for resolving class imbalance. They analyze the behavior of standard pre-processing techniques in a framework called MapReduce framework. Finally, considering the experimental results obtained throughout their work, they discuss on the challenges and future directions for imbalanced Big Data classification. [16] describes an Random Forest algorithm that can handle class imbalance. It states in detail about the random forest methodology. The final result is obtained by combining the weighted results of the RFs. Authors of [13] combine together Random Undersampling (RUS) and Synthetic Minority Oversampling TEchnique (SMOTE) to handle multi class imbalance in a review rating prediction case. Which rids of induced bias and loss of information of the majority class. Research [14] explores various methodologies to handle class imbalance from data based to algorithmic to ensemble methods related to road traffic and vehicle accidents.

Seeing as this is a classification problem, models like SVM, KNN and Decision Trees are what is commonly used. In research [3], KNN as used to categorize the data into 3 different departments. The attrition was based on 3 categories namely: Performance-SVM, Salary-Logistic Regression, Education-Logistic Regression. From each of the three categories, the best employee was predicted and then the attrition for all these employees was predicted using Decision Tree. Apart from them, Random Forest [1, 2, 4, 8, 10, 12], XGBoost [1, 6, 12], Naive Bayes [8] were also used. [15] details various data pre-processing steps including sampling techniques and Principal Component Analysis (PCA) after which XGB classifier is applied on the data to predict employee stress level, which outperforms other models by about ten times. Research quotes major performance metrics used as Accuracy, F1-Score, Precision and Recall, with an added MSE [8]. Previous researches provide different methodologies and metrics to tackle the attrition problem. They do encourage us to focus on various parts of the process from feature selection to handling class imbalance, selection of algorithm and choice of performance metrics. The proposed methodology is to combine different class imbalance approaches to efficiently measure attrition.

3 Employee Attrition Prediction: A Case Study

A. Dataset Description: The dataset used is 'HR Analytics Case Study' which is taken from 'Kaggle'. The dataset has 4411 records and 32 attributes. The overall dataset is combination of 3 datasets:

- Employee Survey Data (employee survey data.csv)
- General Data (general data.csv)
- Manager Survey Data (manager survey data.csv)

The 32 attributes of the dataset contain various related information that can be used for analysis. The attributes help us understand the thoughts and work-life satisfaction of an employee. The table below contains all the attributes present in the whole dataset and its brief explanation (Table 1).

Table 1. Dataset description

Attribute	Description
Age	Age of the employee
Attrition (Target)	Whether employee left in the previous year or not
BusinessTravel	How frequently employee traveled for business purposes in the last year
Department	Department in company
DistanceFromHome	Distance from home in kms
Education	Education Level from 1 to 5 1 'Below College' 2 'College' 3 'Bachelor' 4 'Master' 5 'Doctor'
EducationField	Field of education
EmployeeCount	Employee Count
EmployeeNumber	Employee number/id
EnvironmentSatisfaction	Work Environment Satisfaction Level from 1 to 4 1 'Low' 2 'Medium' 3 'High' 4 'Very High'
Gender	Gender of the employee
JobInvolement	Job Involvement Level from 1 to 4: 1 'Low' 2 'Medium' 3 'High' 4 'Very High'
JobLevel	Job level at company on a scale of 1 to 5
JobRole	Name of job role in company
JobSatisfaction	Job Satisfaction Level from 1 to 4: 1 'Low' 2 'Medium' 3 'High' 4 'Very High'
MaritalStatus	Marital status of the employee
MonthlyIncome	Monthly income in rupees per month
NumCompaniesWorked	Total number of companies the employee has worked for
Over18	Whether the employee is above 18 years of age or not
PercentSalaryHike	Percent salary hike for last year

(continued)

Table 1. (*continued*)

Attribute	Description
PerformanceRating	Performance rating for last year from 1 to 4: 1 'Low' 2 'Good' 3 'Excellent' 4 'Outstanding'

4 Data Preprocessing

A. Remove Unnecessary Attributes: 'EmployeeID' as it does not contribute to determine attrition. The attributes 'EmployeeCount', 'Over18' and 'StandardHours' have only 1 value for all records hence these are removed.

B. Observe the extent of Class Imbalance: From Fig. 1, it is seen that the class imbalance is with a distribution of 711 'Yes' and 3699 'No' and ratio 5.2:1.

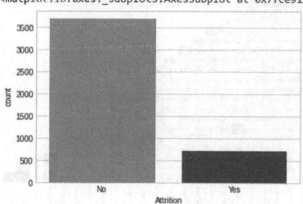

Fig. 1. Distribution of 'Yes' and 'No' values in target values

C. Handle Missing Values: Five attributes are found to have null values. With respect to the dominant majority of a value and skewness of the distribution the missing values are filled.

D. Encode Categorical Attributes: The 'One Hot Encoding' technique is applied to encode the categorical attributes, i.e. 'BusinessTravel', 'Gender' and 'MaritalStatus'.

E. Scale Down Large Valued Attributes: It is observed that 'MonthlyIncome' attribute contains large values hence it is scaled down using Min–Max Standardisation.

5 Exploratory Data Analysis

To look for useful information on the distribution of values, count plots are plotted for every attribute. The following are some of the inferences:

- Most employees rarely travel and very few don't travel at all.
- Most employee records come from the Research department.
- There are more married employees than single or divorced.
- Most employees are male

Another representation known to identify relationships between attributes and the target variable, the Correlation Heatmap. The color gradation flows from red to blue indicates high positive to high negative correlation. The middle point conveys that there is no relationship between the attributes (Fig. 2).

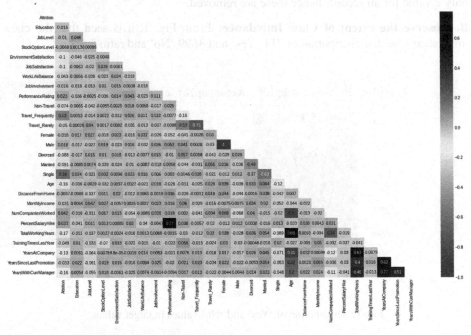

Fig. 2. Correlation Heatmap

The following are some useful observations from the Correlation Heatmap.

- 'PerformanceRating' is positively correlated to 'PercentSalaryHike' by +0.77 so employees with higher salary perform better.
- 'YearsAtCompany' is positively correlated to 'YearsWithCurrManager' by +0.77, that if employees tend to stay more number of years within the company then they would probably be under their current manager.

- 'YearsWithCurrManager' and 'YearsAtCompany' positively correlated to 'YearsSinceLastPromotion' it shows absence of professional growth.

6 Feature Selection

A. Filter Method: The filter method ranks each feature-based uni variate methods like chi square, ANOVA for classification and then selects the highest-ranking features.

(1) *Select K Best Using Chi Square*: Select K best filters the top k values with highest score based on Chi square value. The top ten features using this method are:

MonthlyIncome, TotalWorkingYears, Years at company, Years with current manager, Age, Marital status, Years since last promotion, Job satisfaction, Number of companies worked, Business travel.

(2) *Using ANOVA-F*: ANOVA also known as Analysis of Variance, is a parametric statistical hypothesis test for determining whether the average or mean value of a given dataset come from the same distribution or not. The top features using this method are represented in Fig. 3(a).

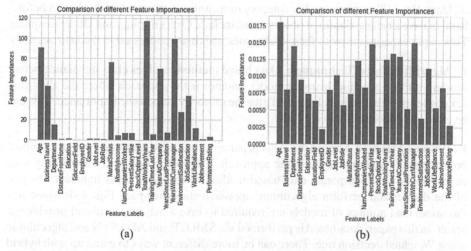

Fig. 3. (a) Feature Selection using ANOVA-f (b) Feature Selection using ExtraTreeClassifier

B. Wrapper Method: The wrapper method uses a subset of attributes and applies models like Recursive Feature Elimination on it and based on the inferences, features can be added or removed from the subset. Recursive Feature Elimination constructs the next model with the remaining features until all the features are utilized and ranks them in order of elimination. The top features from these methods are:

Age, Distance From Home, Job Role, Monthly Income, Num Companies Worked, Percent Salary Hike, Total Working Year, Years Since Last Promotion, Environment Satisfaction, Job Involvement.

C. Ensemble Method: In Extra Tree Classifier the features are selected using the tree-based supervised models and uses multiple tree models constructed randomly from the dataset and sorts out the features that have been most voted for. The top ten features using this method are represented in Fig. 3(b).

Comparing all the different types of feature selection methods and considering the features which have occurred the most amongst all the different methods, the top features selected are:

Age, NumCompaniesWorked, TotalWorkingYears, YearsSinceLastPromotion, EnvironmentSatisfaction, YearsWithCurrManager, MaritalStatus, MonthlyIncome, WorkLifeBalance, YearsAtCompany, JobSatisfaction.

7   Analysis of Resolving Class Imbalance

Class imbalance makes it harder for the algorithm to classify a minority class due to its barely noticeable datapoints. Leading to a disproportionate ratio of observations, that gives an undesirable overfitted model. We can resolve the imbalance with:

A. Data-Based Techniques: The category of techniques that modify the dataset beforehand to have a balanced ratio of observations include Oversampling and Undersampling. This paper experiments with these techniques which are addressed in the later sections.

B. Algorithm Based Techniques: Data based method resolves class imbalance before sending it to the machine learning model, another is the approach where the algorithm acknowledges the data imbalance and give the minority class some importance. This can be called as Algorithm-based.

C. Proposed Hybrid Approach: This paper aims to investigate a new way to combine both the data-based and algorithmic approach. The method in use is, a majority vote ensemble we can incorporate data-based method of resolving class imbalance to part of the dataset and perform algorithmic approach the other part, Fig. 4 demonstrates the same. Odd number of models are required to take a majority vote and provide the result. In this paper, data-based is performed via SMOTE and ADASYN and algorithmic using Weighted decision tree. There can be more different ways to come up with hybrid methods. It is open to future improvisation.

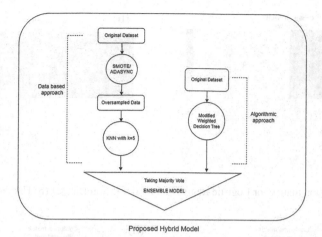

Proposed Hybrid Model

Fig. 4. Architecture of proposed hybrid model

8 Machine Learning Models

The problem statement maps to a classification problem and hence the models used for this research are classification algorithms like Logistic Regression, KNN-3, KNN-5.

A. Before Handling Class Imbalance: The dataset is an imbalanced dataset with ratio 5.2:1. The Logistic Regression model applied on the imbalanced dataset gave a very low F1 score and Recall rate of 22.8 and 13.7% respectively. The model clearly does not perform well and hence the dataset should be balanced to apply various ML Models and test again.

B. After Handling Class Imbalance by Data Based Method – Undersampling: The class imbalance has been reduced by using undersampling technique called NearMiss. The machine learning algorithms applied after handling the imbalance are classification algorithms namely, K-nearest Neighbors using k = 3 and k = 5. and Logistic Regression.

(1) *Using Logistic Regression*: The model has a very low F1 score and precision rate of 34.8 and 23.7% respectively. The accuracy is 60.8 and Recall is 64.8%.
(2) *Using KNN-5*: K value indicates the count of the nearest neighbors. Here, the k is set to 5. The performance metrics for the model are Accuracy: 60.66, F1 Score: 33.65, Recall score: 60.68, Precision score: 23.28. The F1 and Precession score are low (Figs. 5, 6 and 7).

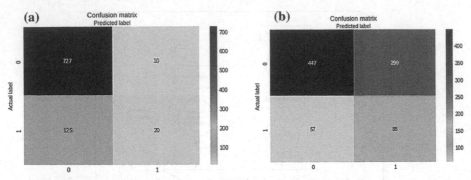

Fig. 5. Confusion matrix for Logistic Regression model (a) on Dataset (b) Undersampled Data

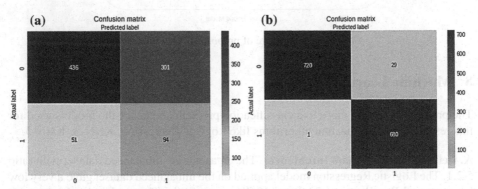

Fig. 6. Confusion matrix for (a) KNN-5 on undersampled data (b) KNN-3 on ADASYN oversampled Data

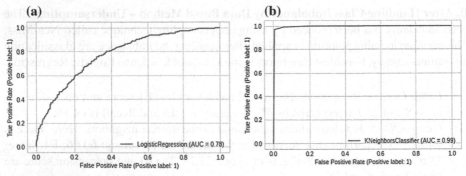

Fig. 7. ROC-AUC curve for (a) Logistic Regression model on SMOTE data (b) KNN-5 on ADASYN Data

C. After Handling Class Imbalance by Data Based Method Called Oversampling:
After the databased approach was applied to balance the dataset, the machine learning algorithms applied were, Logistic Regression, K-nearest Neighbors with k = 5.

(1) *Using Logistic Regression:* was applied on SMOTE balanced data. The performance metric scores are Accuracy 70.00, F1 Score: 70.01, Recall score: 70.09, Precision score: 69.90 An ideal ROC is a one which is an inverted 'L' shape.

(2) *Using KNN-5:* The KNN algorithm was also applied after balancing the dataset using SMOTE oversampling technique. The performance metrics are Accuracy: 92.50, F1 Score: 92.99, Recall score: 99.72, Precision score: 87.11 The ROC curve for the model is a very accurate curve.

(3) *Using Logistic Regression:* Logistic Regression was created to test the dataset after handling the data imbalance using ADASYN oversampling technique The performance metrics for Logistic Regression are Accuracy: 67.48, F1 Score: 66.47, Recall score: 67.69, Precision score: 65.29

(4) *Using KNN-5:* The performance metrics for KNN with 5 neighbors are Accuracy: 91.40, F1 Score: 91.70, Recall score: 99.85, Precision score: 84.78.

(5) *Using KNN-3:* The performance metrics for KNN with 3 neighbors are Accuracy: 97.97, F1 Score: 97.91, Recall score: 100, Precision score: 95.91

D. After Handling Class Imbalance by Algorithmic Method Weighted Decision Tree. (1) *Weighted Decision Tree:* Weighted Decision Tree is a classification algorithm that is used to create a training model that can be used to predict the class or value of the target variable by learning simple decision rules inferred from the training/known data. The performance metrics are Accuracy: 97.62, F1 Score: 93.33, Recall score: 90.74, Precision score: 96.07 (Fig. 8).

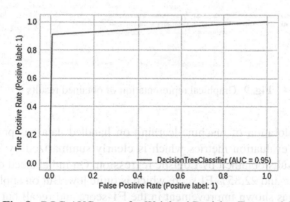

Fig. 8. ROC-AUC curve for Weighted Decision Tree model

E. After Handling Class Imbalance by Proposed Hybrid method the Voting Classifier. A Voting Classifier is a model that combines various machine learning algorithms like an ensemble and gives the average probability or highest of all the chosen class as the output. The voting classifier used in this soft, which outputs the average probability of class prediction from all the models are taken and that is the final result. The models used in the voting classifier are Decision Tree and KNN. The performance metrics are Accuracy: 98.4%, F1 Score: 98.43%, Recall score: 97.69%, Precision score: 99.17%.

9 Results and Analysis

This paper provides a clear illustration that our proposed machine learning model achieves to predict whether an employee will leave the company or not, quite accurately, up to 98.4%. It also proves a good recall rate of 98.43%, precision of 97.69% and F1-score of 99.17%. The dataset that has many challenges such as large number of features and significant presence of class imbalance has been handled by giving great attention. The paper eliminates many nonessential dimensions and uses only 11 most significant parameters which we have selected by carefully performing four techniques of feature selection. These contributing factors are the areas that the company must focus on to reduce the probability of attrition (Fig. 9).

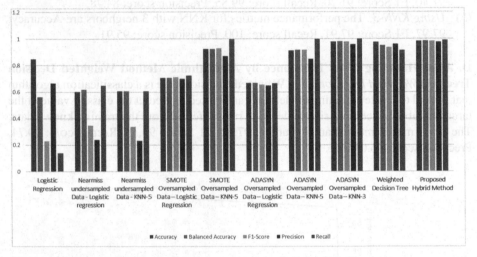

Fig. 9 Graphical representation of obtained results

Intensive exploration of machine learning on handled data has provided multiple vital results and evaluation metrics which is clearly summarized by Table 2. Let us first cover the results obtained by Logistic regression. On imbalanced data it results in 13.79% recall rate and 22.85% F1 score which is quite low. But on applying Near-miss to balance data has shown improvement in the F1-score and recall but by just a small amount. The recall rate increases with ADASYN and SMOTE over sampling by 1.69% and 7.71%.

Table 2. Performance results

Model	Accuracy	Balanced Accuracy	F1-Score	Precision	Recall
Logistic Regression	0.8469	0.5622	0.2285	0.6666	0.1379
Nearmiss undersampled Data - Logistic regression	0.6009	0.6199	0.3481	0.2379	0.6482
Nearmiss undersampled Data - KNN-5	0.6066	0.6067	0.3365	0.2328	0.6068
SMOTE Oversampled Data – Logistic Regression	0.7081	0.7081	0.7127	0.7006	0.7253
SMOTE Oversampled Data – KNN-5	0.9243	0.9244	0.9293	0.8701	0.9972
ADASYN Oversampled Data – Logistic Regression	0.6692	0.6690	0.6569	0.6489	0.6651
ADASYN Oversampled Data – KNN-5	0.9133	0.9172	0.9164	0.8468	0.9985
ADASYN Oversampled Data – KNN-3	0.9797	0.9806	0.9791	0.9591	1.0
Weighted Decision Tree	0.9773	0.9526	0.9367	0.9610	0.9135
Proposed Hybrid Method	0.9845	0.9865	0.9843	0.9769	0.9917

The next applied nearest neighbor model yields highly greater results compared to logistic regression in all cases of handled data with highest recall on ADASYN over sampled data with 99.85%. The algorithmic approach with weighted Decision tree also yields a good score with 0.9773 accuracy, 0.9135 recall and 0.9591 precision.

An amazing result of 1.0 recall rate is observed on applying KNN model with 3 neighbors on ADASYN over sampled data. The proposed hybrid approach combining KNN and Decision tree yields an extremely good result in all the evaluation metrics - 0.984 accuracy with 0.9917 recall and 0.9865 F1-Score. The same can be observed in the ROC-AUC curves for the models.

A. Conclusion

This work was focused on using machine learning algorithm to accurately predict whether an employee has a chance of leaving the company or not. Also, to identify the factors that could contribute to the employee leaving the company. The paper also aimed to find a way to reduce the dominant imbalance in classification. In the process of resolving class imbalance, we came up with a proposed idea of incorporating both data-based and algorithmic method of handling imbalance called a Voting classifier. We found about 11 most important factors such as age, number of companies they worked for, etc. that impacted the occurrence of attrition. We could gain some useful inferences about how it takes more years to get a promotion, how employees who are paid more seem to be retain in their job, etc.

In the process of developing a predicting model, we have experimented in various ways and got results. The top 2 models that performs the best for the problem statement are KNN-3 after handling data imbalance using ADASYN Oversampling technique with an accuracy of 97.97%, F1-Score of 97.91%, precision 95.91% and recall rate of 100% and Weighted Decision Tree with an accuracy of 97.73%, F1-Score of 96.67%, precision of 96.10% and recall rate of 91.35%. The hybrid model provides an accuracy of 98.4%, F1-Score of 98.43%, precision of 97.69% and recall rate of 99.17%.

References

1. Alduayj, S.S., Rajpoot, K.: Predicting employee attrition using machine learning, pp. 93–98 (2018)
2. Ali, A., Shamsuddin, S.M.H., Ralescu, A.L.: Classification with class imbalance problem: a review (2015)
3. Dutta, S., Bandyopadhyay, S.: Employee attrition prediction using neural network cross validation method. Int. J. Comm. Manag. 6, 80–85 (2020)
4. Fallucchi, F., Coladangelo, M., Giuliano, R., William DeLuca, E.: Predicting employee attrition using machine learning techniques. Computers 9(4) (2020)
5. Fernandez, A., del Río, S., Chawla, N., Herrera, F.: An insight into imbalanced big data classification: outcomes and challenges". Complex Intell. Syst. 3, 105–120 (2017)
6. Jain, P.K., Jain, M., Pamula, R.: "Explaining and predicting employees' attrition a machine learning approach. SN Appl. Sci. 2, 757 (2020)
7. Jain, R., Nayyar, A.: Predicting employee attrition using XGBoost machine learning approach, pp. 113–120 (2018)
8. Karande, S., Shyamala, L.: Prediction of employee turnover using ensemble learning, pp. 319–327 (2019)
9. Mhatre, A., Mahalingam, A., Narayanan, M., Nair, A., Jaju, S.: Predicting employee attrition along with identifying high risk employees using big data and machine learning, pp. 269–276 (2020)
10. Patel, A., Pardeshi, N., Patil, S., Sutar, S., Sadafule, R., Bhat, S.: Employee attrition predictive model using machine learning 07(05) (2020)
11. Shankar, R.S., Rajanikanth, J., Sivaramaraju, V., Murthy, K.: Prediction of employee attrition using datamining, pp. 1–8 (2018)
12. Yahia, N.B., Hlel, J., Colomo, R.: From big data to deep data to support people analytics for employee attrition prediction. IEEE Access 9, 60447–60458 (2021)
13. Mahadevan, A., Arock, M.: A class imbalance-aware review rating prediction using hybrid sampling and ensemble learning. Multimed. Tools Appl. 80, 6911–6938 (2021)
14. Prathilothamai, M., Indra Kumar, V., Anjali Ragupathi, Aradhana, J.: Analysis of techniques to handle class imbalance in road traffic prediction. Int. J. Adv. Sci. Technol. 29(05), 7549–7567 (2020)
15. Garlapati, A., Krishna, D.R., Garlapati, K., Narayanan, G.: Predicting employees under stress for pre-emptive remediation using machine learning algorithm, pp. 315–319 (2020). https://doi.org/10.1109/RTEICT49044.2020.9315726
16. Jose, C., Gopakumar, G.: An improved random forest algorithm for classification in an imbalanced dataset. In: 2019 URSI Asia-Pacific Radio Science Conference (AP-RASC), pp. 1–4 (2019). https://doi.org/10.23919/URSIAPRASC.2019.8738232.

17. Ketha, S., Balakrishna, P., Ravi, V., Dr. Soman K.P.: Deep learning based frameworks for handling imbalance in DGA, Email, and URL data analysis. In: Communications in Computer and Information Science (2020)
18. Vinayakumar, R., Soman, K.P., Poornachandran, P.: DeepDGAMINet: cost-sensitive deep learning based framework for handling multiclass imbalanced DGA detection. In: Gupta, B., Perez, G., Agrawal, D., Gupta, D. (eds.) Handbook of Computer Networks and Cyber Security. Springer, Cham (2020). https://doi.org/10.1007/978-3-030-22277-2-37
19. Mohammed Harun Babu, R., Vinayakumar, R., Soman, K.P.: CostSensitive long short-term memory for imbalanced DGA family categorization. In: Sengodan, T., Murugappan, M., Misra, S. (eds.) Advances in Electrical and Computer Technologies. Lecture Notes in Electrical Engineering, vol. 672. Springer, Singapore (2020). https://doi.org/10.1007/978-981-15-5558-9-49.

Password Generation Based on Song Lyrics and Its Management

E. Sai Kishan, M. Hemchudaesh$^{(\boxtimes)}$, B. Gowri Shankar, B. Sai Brahadeesh, and K. P. Jevitha

Department of Computer Science and Engineering, Amrita School of Computing, Coimbatore, Amrita Vishwa Vidyapeetham, Coimbatore, India
{cb.en.u4cse18048,cb.en.u4cse18120,cb.en.u4cse18225, cb.en.u4cse18247}@cb.students.amrita.edu, kp_jevitha@cb.amrita.edu

Abstract. This paper focuses on creating a Password Manager which is secure and *easy-to-remember*. Password managers prove efficient in generating and storing passwords but they were never able to generate passwords that people could remember easily. Our password manager solves this by creating passwords using song lyrics. Instead of generating passwords randomly, our tool will create passwords from the lyrics of the song selected by the user. The passwords generated are checked thoroughly for validity, by using the *haveibeenpwned* API(data breach checks) and are made secure by making combinations of letters, numbers special characters of the password and suggesting the best secure password from the list.

Keywords: Password · Password Manager(PM) · Master key · Hint key · Haveibeenpwned · Song lyrics

1 Introduction

The number of websites requiring a password has been increasingly growing across the years, making people resort to password reuse due to lack of remembrance. As a frequent user of the internet, one would have to use and remember at least 15–20 passwords daily. Starting from your email account, social media, bank account and shopping portals, every website requires a password and this makes it extremely difficult to remember them. According to [2] many end up writing it on paper or simply store them as a file on their computer. That was the time when password managers came into existence.

Although password managers prove efficient in generating and storing passwords, they never were able to create passwords that people could remember easily. As we have seen in [3], compromising security over remembrance is also not a feasible option. Hence, PMs which give importance to security as well as remembering capability is the need of the hour. Our password manager solves this problem using a simple method, creating passwords using song lyrics. Instead of generating passwords randomly, our tool will create passwords from the lyrics of the song selected by the user.

M. Singh et al. (Eds.): ICACDS 2022, CCIS 1614, pp. 278–290, 2022.
https://doi.org/10.1007/978-3-031-12641-3_23

2   Literature Survey

In [1], a high secure scheme of how to manage complex user passwords without the lack of user friendliness was elaborate. This was done because users were relying heavily on their services on internet and due to their remembering capabilities, had to resort to password reuse.

In [2], people with more than 100 accounts used online PM applications which generates random passwords and stores them in the application itself. Many of them initially used traditional methods for storing passwords(files, papers etc.) but due to various security issues, they shifted to PMs later as the chances of passwords being compromised were increasing rapidly.

In [3], it was found that most people reuse passwords in both exact and partial form. It was extremely rampant because it was easy to remember. People reuse approximately 80% of their passwords across domains. Most use exact, partial and some used simpler strategies to reuse their passwords. It was found that there was no statistically significant effect of the presence of a PM. Only 19 participants had installed PMs. Passwords used in government websites tend to be reused less because users consider government website more important in terms of security.

In [4], various leaked passwords were analyzed. It was found that in Hash-Based login systems, the password is hashed and stored as a code in the database, which for checking will be retrieved back by using the hash index in the database. But it can often be guessed, forgotten, or revealed. To overcome this weakness, a stronger password should be created or additional authentication factors be used, such as physical token, digital certificates, one-time access code, etc. Dictionary, Brute Force and MITM attack are attacks on passwords.

In [5], a client-side PM application was used to prevent insider attacks, remote exploits of weakly secured sites, keylogging on public terminals and websites spoofing. Along with the existing features, this delivers usability, transportability and increased security as well. To reduce computation time, they implemented client-side caching in which the password generation will only be slow for the first time compared to subsequent attempts to generate.

In [6], usage of secured password encrypting algorithms such as PBKDF2, bcrypt and scrypt provides more security compared to modern hash algorithms since they will be slower which in turn results in increased time for cryptoanalysis. Using personal details while creating passwords can make them more prone to easier attacks(rainbow attacks and birthday attack).

In [7], it was understood that there was no perfect PM, and every product tries to balance between providing users with security and convenience. From analyzing the recently leaked passwords, users tend to use weak passwords or reuse passwords on different sites.

In [8], usage of PM in the online mode was preferred over USB, followed by Phone. The online PM was the last choice for non-technical people, who mostly preferred the phone manager, technical people were more inclined towards the USB manager in comparison to the online manager.

In [9], critical vulnerabilities like bookmarklet vulnerability, web vulnerability (Cross-Site Scripting (XSS), Cross-Site Request Forgery (CSRF)), authorization and user interface vulnerabilities was found in all the PMs. Two of the three PMs that support credential sharing both mistake authentication for authorization. In [10], it was found that a good PM would prioritize security over ease of use. While a long password may be annoying to the user, the master password must be strong. Every step should be automated as possible to increase usability. In [11], it was found how partial password authentication techniques were very important and their role in e-banking. The security side of them is not discussed much in academics. Their proposed scheme was based on the 2DNF formula.

In [12], more about partial password authentication and its techniques were learnt. They played a very important role in cryptography. This concept involved that the user does not give his full password. Due to this, guessing attack becomes less. Since number of attempts to break password increases, dictionary attack also becomes less.

In [13], passwords containing user data in some way or the other increases the possibility of it being breached or stolen. This is done by the user in order to remember the password. To find whether users are using their personal data, they propose segmentation algorithms which help to find user data, and find the strength of a password based on it.

3   Proposed System

Our algorithm (Fig. 1) generates password randomly. Users have option to set their password size range from 10(minimum) to 15. A special character set is initially given($\sim$,!,@,#,$,%,&,*,-,_,+,.,?,/) to be taken in case the user does not have a preference. From the lyrics of the password extracted in the website, two words at random are chosen, along with numbers(ranging from 100–10000) and two special characters either from the preferred set or the given set is taken and combined to form a password. The lyrics of the password should be in English but the song can be in any language.

This is one combination of a password taken from the lyrics. Like this, the five elements of the password (lyric words(2), number and special characters(2)) are used to form combinations. Totally 120 combinations are possible. In case of a user defined lyric input, the number of combinations would be 24 passwords. The combinations are then fed into the *haveibeenpwned* API for data-breach check. The best password among them are selected and given back to the user. The user is prompted for the number of passwords to be selected from a song (3, 5, 7). The above process of selecting a single password from 120 is repeated for the number of password suggestions the user had suggested. So if a user had chosen three passwords, the 120 combinations would have been repeated for three times in order to choose three passwords. So there would be 360 combinations.

```
def generatepassword(self):
    self.url_value=self.url.get()
    self.username.value=self.username.get()
    if(self.isValidURL(self.url_value)==True and self.username.value='' and self.phisingdetection(self.url_value)==1):
        row_extract = self.cur.execute(" SELECT song_title FROM user_settings WHERE emailid ?", (self.customer,)).fetch
        print(row_extract)
        songlist_add=row_extract[0]
        l_temp=list(songlist_add.split(','))
        self.selectedsong_title=random.choice(l_temp)
        print(self.selectedsong_title)
        row_extract = self.cur.execute(" SELECT prefer_no,Random_no,specialchar,random_special,suggestion_no FROM user
        print(row_extract)
        if(row_extract[1]==0):
            numberprefer=list(row_extract[0].split(','))
        if(row_extract[3]==0):
            specialprefer=list(row_extract[2].split(','))
        counter=row_extract[4]
        song_array=self.song_lyric_extractor(self.selectedsong_title)
        self.all_pass_generate=[]
        while(counter>0):
            songword=[]
            song_taker=random.choice(song_array)
            song_taker1=random.choice(song_array)
            song_taker=song_taker+song_taker1
            songword.append(song_taker)
            if(row_extract[1]==0):
                songword.append(random.choice(numberprefer))
            else:
                songword.append(random.choice(range(100, 10000)))
            if(row_extract[3]==0):
                songword.append(random.choice(specialprefer))
            else:
                songword.append(random.choice(self.special_character_set))
                songword.append(random.choice(self.special_character_set))
            print(songword)
            permutations = list(itertools.permutations(songword))
            permutation_arr=[''.join(permutation) for permutation in permutations]
            print(permutation_arr)
            self.passwordbreach_testing(permutation_arr,songword)
            self.all_pass_generate.append(random.choice(permutation_arr))
            counter-=1
    print('All password Generated')
```

Fig. 1. Implementation of password generation algorithm

This ensures different combinations of the passwords to be explored and the selected passwords are checked for **validation** properly (data-breach check).

The process (Fig. 2) starts with signup/signin process. The signup process involves creating the account with a master key (main key to retrieve all passwords). Then it goes to the settings page where preferences are given according to the users choice (Number, Special Character, Number of Songs and Number of password Preference). Afterwards, it navigates to the song selection page where lyrics are extracted from *lyric_extractor* and kept in the database.

The Search Bar in our Password Manager is implemented using Python. The songs searched in the search bar are used by *lyric_extractor* which feeds in the song to Google custom search engine which searches for the song in different websites using API calls. The song details(ID, Artist Name, Album Name, Title) are returned if the song is present. The engine stops searching if the song is found in one website. There is also a feature called getSongByID(ID) which takes in the ID number of the song stored in the database and returns the song lyrics as a single string. Also the song title is extracted and added to our database which can be later used for hint suggestion purposes.

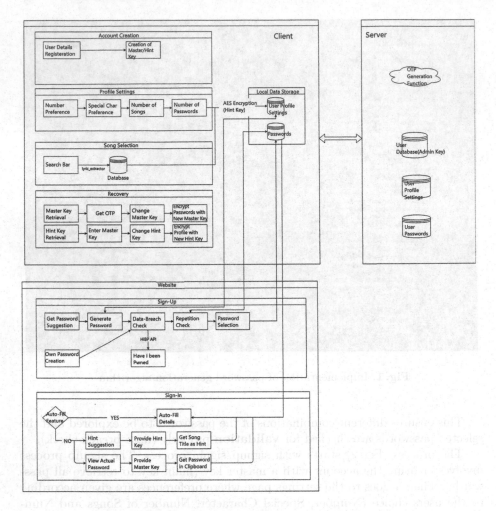

Fig. 2. Architecture diagram - work flow

From lyrics of the song preferred which has been extracted by the getSong-ByID(ID) function, selection of 2 to 3 different words from it randomly will take place along with the addition of some special characters and numbers to generate our password. The number of passwords suggested, special characters used and all other features used to create the password can be fixed and altered using password settings.

3.1 Password Manager Features

Using **Auto-Fill** feature, the passwords are automatically filled with a pop-up prompt given to user who decides to use the feature or not to. The **Auto-Grab** feature helps to capture new passwords entered in websites to be automatically stored in PM.

Our password manager allows the user to initially set a master-key. All the passwords stored in the database by the user is encrypted/decrypted using this. In addition to that, a hint is also provided to the user (pop-up button) which will help the person to use the hint. The hint provided here is the name of the song with which the user has set the password. When the user fails to remember both the Master Key and Hint Key, a OTP (One Time Password) is sent to the user's phone number. The phone number of the user is retrieved by checking his details from the cloud database). Values for one-time passwords are generated using the Hashed Message Authentication Code (HMAC) algorithm and a moving factor, such as time-based information (TOTP). After OTP is sent, the user needs to enter the correct OTP within a stipulated buffer time till which it is valid. Upon entering correctly, the user will be given an option to change the Master Key. Changing it will now result in the encryption and decryption rules to be changed.

Users have freedom to choose the special characters and numbers of their choices. They can choose the list of songs from which words are chosen from lyrics and password will be generated eventually. The minimum number of songs selected must be 4, the default is 5 and can be a maximum of 8 songs.

3.2 Security Features

The APIv2 allows the list of pwned accounts (email addresses and usernames) to be searched via a RESTful service. The API is used to return a list of all breaches a particular account has been involved in. The API takes a single parameter which is the account to be searched for. The account is not case sensitive and will be trimmed of leading or trailing white spaces. The account should always be URL encoded. The API should called by the URL [14]

Encryption- AES 256 encryption is virtually impenetrable using brute-force methods since it is used to encrypt our passwords and store it the local database(trusted device) as well as in the cloud. Web Crypto API is used to encrypt passwords using AES-GCM mode. The encrypt() method of the Subtle-Crypto [15] interface encrypts data. It takes as its arguments a key to encrypt with, some algorithm-specific parameters, and the data to encrypt (also known as "plaintext"). It returns a Promise which will be fulfilled with the encrypted data (also known as "ciphertext").

The PM will check for password repetition within the Local Database present in user's trusted device or the cloud Database, which can be used for suggesting a good password.

Decryption- The decryptor will check the song chosen by the user and decrypts all password for that song by using a master key(AES Decryption). This decryption is used to check for password repetiton and for data-breached passwords. Web Crypto API is used to decrypt passwords using AES-GCM mode. One major difference between this mode and the others is that GCM is an "authenticated" mode, which means that it includes checks that the ciphertext has not been modified by an attacker. The decrypt() method of the SubtleCrypto interface [15] decrypts some encrypted data. It takes as arguments a key to decrypt with, some optional extra parameters and the data to decrypt (also known as "ciphertext"). It returns a promise which will be fulfilled with the decrypted data (also known as "plaintext").

4   Implementation and Results

We have used Python, Tkinter module [16] (python GUI), SQLite3 [17] for GUI purposes, *haveibeenpwned* API for checking passwords involved in data breach, lyric_extractor along with Google Custom Search Engine(song lyric extraction) to implement this project. Figure 3 represents the login page for our PM. The input validation is done for both fields username and password respectively. The user will be logged in or made to stay on the same page depending on the validation output of the password input.

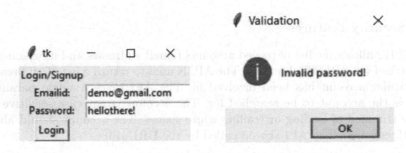

Fig. 3. Login and validation of user details

If the user logged in is a old user, he will be directly taken to password generation page. If the user logged in is a new user, he will be taken to the Settings or Customization page (refer Fig. 4). The page allows the user to set up the master key and the hint key. There are preferences given for numbers (100–10000) and special characters (min 2). The user can also click the random checkbox below to allow the software to automatically select any number or special character during password generation. The number of password suggestions for a particular website option is also present. The user is logged in after setup and taken to the music selection page and password generation page.

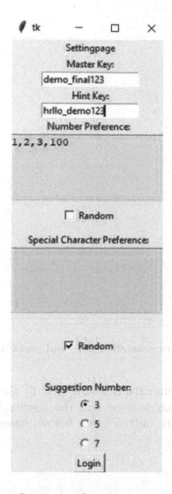

Fig. 4. Settings and customization page

The search bar is implemented using custom search engine (Fig.5), where the song and its lyrics is searched in the list of user given websites using API and the first match will fetch the user the song lyrics from that website. No other search is done after the first match in order to reduce time. This page can also be accessed from the 'Add Song' option present in the Password Generation Page. The user can also delete selected songs and save the current configuration of songs present in the database. There will be a display window which shows the author and song name for each corresponding song.

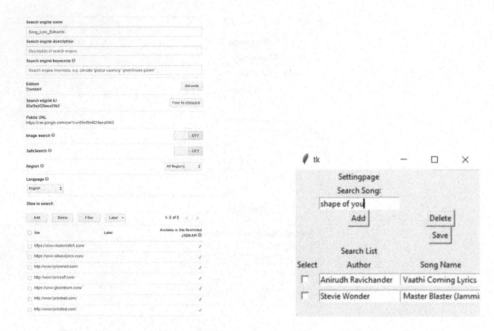

Fig. 5. Password generation page and passwords generated

In the **Password Generation Page** (Fig. 6) the user can put the URL where the account creation is done and the username used for the account. Upon clicking the generate button given below, password suggestions will be shown to the user.

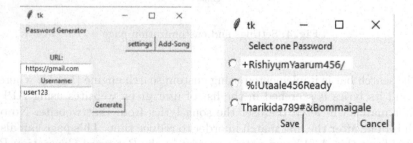

Fig. 6. Password generation page and passwords generated

4.1 Testing

Password Authentication and Validation testing is done using the haveibeen-pwned API (Fig. 7), which checks if the password given is one among data breached passwords which have been stored by the API. Using this, the passwords are generated with proper authenticity.

When a user selects any one of the passwords, the software will create a set of passwords using the selected password (using permutation and combination) and will feed it to the haveibeenpwned API. The results generated will be stored in a text file. The password with the least number of data-breaches will be taken.

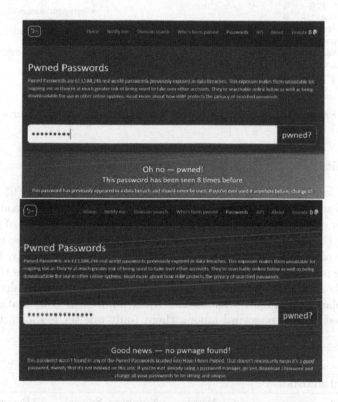

Fig. 7. haveibeenpwned API - data breached and non breached password

Figure 8 shows us the output from the API assuming that the user has selected the choice 'goingchance126~*'. None of the set of passwords were found to be occuring in the API, which makes the passwords generated in our software more reliable and increases the degree of validation.

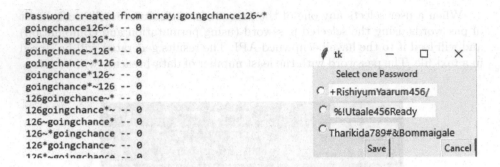

```
Password created from array:goingchance126~*
goingchance126~* -- 0
goingchance126*~ -- 0
goingchance~126* -- 0
goingchance~*126 -- 0
goingchance*126~ -- 0
goingchance*~126 -- 0
126goingchance~* -- 0
126goingchance*~ -- 0
126~goingchance* -- 0
126~*goingchance -- 0
126*goingchance~ -- 0
126*~goingchance -- 0
```

Fig. 8. Passwords generated for a URL and suggested passwords for a song

4.2   Results

As already seen in Fig. 8, almost all of the passwords generated (95–97%) are not data-breached which strengthens our algorithm. The number of combinations produced for each password is 5!(120), and it increases with number of passwords required by the user. For example, if the user requires two or three passwords, the number of combinations generated are 240 and 360 respectively. From that two or three passwords are selected Above example (Fig. 8) shows us the number of passwords (in this case three). Generated and selected by the algorithm for one particular song. Figure 9 shows the total number of combinations which are generated for one password among them (only one set is shown here as an example).

Password statistics				
Password	Words involved	Numbers involved	Special characters involved	Passwords generated
+RishiyumYaarum456/	2	3	2	120
%!Utaale456Ready	2	3	2	120
Tharikida789#&Bommaigale	2	3	2	120

Fig. 9. Total combinations for a song

5 Summary

The passwords generated by the manager are always easy to remember, since they are generated from song lyrics with the combination of numbers and special characters. The passwords are populated for a lot of combinations (Fig. 9) and are also validated and checked for data-breach in the haveibeenpwned API. Above 95% of our passwords haven't been occurring even once in the data-breached passwords which makes our software even more better.

Research Gaps/Future enhancements- The password can be easily guessed when the song is exposed and is an English song by brute force method. But it is difficult in out model as we use more than one word combined with numbers and characters. Dictionary attacks are also possible. Our existing model can allow user to select his own password by splitting the selected password into blocks (character block, number and special character block) and the user now needs to move the blocks to select his preferred combination. The lyrics extracted should be meaningful, in order to do that this model should be extended so that it can use certain machine learning concepts to extract meaningful words from the song lyrics, as well as discarding a song when it is used too much.

References

1. Maruthi, S., Manchikanti, B., Chetan Babu, G., Harini, N.: A secure scheme to manage complex password capable of overcoming human memory limitations. Int. J. Recent Technol. Eng. **7**(5), 137–143 (2019)
2. Luevanos, C., Elizarraras, J., Hirschi, K., Yeh, J.-H.: Analysis on the security and use of password managers. In: 2017 18th International Conference on Parallel and Distributed Computing, Applications and Technologies (PDCAT), pp. 17–24. IEEE (2017)

3. Karole, A., Saxena, N., Christin, N.: A comparative usability evaluation of traditional password managers. In: Rhee, K.-H., Nyang, D.H. (eds.) ICISC 2010. A comparative usability evaluation of traditional password managers, vol. 6829, pp. 233–251. Springer, Heidelberg (2011). https://doi.org/10.1007/978-3-642-24209-0_16

4. Haugum, T., Rygh, L.-C.K.: Design, implementation and analysis of a theft-resistant password manager based on kamouflage architecture. Master's thesis, Universitetet i Agder; University of Agder (2015)

5. Halderman, J.A., Waters, B., Felten, E.W.: A convenient method for securely managing passwords. In: Proceedings of the 14th International Conference on World Wide Web, WWW '05, pp. 471–479. Association for Computing Machinery, New York (2005)

6. Pearman, S., Zhang, S. A., Bauer, L., Christin, N., Cranor, L.F.: Why people (don't) use password managers effectively. In: Fifteenth Symposium on Usable Privacy and Security (SOUPS 2019), pp. 319–338. USENIX Association, Santa Clara (2019)

7. Li, Z., He, W., Akhawe, D., Song, D.: The emperor's new password manager: security analysis of web-based password managers. In: 23rd USENIX Security Symposium (USENIX Security 14), pp. 465–479 (2014)

8. Pham, N., Chow, M.: The security of password managers. Usability News 9(1), 2833–2836 (2016)

9. Pearman, S., et al.: Let's go in for a closer look: Observing passwords in their natural habitat. In: Proceedings of the 2017 ACM SIGSAC Conference on Computer and Communications Security, CCS 2017, pp. 295–310. Association for Computing Machinery, New York (2017)

10. Shancang, L., Romdhani, I., Buchanan, W.: Password pattern and vulnerability analysis for web and mobile applications. ZTE Commun. 32, 1 (2016)

11. Sethumadhavan, Praveen, I.: Partial password authentication using vector decomposition. Int. J. Pure Appl. Math. 118(7), 381–86 (2018)

12. Sreedevi, M., Praveen, I.: Password authentication using vector decomposition. Int. J. Control Theory Appl. 9, 4639–4645 (2016)

13. Sivapriya, K., Deepthi, L.R.: Password strength analyzer using segmentation algorithms. In: 2020 5th International Conference on Communication and Electronics Systems (ICCES), pp. 605–611. IEEE (2020)

14. Api key. Have I Been Pwned. (n.d.). https://haveibeenpwned.com/api/v2/breachedaccount/account

15. SubtleCrypto.decrypt() - web apis: MDN. Web APIs — MDN. (n.d.), from https://developer.mozilla.org/en-US/docs/Web/API/SubtleCrypto/decrypt

16. Tkinter - Python interface to TCL/TK. tkinter - Python interface to Tcl/Tk - Python 3.10.2 documentation. (n.d.), from https://docs.python.org/3/library/tkinter.html

17. SQLITE3 - DB-API 2.0 interface for SQLite databases. sqlite3 - DB-API 2.0 interface for SQLite databases - Python 3.10.2 documentation. (n.d.), from https://docs.python.org/3/library/sqlite3.html

Recognition of Handwritten Gujarati Conjuncts Using the Convolutional Neural Network Architectures: AlexNet, GoogLeNet, Inception V3, and ResNet50

Megha Parikh(✉) and Apurva Desai

Department of Computer Science, Veer Narmad South Gujarat University, Surat, India
meghaparikh4@gmail.com, aadesai@vnsgu.ac.in

Abstract. A methodology for recognizing offline handwritten Gujarati conjuncts has been introduced in the proposed article. This article includes 767 types of frequently utilized conjuncts. The convolutional neural network (CNN) architectures AlexNet, GoogLeNet, Inception V3, as well as ResNet50, are used to recognize handwritten Gujarati conjuncts. The performance of these CNN architectures is systematically evaluated. Gujarati conjuncts are required to segment from the appropriate place before recognition. Conjuncts are segmented into preceding components and succeeding components before recognition. These preceding components and succeeding components can be then individually recognized by the system. 19694 sample images of preceding components of the conjuncts and 28050 samples images of the succeeding components of the conjuncts are utilized for the proposed research work. Maximum accuracy of 93.41% for the preceding components and maximum accuracy of 89.01% for the succeeding components have been achieved by using GoogLeNet.

Keywords: Handwritten Gujarati conjuncts recognition · Recognition of consonant clusters · AlexNet · GoogLeNet · Inception V3 · ResNet50

1 Introduction

Automation of handwritten test segmentation, as well as text recognition, is quite a difficult and challenging task. The high variance and the ambiguity of strokes in handwriting styles across a person, poor quality of the source document due to degradation over time, variable rotation, size, and type of the characters cause significant hurdles while transforming handwritten text into machine-readable text. It's a crucial problem for multiple industries like banking, healthcare, insurance, and many more. Several strategies for recognizing handwritten text have been explored by various scholars over the past several decades. Deep neural network is one of them. It has captured the interests of scholars because of their prominent results in addressing computer vision problems like recognition, classification, object detection [1], etc. A convolutional neural network is one of

M. Singh et al. (Eds.): ICACDS 2022, CCIS 1614, pp. 291–303, 2022.
https://doi.org/10.1007/978-3-031-12641-3_24

the widely accepted deep neural networks. CNN is capable of learning and retrieving features from two-dimensional images [2].

In this research work, the recognition of handwritten Gujarati conjuncts using CNN architectures AlexNet, GoogLeNet, Inception V3, and ResNet50 have been discussed. The Gujarati language is a part of the Indo-Aryan family [30]. According to Articles 344(1) along with 351 of the Indian Constitution, the 8th schedule contains Gujarati as one of the twenty-two official languages of the Republic of India [31]. It is the first language of the people of Gujarat state. Apart from Gujarat, it is also spoken in adjoining union territories of Daman and Diu and Dadra and Nagar Haveli [30]. It is also spoken in countries such as Bangladesh, Fiji, Kenya, Malawi, Mauritius, Singapore, South Africa, the United Kingdom, the United States of America, Zambia, Zimbabwe, and others [32].

This article is organized as follows: the second section includes the introduction to Gujarati conjuncts which is followed by the literature review covering research work carried out by various scholars in the following section. The fourth section comprises methodology and system architecture, which is followed by the data collection in the next section. In the following section data augmentation techniques used for the model have been covered. In the seventh section classification and model training have been discussed followed by experimental results and comparison in the eighth section. The conclusion of the proposed study is outlined in the last section.

2   Gujarati Conjuncts

Gujarati contains 12 vowels, $34 + 2^1$ consonants, and 10 digits [27]. Apart from this, Gujarati contains consonant clusters known as conjuncts. Consonant clusters or conjuncts are created when consonants without the inherent vowel sound are combined [27]. Resultant conjuncts are created based on certain rules [6, 27]. This study covers the conjuncts formed by joining a part of the preceding consonant to the succeeding consonant. 23 out of the 36 consonants contain a right vertical stroke (ખ - kha, ઘ - gha, ન - na, etc.). When these consonants participate as first or middle members of conjuncts, they lose their stroke [27] while connecting with the succeeding consonants. This study includes these conjuncts as well. 767 Gujarati conjuncts are covered in this study. Some of the sample images of handwritten Gujarati conjuncts are covered in Fig. 1.

These conjuncts cannot directly recognized by the system. These conjuncts are required to segment from the appropriate place [6] so that the preceding consonant and succeeding consonant can be retrieved from the conjuncts for recognition. These preceding consonants and succeeding consonants can be further recognized by the system individually to identify the original conjunct.

---

[1] In Gujarati script, consonant clusters ક્ષ (kṣa) as well as જ્ઞ (gña) are considered as basic consonants.

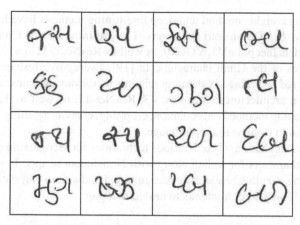

Fig. 1. Sample images of handwritten Gujarati conjuncts

3   Literature Review

Many researchers have contributed their research work for various languages.

Author [3] has used a multi-layered feed-forward neural network to classify Gujarati numerals. Algorithms for recognizing handwritten Gujarati digits as well as characters have been proposed by the author in [4, 5]. In [6], the authors proposed a method for segmenting Gujarati conjuncts into their participating components. The authors have used information of the neighboring pixels for the segmentation of the conjuncts. In [7, 16], a vertical projection profile along with morphological operations are used for retrieving words from the line in Gujarati Script. In [8], the authors have discussed a process that is based on distance transform to identify several zones of handwritten Gujarati words. This can be utilized for retrieving modifiers from the characters. A deep learning approach has been proposed by authors in [9] to recognize Gujarati characters retrieved from the image. Authors have developed OCR in [10], for the uppercase English alphabet. Authors have utilized the convolutional neural network for extracting features of the alphabet along with the Error correcting output code for classification. A convolutional deep model in [11] has been proposed to recognize handwritten Bengali characters. Kernels, as well as local receptive fields, are used for learning the required feature set, and then densely connected layers have been employed for the discrimination task. Authors in [12] have developed a streamlined version of GoogLeNet to recognize handwritten Chinese characters. They have used a model which is 19 layers deep and includes 7.26M parameters. They have attained a new level of recognition accuracy. AlexNet based CNN [13] has been presented to extract features from 44 basic as well as 36 compound Malayalam characters. Three different Convolution Neural Networks (CNN) architectures [14] that are CNN, Modified LeNet CNN as well as Alexnet CNN have been presented to recognize Devanagari Characters. Authors have proposed a modified LCNN by using the activation function Rectified Linear Unit instead of the standard activation function tanh(x). Deep Convolution Neural Network architectures [15] AlexNet, DenseNet 121, DenseNet 201, VGG 11, VGG 16, VGG 18, as well as Inception V3 have been used for feature extractor and recognition of Devanagari characters. VGG16 and DenseNet

[17] with freezing weights method and deep fine-tuning methods have been utilized to recognize handwritten Devanagari characters. The pre-trained Deep Convolution Neural Network models InceptionV3, VGG19, as well as ResNet50 [18] have been utilized to extract features of the Hindi characters. In [19], twenty-five reshaping techniques have been analyzed spanning six datasets relating to various domains, and trained on three well-known architectures: InceptionV3, ResNet-18, as well as 121-Layer deep DenseNet. Reshaping methods interpolation, cropping, containing, tiling, and mirroring has been explored to reshape the input image.

Many researchers have analyzed various techniques for recognizing printed as well as handwritten characters for Indian along with Non-Indian scripts. In our experience and understanding, no other research on recognizing handwritten Gujarati conjuncts has been conducted. This research intends to bridge that gap.

4 Methodology and System Architecture

Convolutional neural networks have made immense progress in image classification, face recognition, object detection, and other computer vision applications [20]. A CNN contains neurons or nodes. These neurons have learnable weights and biases. In CNN architecture, every single neuron does a convolution operation on the received input. A classical CNN contains three types of layers viz. convolution layers, pooling layers, along with fully connected layers. These layers are stacked collectively for constructing a convolutional neural network [21].

One of the key benefits of CNN is it requires very less preprocessing of data compared to other deep learning networks. LeNet, AlexNet, VGG, GoogLeNet, ResNet, SqueezeNet, and EfficientNet are some of the well-known CNN architectures. In the present study, AlexNet, GoogLeNet, Inception V3, and ResNet50 have been used to recognize preceding and succeeding components of the Gujarati conjuncts.

AlexNet
Alex Krizhevsky developed AlexNet along with Ilya Sutskever and Geoffrey Hinton. AlexNet consists of 60M parameters as well as 650,000 nodes, along with eight layers that are five convolutional layers and three fully-connected layers [22]. AlexNet uses the ReLU activation function instead of the standard activation functions tanh and sigmoid. According to the authors, the model can be trained much faster compared to standard activation functions. Authors have also used data augmentation and dropout techniques to reduce overfitting [22].

GoogleNet
GoogLeNet also known as Inception V1, is a 22 layers deep Convolution neural network developed by researchers at Google [23]. It is a convolutional neural network based on the Inception architecture. GoogLeNet consists of seven million parameters which are much less compared to AlexNet, even though GoogLeNet is much wider and deeper [23]. To solve the problem of overfitting, the authors proposed to have different filters

which can operate on the same level. Because of it, the network becomes wider rather than deeper [23].

Inception V3

Inception V3 is a widely-used image recognition model developed by Szegedy, et al. [24]. The model is made up of symmetric as well as asymmetric building blocks. It is a 42 layers deep convolution neural network including convolutions, average pooling, max pooling, concatenations, dropouts, and fully connected layers. Softmax is utilized for calculating loss [24].

ResNet50

ResNet is a successful classic neural network utilized for various computer vision tasks [25]. Here 50 represents the number of neural network layers a network has. ResNet learns residual functions with reference to the layer inputs. This was the first network to introduce skip connections. It uses 3-layer bottleneck residual block design to ensure the accuracy of the system and lesser training time which directly contribute to the performance of the system [25].

The features of the CNN architectures used in the proposed study to recognize preceding and succeeding components of the Gujarati conjuncts are described in Table 1.

We have developed models using this CNN architecture. For the unbiased comparison between the models, we have tried to standardize the hyper-parameters across the models. The hyper-parameters described in Table 2 are used to develop models.

Table 1. Features of the utilized CNN architectures

CNN architecture	Depth	Parameters	Image input size
AlexNet [22, 26]	8	60 M	227 × 227
GoogLeNet [23, 26]	22	7 M	224 × 224
Inception V3 [24, 26]	42	23.9 M	299 × 229
ResNet 50 [25, 26]	50	25. 6M	224 × 224

Table 2. Hyper-parameters used for the models

Hyper-parameters	Value
Optimization algorithm	Adam
Initial learning rate	1.0000×10^{-3}
Epochs	20
Batch size	128

5 Data Collection

For our research work as described in [6], we have created a database containing hand-written Gujarati conjuncts. This study covers 767 types of Gujarati conjuncts, which have been collected from writers of different age. These collected sample images are scanned by using a flatbed scanner that has a resolution of 300 dpi. These sample images are stored in jpg file format. Morphological operations clean, open, thicken, along with close have been used to preprocess the image. After completing pre-processing, conjuncts are segmented into the preceding components and the succeeding components of the conjuncts using the algorithm proposed in [6]. These images of the preceding components and the succeeding components then can be individually recognized by the proposed system. 19694 sample images of preceding components and 28050 sample images of succeeding components have been used for the models. This research study covers 25 different types of consonant characters which can participate as preceding components of the conjuncts. It also covers 33 different types of consonant characters which can participate as succeeding components of the conjuncts.

6 Data Augmentation

Data augmentation is a method utilized for extending the size of the training dataset by applying some modifications like rotation, position shifting, zooming, shearing, and so on to the existing data for creating new data [28]. It artificially creates variation in the existing images in order to expand the dataset's size. This is required for the model so that it can generalize well on new data. Due to this technique, different images are generated randomly in every epoch. These randomized methods introduce some diversity to the training images that assist the models in learning the properties of the alphabet during the training stage. For this research work rescaling, rotation, shear, and zooming technique has been applied to the images. The data augmentation technique utilized for this research study is described in Table 3.

Table 3. Data augmentation techniques utilized for this research study

Data augmentation technique	Value
Rescale	1./255
Rotation	40
Shear	0.2
Zoom	0.4

7 Classification and Model Training

Data is divided into two sets: a training set as well as a validation set. The training set is utilized to train the model while the validation set is utilized for validating the accuracy of the model on unknown data. 80% of the total dataset is utilized for training purpose while the remaining 20% of the dataset is utilized for validation purpose. As discussed, there are 19694 sample images of preceding components and 28050 sample images of succeeding components available for the model. 15756 sample images of the preceding components and 22440 sample images of the succeeding components are reserved for the training purpose, while 3938 sample images of the preceding components and 5610 sample images of the succeeding components are reserved for the validation purpose. Data augmentation techniques discussed above are applied to the training set.

The training process was executed for 20 epochs for all the models. The total number of steps in each epoch, the time required to execute each step in an epoch, and the average time required to train each epoch are described in Table 4 and Table 5 for the preceding components and the succeeding components of the conjuncts respectively.

Table 4. Training the model (for the preceding components of the conjuncts)

CNN architecture	A	B	C
AlexNet	123	3 s/step	339–387 s
GoogLeNet	123	3–4 s/step	382–460 s
Inception V3	123	12–13 s/step	1536–1644 s
ResNet50	123	10 s/step	1179–1239 s

Table 5. Training the model (for the succeeding components of the conjuncts)

CNN architecture	A	B	C
AlexNet	175	3 s/step	515–550 s
GoogLeNet	175	3 s/step	557–578 s
Inception V3	175	12–17 s/step	2023–3061 s
ResNet50	175	8 s/step	1454–1478 s

Where,
A: Total number of steps in each epoch
B: Time required to execute each step in an epoch
C: Time required to train each epoch
This time can vary for the initial epochs.

The number of nodes in the classification output layer is the same as the number of classes. So there are 25 classes included for the preceding components whereas for the succeeding components 33 possible classes are included. These models are capable of learning features automatically during the training stage. So no further feature extraction technique is required for retrieving the features of the images. Because this is a multiclass classification problem, the softmax layer is used as the last layer for all the models. It transforms a previous layer's output into a vector of probabilities. The model then selects the class having the maximum probability as its class prediction. This decides the preceding component and the succeeding component of the conjunct.

8   Experimental Results and Comparisons

In the present research study, an assessment of CNN models for classifying the preceding components and the succeeding components of the conjuncts has been performed. The purpose of the proposed study is to compare the CNN models by analyzing the accuracy, precision, sensitivity, along with f1-score by fine-tuning.

Performance measures in terms of the training, as well as the validation accuracy of this experimental study for the preceding components of the conjuncts, have been described in Table 6. Figure 2 represents the validation accuracy of the preceding components of the conjuncts.

Table 6. Performance measure in terms of accuracy of this experimental study for the preceding components of the conjuncts

CNN architecture	Training accuracy	Validation accuracy
AlexNet	91.92%	89.11%
GoogLeNet	93.01%	93.41%
Inception V3	92.95%	89.27%
ResNet50	91.76%	87.45%

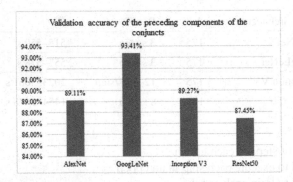

Fig. 2. Validation accuracy of the preceding components of the conjuncts

Each model discussed above presented statistically significant performance. It is observed from Table 6 and Fig. 2 that ResNet50 showed the lowest performance of 87.45% while training for the preceding components of the conjuncts. While for the preceding components of the conjuncts GoogLeNet reported the highest accuracy of 93.41% followed by Inception V3.

Performance measures in terms of the training, as well as the validation accuracy of this experimental study for the succeeding components of the conjuncts, have been described in Table 7. Figure 3 represents the validation accuracy of the succeeding components of the Gujarati conjuncts.

Table 7. Performance measure in terms of accuracy of this experimental study for the succeeding components of the conjuncts

CNN architecture	Training accuracy	Validation accuracy
AlexNet	89.86%	81.49%
GoogLeNet	90.79%	89.01%
Inception V3	91.88%	90.32%
ResNet50	90.83%	83.72%

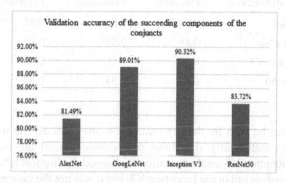

Fig. 3. Validation accuracy of the succeeding components of the conjuncts

For the succeeding components of the conjuncts, as depicted in Table 7 and Fig. 3 the lowest accuracy of 81.49% was obtained by the AlexNet, and the highest accuracy of 90.32% was reported by the Inception V3 which is close to the accuracy achieved by GoogLeNet i.e. 89.01%.

Finally, in the precision, sensitivity, along with f1-score evaluation, the ResNet50 performance was 0.87, 0.86, and 0.86, correspondingly, whereas in GoogleNet, which achieved the highest accuracy in the prior metrics with 0.93, 0.93, and 0.93 correspondingly for the preceding components of the conjuncts. The reported values for precision, sensitivity, and f1-score for the preceding components are depicted in Table 8.

For the succeeding components of the conjuncts, the highest performance was observed by Inception V3 with 0.90, 0.90, and 0.90 in the precision, sensitivity, along with f1-score evaluation while the lowest performance was reported by AlexNet with 0.84, 0.81, and 0.81 correspondingly. Accordingly, to the measurements, the Inception V3 implementation achieved the highest performance which is almost similar to the performance of the GoogLeNet. The reported values for precision, sensitivity, along with f1-score for the succeeding components are displayed in Table 9.

Table 8. Performance of the models for the evaluation of the preceding components of the conjuncts

CNN architecture	Precision	Sensitivity	F1-score
AlexNet	0.89	0.88	0.88
GoogLeNet	0.93	0.93	0.93
Inception V3	0.90	0.89	0.88
ResNet50	0.87	0.86	0.86

Table 9. Performance of the models for the evaluation of the succeeding components of the conjuncts

CNN architecture	Precision	Sensitivity	F1-score
AlexNet	0.84	0.81	0.81
GoogLeNet	0.89	0.89	0.89
Inception V3	0.90	0.90	0.90
ResNet50	0.85	0.84	0.84

Certainly, it can be seen by the performance matrix that all the models discussed above have shown statistically significant performance in all the measures, but the GoogleNet and Inception V3 implementations achieved the highest results. Some cases of overfitting were observed in the Inception V3, but it was not the case with GoogLeNet. The processing time to complete the classification process of the GoogLeNet is far better than that of Inception V3. By considering all the measurements, it appears that GoogLeNet outperforms Inception V3. In addition, AlexNet performed best based on the processing time each model took to complete the classification process. This means it is more efficient than other architectures. But the accuracy achieved by the model was less compared to other models. So in this case, a model having better accuracy has been selected over a model having better processing efficiency.

9   Conclusion

In the presented study, we analyzed the performance of several popular deep convolutional neural networks architectures for recognizing frequently used handwritten Gujarati conjuncts. The experimental finding shows that GoogLeNet is the best performer in classifying preceding and succeeding components of the Gujarati conjuncts. On the other hand, AlexNet and ResNet50 show the lowest performance compared to other architectures. We achieved a recognition rate of 93.41% and 89.01% for the preceding and the succeeding components of the Gujarati conjuncts respectively using GoogLeNet. The presented study is expected to make some contribution in the field of character recognition for the Gujarati script using a convolution neural network.

Acknowledgments. The authors acknowledge the support of the University Grants Commission (UGC), New Delhi, for this research work through project file no.: F.3–12/2018/ DRS-II (SAP-II).

References

1. Bai, J., Chen, Z., Feng, B., Xu, B.: Image character recognition using deep convolutional neural network learned from different languages. In: 2014 IEEE International Conference on Image Processing (ICIP), pp. 2560–2564 (2014). https://doi.org/10.1109/ICIP.2014.7025518

2. Maitra, D.S., Bhattacharya, U., Parui, S.K.: CNN based common approach to handwritten character recognition of multiple scripts. In: 2015 13th International Conference on Document Analysis and Recognition (ICDAR), pp. 1021–1025 (2015). https://doi.org/10.1109/ICDAR.2015.7333916

3. Desai, A.A.: Gujarati handwritten numeral optical character reorganization through neural network. Pattern Recogn. **43**(7), 2582–2589 (2010). https://doi.org/10.1016/j.patcog.2010.01.008

4. Desai, A.A.: Support vector machine for identification of handwritten Gujarati alphabets using hybrid feature space. CSI Transactions on ICT **2**(4), 235–241 (2015). https://doi.org/10.1007/s40012-014-0059-z

5. Desai, A.A.: Handwritten Gujarati numeral optical character recognition using hybrid feature extraction technique. In: Proceedings of the 2010 International Conference on Image Processing, Computer Vision, Pattern Recognition, vol. 2, pp. 733–739 (2010)

6. Parikh, M., Desai, A.A.: Segmentation of Frequently Used Handwritten Gujarati Conjunctive Alphabet. In: 2019 5th International Conference On Computing, Communication, Control And Automation (ICCUBEA), pp. 1–6 (2019). https://doi.org/10.1109/ICCUBEA47591.2019.9128510

7. Patel, C., Desai, A.A.: Segmentation of text lines into words for Gujarati handwritten text. In: 2010 International Conference on Signal and Image Processing, pp. 130–134 (2010). https://doi.org/10.1109/ICSIP.2010.5697455

8. Patel, C., Desai, A.A.: Zone Identification for Gujarati Handwritten Word. In: 2011 Second International Conference on Emerging Applications of Information Technology, pp. 194–197 (2011). https://doi.org/10.1109/EAIT.2011.47

9. Shukla, D., Desai, A.: Extraction and recognition of handwritten Gujarati characters and numerals from images using deep learning. In: Thakkar, F., Saha, G., Shahnaz, C., Hu, Y.-C. (eds.) Proceedings of the International e-Conference on Intelligent Systems and Signal Processing. AISC, vol. 1370, pp. 657–669. Springer, Singapore (2022). https://doi.org/10.1007/978-981-16-2123-9_51

10. Bora, M.B., Daimary, D., Amitab, K., Kandar, D.: Handwritten character recognition from images using CNN-ECOC. Proced. Comput. Sci. **167**, 2403–2409 (2020). https://doi.org/10.1016/j.procs.2020.03.293, ISSN 1877–0509

11. Purkaystha, B., Datta, T., Islam, M.S.: Bengali handwritten character recognition using deep convolutional neural network. In: 2017 20th International Conference of Computer and Information Technology (ICCIT), pp. 1–5 (2017). https://doi.org/10.1109/ICCITECHN.2017.8281853

12. Zhong, Z., Jin, L., Xie, Z.: High performance offline handwritten Chinese character recognition using GoogLeNet and directional feature maps. In: 2015 13th International Conference on Document Analysis and Recognition (ICDAR), pp. 846–850 (2015). https://doi.org/10.1109/ICDAR.2015.7333881

13. James, A., Manjusha, J., Saravanan, C.: Malayalam handwritten character recognition using AlexNet based architecture. Indonesian J. Elec. Eng. Inf. (IJEEI), **6**(4), 393–400, ISSN: 2089–3272. (2018). https://doi.org/10.11591/ijeei.v6i1.518
14. Prashanth, D.S., Mehta, R.V.K., Ramana, K., Bhaskar, V.: Handwritten devanagari character recognition using modified Lenet and Alexnet convolution neural networks. Wireless Pers. Commun. **122**(1), 349–378 (2021). https://doi.org/10.1007/s11277-021-08903-4
15. Aneja, N., Aneja, S.: Transfer Learning using CNN for Handwritten devanagari character recognition. In: 2019 1st International Conference on Advances in Information Technology (ICAIT), pp. 293–296 (2019). https://doi.org/10.1109/ICAIT47043.2019.8987286
16. Patel, C., Desai, A.: Extraction of characters and modifiers from handwritten Gujarati words. Int. J. Comput. Appli. **73**(3), 7–12 (2013). https://doi.org/10.5120/12719-9541
17. Bhati, G.S., Garg, A.R.: Handwritten devanagari character recognition using CNN with transfer learning. In: Sharma, H., Saraswat, M., Yadav, A., Kim, J.H., Bansal, J.C. (eds.) CIS 2020. AISC, vol. 1335, pp. 269–279. Springer, Singapore (2021). https://doi.org/10.1007/978-981-33-6984-9_22
18. Rajpal, D., Garg, A.R., Mahela, O.P., Alhelou, H.H., Siano, P.: A fusion-based hybrid-feature approach for recognition of unconstrained offline handwritten Hindi characters. Future Internet. **13**(9), 239 (2021). https://doi.org/10.3390/fi13090239
19. Ghosh, S., Das, N., Nasipuri, M.: Reshaping inputs for convolutional neural network: Some common and uncommon methods. Pattern Recog. **93**, 79–94 (2019). https://doi.org/10.1016/j.patcog.2019.04.009, ISSN 0031–3203
20. Huang, S.-C., Le, T.-H.: Chapter 8 - convolutional neural network architectures. In: Huang, S.-Le, T.-H., (eds.) Principles and Labs for Deep Learning, pp. 201–217. Academic Press, ISBN 9780323901987 (2021). https://doi.org/10.1016/B978-0-323-90198-7.00001-X
21. Gupta, V., Sachdeva, S., Dohare, N.: Chapter 8 - Deep similarity learning for disease prediction. In: Piuri, V., Raj, S., Genovese, A., Srivastava, R., (eds.) Hybrid Computational Intelligence for Pattern Analysis, Trends in Deep Learning Methodologies, pp. 183–206. Academic Press (2021). https://doi.org/10.1016/B978-0-12-822226-3.00008-8, ISBN 9780128222263
22. Krizhevsky, A., Sutskever, I., Hinton, G.E.: Imagenet classification with deep convolutional neural networks. Neural Inf. Process. Syst. **25**, 1097–1105 (2012). https://doi.org/10.1145/3065386
23. Szegedy, C., et al.: Going deeper with convolutions. In: 2015 IEEE Conference on Computer Vision and Pattern Recognition (CVPR), pp. 1–9 (2015). https://doi.org/10.1109/CVPR.2015.7298594
24. Szegedy, C., Vanhoucke, V., Ioffe, S., Shlens, J., Wojna, Z.: Rethinking the inception architecture for computer vision. In: 2016 IEEE Conference on Computer Vision and Pattern Recognition (CVPR), pp. 2818–2826 (2016). https://doi.org/10.1109/CVPR.2016.308
25. He, K., Zhang, X., Ren, S., Sun, J.: Deep residual learning for image recognition. In: 2016 IEEE Conference on Computer Vision and Pattern Recognition (CVPR), pp. 770–778 (2016). https://doi.org/10.1109/CVPR.2016.90
26. Khan, A., Sohail, A., Zahoora, U., Qureshi, A.S.: A survey of the recent architectures of deep convolutional neural networks. Artif. Intell. Rev. **53**(8), 5455–5516 (2020). https://doi.org/10.1007/s10462-020-09825-6
27. Govindaraju, V., Setlur, S.: Guide to OCR for Indic Scripts: Document Recognition and Retrieval, 1st edn. Springer Publishing Company, Incorporated (2009)
28. Wikipedia contributors. Data augmentation. Wikipedia, The Free Encyclopedia. February 10, 2022, 21:55 UTC. Available at: https://en.wikipedia.org/w/index.php?title=Data_augmentation&oldid+1071104120, (Accessed 16 February 2022)
29. Wikipedia contributors. Gujarati script. Wikipedia, The Free Encyclopedia. https://en.wikipedia.org/w/index.php?title=Gujarati_script&oldid=1079209121, (Accessed 16 December 2021)

30. Indian   Mirror.   https://www.indianmirror.com/languages/gujarati-language.html, (Accessed 16 December 2021)
31. Wikipedia contributors. Eighth Schedule to the Constitution of India. Wikipedia, The Free Encyclopedia. 27 October 2021, 17:50 UTC. https://en.wikipedia.org/w/index.php?title= Eighth_Schedule_to_the_Constitution_of_India&oldid=1052154181, (Accessed 16 December 2021)
32. Omniglot. http://www.omniglot.com/writing/gujarati.htm, (Accessed 16 December 2021)

Evolving Spiking Neural Network as a Classifier: An Experimental Review

M. Saravanan[1(✉)], Annushree Bablani[2], and Navyasai Rangisetty[2]

[1] Ericsson India Global Services Pvt. Ltd., Perungudi, India
m.saravanan@ericsson.com
[2] Indian Institute of Information Technology, Sri City, India
{annushree.bablani,navyasai.r17}@iiits.in

Abstract. The brain-inspired Spiking Neural Networks (SNNs) are considered as the third generation of neural networks for AI applications. The spiking neural network has been proved very efficient to predict and classify data with spatial and temporal information in the way of modeling the behavior and learning potential of the brain. The aim is to understand the working of the Evolving SNN (ESNN) as a classifier and how it is different from the existing neural models. Besides exploring the existing ESNN architecture, the results have been generated by tuning the various parameters of the ESNN model which may be contributing to provide better prediction accuracies. The tuned ESNN model is applied to various datasets and compared with the existing second-generation neural network model like LSTM. The results show comparable improvement in the classification accuracy using ESNN which concludes that the ESNN and its variants are the beginning of a new era of Neural Networks.

Keywords: Evolving spiking neural network · Reservoir · Gaussian receptive field · Spatial-temporal data

1 Introduction

In the past few years, Neural Networks have evolved a great deal in solving a lot of computationally complex problems. Being a decade-old algorithm with such flexibility to evolve into various generations, the increasing complexity of the problem has made neural networks, one of the most used algorithms by researchers. With this emergence, a new type of model, SNN has been introduced [1], which is computationally more powerful than the previous generations. In the conventional Artificial Neural Network, the neurons are continuous data variables, here in SNN (aka, Third Generation Neural Network), the neurons receive and transmit information in the form of spike trains (inspired from the biological brain).

One of the SNN architectures, called Evolving Spiking Neural Network, has been introduced in [2] which makes SNN more adaptive and faster by incrementally merging the evoked neurons to capture the pattern in the given problem. ESNN takes this property from Evolving Connectionist Systems (ECOS) [3], where the new neurons, new features, new connections are generated during the execution of the system. This is a dynamic model, which adapts or evolves with time. In this paper, we are trying to explore the evolving nature of ESNN as a classifier using three case study applications to understand its effectiveness. Because of its dynamic and evolving nature, ESNN can capture data as and when available without the requirement of retraining the network. The ESNN learns the mapping from signal data vector to specified class label and hence this makes it a unique classifier.

To improve the performance of models on spatiotemporal data, an extension to ESNN was proposed in [4] called Recurrent ESNN, which follows the principle of probabilistic Liquid State Machine, by adding a layer to the architecture. This additional layer acts as the reservoir which transforms the input pattern into a single high-dimensional network. The high-dimensional network then can be trained using ESNN. The ESNN model is illustrated in this paper and applied on three different applications: 1) Mobility Dataset 2) 5G Data and 3) EEG data for emotion Recognition and these datasets exhibit spatiotemporal features as a common factor. The reason for choosing two communication datasets is because of their dynamic behavior and drift caused during various times and locations. There is no specific pattern in these changes. Also, we have used EEG data which is biological and highly dynamic data. In EEG data there underlies a pattern but this pattern changes from human to human, like communication data which changes with the type of user. The aim is to produce a more accurate, energy-efficient, and fast learning system to achieve better classification results on spatiotemporal datasets. To achieve that instead of learning from all the fired neurons we have selected the best k, neurons as the input to the ESNN. The detailed methodology and results achieved are explained in the rest of the paper.

2 Related Works

Researchers have applied ESNN to achieve better results for the limited size neuron repository [5]. The model presented in [5] classifies input after one presentation. These models are good for online learning, where the input changes dynamically and the model can classify the data without the requirement of retraining. The model followed a sliding window approach where the input samples within a particular window were encoded. In another work, the classification performance of ESNN was improved by selecting an optimal set of features using the wrapper and quantum-inspired evolutionary optimization approaches [7]. Using this combination relevant features were identified and finally, an optimal parameter setting was achieved giving better classification results. Parameter optimization is a very crucial part of any kind of network model and to achieve that like previous work using a quantum inspired SNN with Particle Swarm Optimization (PSO) approach was proposed [8]. This quantum inspired PSO (QiPSO) was designed for search in binary spaces, and it has shown comparable results to the formal QiSNN. Later Dynamic QiPSO [9], was proposed where the search strategy was performed in both binary and continuous search spaces.

ESNN has been implemented by several researchers in a variety of applications. The ESNN classifier has been used for Spatial and Spectro-temporal pattern recognition problems. Various models with a single layer or with a layer of the reservoir have been applied. In [9, 10] dynamic updates in the input neurons are added to the conventional Leaky Integrate-and-Fire (LIF) Model and updating of the weights was not only dependent on rank order learning, but on the time of the following incoming spikes to the post-synaptic neuron. This model was referred to as Dynamic ESNN, and they used a single layer of neurons [9,10]. When the size of the input neuron increases (temporal pattern), a single layer may not be sufficient and efficient to learn these patterns. So, a reservoir based ESNN has been proposed in [6]. Initially, the input is converted to a spike train using an encoding scheme and then these input spikes are passed through a filter to collect the temporal features. This filter acts as a liquid state or reservoir and ESNN acts as the readout or layer. The reservoir is constructed of LIF neurons with exponential synaptic currents. Here, the RESNN has been utilized to model Spatial-temporal patterns. The study used the real-world sign-language dataset LIBRAS on this model and it achieved encouraging classification results [6].

ESNN and its improved architectures have been used in applications like image recognition [11–13], speaker authentication system [14], audio-visual pattern recognition system [15], taste recognition [16, 17], sign language [6], object recognition [9], EEG pattern analysis [10] and many others. In this paper, we have attempted to understand how ESNN works as a classification approach and compared it with the earlier NN approaches. As ESNN has shown promising performance in earlier studies also, it motivated us to try to propose an ESNN by selecting the best-fired neurons and applying the model on three different use cases. Here, additionally we tried to tune the ESNN parameters suitably to get a best performance from the classifier. Detailed methodology is explained in the subsequent section.

3 Evolving Methodology

The data classification is based on prior knowledge or statistical information extracted from the given data. The data is usually classified as measurements or observations, defining points in an appropriate multidimensional space. Classification is dependent on the type of learning procedure that generates the output value based on template matching, statistical classification, syntactic or structural matching, and neural networks.

ESNN being a type of neural network classifies the data by creating spikes that contains the temporal information of data and hence are very powerful in discriminating the information. The implementation of the ESNN initially begins by creating an empty neuron repository and a new output neuron is generated and added to the empty repository. For every input sample, the numerical data is converted into a trail of spikes, this is called *encoding*. To perform encoding on numerical data, *Gaussian Receptive Field (GRF) population* [5] encoding scheme has been used. Every feature input is distributed over several neurons (called Gaussian Receptive Field Neurons or GRFN) and each of these neurons is fired only once during the time interval (T). After encoding each input feature is represented in the form spatiotemporal spike pattern. The center μ_j and the width σ_j of each GRF presynaptic neuron are computed as [5]:

$$\mu_j = I^n_{min} + \frac{2j - 3}{2}\left(\frac{I^n_{max} - I^n_{min}}{N - 2}\right) \tag{1}$$

$$\sigma_j = \frac{1}{\beta} \left(\frac{I_{max}^n - I_{min}^n}{N - 2} \right) \tag{2}$$

Here I_{max}^n and I_{min}^n are maximum and minimum values of nth feature in given window size, N is the number of receptive fields (GRF neurons for each feature), β is a parameter in [1, 2]. The output of neuron j is defined as

$$out_j = exp\left(\frac{(x - \mu_j)^2}{2\sigma_j^2} \right) \tag{3}$$

where x is the input value. The firing time of each presynaptic neuron j is defined as $T_j = T(1 - out_j)$, where T is the simulation time or spike interval.

The advantage of using the GRF encoding scheme over the mostly used Poisson encoding scheme is the time taken for encoding. In Poisson encoding, the encoder waits for all spikes to get fired and then encodes it, whereas in GRF each spike as it is fired is encoded. GRF does not wait for a full spike train, hence reducing the time of encoding in the model.

Here LIF model [1] is used to create the initial neurons. Each neuron fires at most once and a neuron fire only when its Postsynaptic Potential (PSP) reaches its threshold value. PSP of the ith neuron is defined as:

$$PSP_i = \begin{cases} 0. if\ fired \\ \sum_j w_{ji} \cdot mod^{order(j)} . Otherwise \end{cases} \tag{4}$$

where w_{ji} represents the weight of the synaptic connection between presynaptic neuron j to output neuron i, mod is the modulation factor in [0, 1], and $order(j)$ defines the rank of the presynaptic neurons spike. The first rank is assigned as 0 and subsequently, the rank is increased by 1 based on the firing time of each presynaptic neuron.

Firstly, the model creates an empty repository for output neurons. For each pattern that belongs to the same given class, a new output neuron is created and connected to all presynaptic neurons in the previous layer, and weights are assigned using rank order. The weight w_{ji} is calculated as

$$w_{ji} = mod^{order(j)} \tag{5}$$

A numerical threshold γ_i is set for the newly created output neuron as the fraction c in (0, 1) of its maximum postsynaptic potential $PSP_{max,i}$ as

$$\gamma_i = PSP_{max,i} \cdot c \tag{6}$$

The weight vector of a newly created output neuron is then compared with the already present output neurons in the repository. If the Euclidean distance between the newly created output neuron weight vector and that of any of the already trained output neurons is smaller than a similarity parameter (SIM), they are considered to be similar. As a result, their thresholds and weight vectors are merged according to

$$w_{ji} = \frac{w_{new} + (w_{ji} \cdot M)}{M + 1} \tag{7}$$

$$\gamma_{ji} = \frac{\gamma_{new} + (\gamma_{ji} \cdot M)}{M + 1} \tag{8}$$

where M is the number of previous merges of similar neurons through the learning history of the ESNN. After merging, the weight vector of the newly created output neuron is discarded, and the new pattern is presented to the model. If none of the already trained neurons in the repository is found to be similar (as per the SIM parameter) to the newly produced output neuron, then it is added to the repository.

The testing phase is carried out by propagating the spikes that encode the test sample to all trained output neurons. The class label for the test sample is assigned according to the class label of the output neuron which has fired first after reaching its threshold value γ_i. The methodology explained above is given as a flow diagram in Fig. 1.

In this paper, the existing methodology is improved by tuning the ESNN parameters c, mod, and SIM. The parameter c determines that how many fractions of PSP will be used as the threshold value. The parameter mod is the modulation factor, and it determines the importance of the order of the first spike, and the parameter SIM, which measures the similarity between two neurons, and on basis of SIM, a neuron is either added to a repository or is replaced the already available similar neuron in a repository.

The synapses are dynamic; their values change over the timescale of training. To have that variability in the ESNN model the parameters c, mod, and SIM are utilized. ESNN model unlike any other neural network is quite sensitive to its parameters and they play a major role in defining the accuracy of the model. Determining an optimal set of parameters is crucial and challenging hence an attempt has been made to tune these parameters with various datasets and results are presented in the next section. All these variables lie in the range of (0,1) and varying them and tuning according to the model will improve the efficiency.

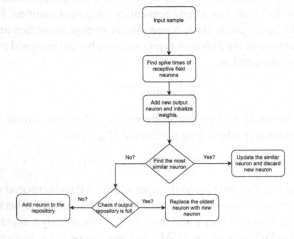

Fig. 1. ESNN workflow diagram

4 Dataset Description

Here, we first implement the ESNN for the classification tasks and to understand how the parameter tuning in ESNN can impact three different types of communication datasets, which will be analyzed. The motivation to use these three datasets is that the datasets are represented using their temporal information and ESNN's spikes manage the temporal information efficiently, which is expected to generate a better classification.

The first dataset is the telecom dataset [18] with 42 eNode B's and 10,000 UE's (unique id for each user mobile) for 5000 min. The following multiple mobility patterns were simulated: 1) Work Professional: Person travels from fixed home location to office location and comes back to home. 2) Sales Professional: Person travels from fixed home location and travels to 10 random locations. 3) Random waypoint: Person travels from random home location to random destinations until the day finishes off. In the simulator, one step corresponds to one minute, and also to introduce concept drift we switched off 8 of the eNode B's at a time of 3000 min and woke them up after 4000 min. For further details of the dataset, the reader is suggested to go through [18]. The second dataset utilized is a 5G dataset [19, 20] which identifies the incoming connection request and assigns it to the most optimal slice based on important KPIs. These KPIs can be captured from control packets between the UE and network. It has multiple types of input devices which include smartphones, general IoT, AR-VR, Industry 4.0 traffic, e911 or public safety communication, healthcare, smart city, or smart homes traffic, etc. It can even capture an unknown device requesting access to one or multiple services. Here, they have UE category values defined to them and the network also allocates a pre-defined QoS class identifier (QCI) value to each service request. In 5G, the packet delay budget and the packet loss rate are an integral part of the 5QI (5G QoS Identifier). DeepSlice will also observe what time and day of the week the request is received in the system [19].

To understand the variability in the results the third dataset utilized is a multimodal electroencephalogram (EEG) data set for the analysis of human affective states [21]. The EEG and peripheral physiological signals of 32 participants were recorded as each watched 40 one-minute-long excerpts of music videos. Participants rated each video in terms of the levels of arousal, valence, like/dislike, dominance, and familiarity. All three datasets were included for an experimental study to perform classification based on our proposed technique.

5 Results

The experimental results were performed using Intel(R) Xeon(R) CPU Model Name with 2.30 GHz frequency having 2 number of CPU Cores. The CPU Family Haswell with RAM 12 GB (upgrade to 26.75 GB) has been utilized for the implementation of the model on the three datasets (Sect. 4). From Eqs. 4 to 6, updating the weight and adding the right neuron in the repository using ESNN depends on parameters like c, mod, and SIM considered in the study.

To understand the best of the model, we performed various experiments. An attempt has been made to tune these parameters such that the accuracy of the model is improved.

We have applied the model and tuned these parameters c, mod, and SIM for the telecom dataset. The results of the same are given in Fig. 2. The variations in the accuracy with the varied value of parameters are shown. We can see that from Fig. 2(a), if the mod has lower values ranging in between 0.3 to 0.4 it gives good accuracy and provides comparable accuracy with values ranging between 0.6 to 0.7. This suggests that the modulation factors involved in updating the weight cannot be very less or exceedingly high. A moderate modulating factor achieves a weight that finally improves the network and achieves better accuracy. If we look at the graph of parameter c in Fig. 2(b), we can conclude that if the fraction lies in the range of 0.5 to 0.8 will be proportional to the threshold applied at the output neuron that will give better results. For Fig. 2(c) we can specify that the parameter SIM which tells us about the similarity between the two output neurons generated, should be less, less the similarity, better the learning, more accurately we can classify the data.

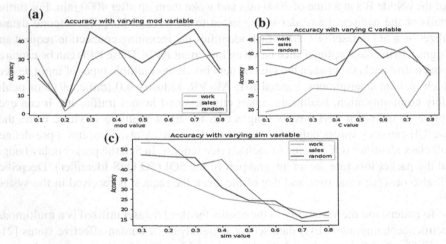

Fig. 2. ESNN Parameters *mod, c, SIM* in range (0,1) are tuned. (a) Accuracy with varying *mod*. (b) Accuracy with varying *C*. (c) Accuracy with varying *SIM*

The ESNN model results have been compared with Long Short-Term Memory (LSTM) [22] and the conventional SNN to demonstrate the improvement in the performance of the system at various levels using different parameters. The LSTM also belongs to the family of Neural Network, where instead of having only forward connections it has feedback connections. The LSTM has been developed to be used on time series data and to solve the problem of vanishing gradient descent. In LSTM, each block is made up of memory cells with input, forget, and output gates. The gates allow the memory cells to store and access information for longer periods to improve performance. The training time taken by the telecom dataset to perform LSTM is 208.57 s for the work-professional dataset, 420.52 for the sales-professional dataset, and 323.75 s for the random waypoint. Whereas, the ESNN is much faster taking 12.88 s for the work-professional dataset, 27.43 for the sales-professional dataset, and 25.09 s for the random waypoint. ESNN takes comparatively very less time than LSTM to complete the training. LSTM requires more time due to its architecture, whereas ESNN completes this task faster. Not only faster training times, but also better accuracy on the dataset has been achieved using ESNN.

A comparison of the accuracy of the ESNN model with the LSTM model is given in Table 1. Not only training and the testing of the similar variant of mobility dataset, but we have also experimented with cross-training and testing within the mobility dataset, to understand the working of the model. For example, the ESNN model is trained with work professional telecom data and tested on sales professional telecom data and vice versa. The aim is to reach a single universal model which can perform well on similar datasets collected at or for different scenarios.

Table 1. Accuracy of the ESNN model, compared with LSTM.

Training	Dataset	work-professional	Sales-professional	random waypoint
Testing with ESNN	Work-professional	99.31	73.42	73.09
	Sales-professional	99.25	99.95	99.63
	random waypoint	99.27	99.97	99.97
Testing with LSTM	Work-professional	43.95	8.52	9.23
	sales-professional	35.56	65.34	63.75
	random waypoint	34.53	65.81	66.14

A comparative analysis has also been performed by considering the best 5 or top 5 results while testing. In the telecom dataset, there is a possibility that if more than one service devices are present in that location then our mobile can connect to any of the service devices. Since the output can be any of the service devices, we need to consider more than one output neuron, so we are using *top k* to overcome this problem. Figure 3 shows the results achieved using various combinations of training and testing models. A comparison is performed using LSTM and ESNN approaches. While testing using LSTM, a prediction on the probability of each class using 'model.predict_proba' is made and we have considered top-5 classes (classes with top-5 probabilities) to check for true class. If the true class is present in these 5 classes, then the prediction is considered true. However, while testing with ESNN we considered the first 5 neurons that are fired (whose PSP value reached threshold value γi). If a true is present in any of the classes of these 5 neurons, our prediction is true. More experimental analysis has been performed on the telecom dataset. As shown in Fig. 3, the y-axis represents the accuracy achieved by each model. Here,

- Model-1 represents training with work-professional data and testing on work-professional data.
- Model-2 represents training with work-professional data and testing on sales-professional data.

- Model-3 represents training with work-professional data and testing on random-waypoint data.
- Model-4 represents training with sales-professional data and testing on work-professional data.
- Model-5 training with sales-professional data and testing on sales-professional data.
- Model-6 represents training with sales-professional data and testing on random-waypoint data.

From Fig. 3 we can conclude that using top-5 max fired PSP neurons for testing using ESNN (in blue line) achieves the highest results among all types of models. Considering top-5 LSTM (in yellow line) probabilities for testing represents a comparable result for all types of models. While comparing between various models, model -5 and model-6 give us good accuracy results. This suggests that when trained with one type of data and tested with the same data gives us better accuracy.

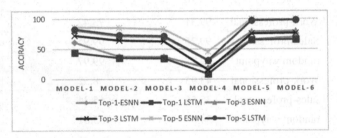

Fig. 3. Accuracy achieved on different combinations of the model whilst considering the testing on top-k probabilities.

However, model-6 where training is performed on sales professional data and tested on random waypoint has also achieved better accuracy. When you compare it with the result of model-3, model-6 shows promising results that is because the sales professional data has data of various locations a salesman visits whereas in the work professional dataset a specific set of locations are only captured. When these datasets are tested with a random waypoint where locations are random hence the model trained on random locations of sales professionals achieves better results.

We have also performed experiments and calculated the accuracy of the model by initially training it on sales professional data and testing it on work professional data. Then we add a few samples from work professional data to sales professional data and trained it again and tested on work professional. We keep on adding the samples of work professionals and train the model and then test. The results achieved are shown in Fig. 4(a) for both LSTM and ESNN models. A similar experiment is performed by adding samples of work professional dataset on random waypoint dataset and the results are shown in Fig. 4(b). In Fig. 4, the y-axis gives the accuracy achieved and the x-axis represents the number of work professional datasets added while training the model. An attempt is made to compare the improved ESNN model with the state-of-the-art neural network and it is shown in Table 2. On comparing ANN, SNN, and RESNN with our improved ESNN, it is explicit that the ESNN outperforms all other approaches. The experiment results have also been computed by training the 5G data (second dataset) using LSTM an accuracy of 96.14% has been achieved in 289.05 s, training with RESNN an accuracy of 82.89% has been achieved with a training time of 588.31 s. Whereas on using the ESNN model, an accuracy of 100% is achieved with a training time of 89.35 s.

Table 2. Accuracy and training time for ANN (top-5), SNN (top-5), ESNN and RESNN when training on work-professional and testing on work-professional dataset.

Model (top-5)	Training time	Accuracy
ANN	208.57	83.27
SNN	222.55	83.87
ESNN	12.88	99.31
RESNN	80.28	99.05

We can compare and say that ESNN has not only shown efficient performance in terms of accuracy but also in terms of training time. We have also used the spatiotemporal data i.e., EEG data for the implementation in this paper. The EEG data used is recorded while presenting a set of music video stimuli to the users [21]. EEG data were recorded from 32 channels or electrodes placed on the scalp following the 10–20 international placement protocol. The total recording was done for 2 min where the stimuli were presented to the user on a PC screen. Before performing experiments, the participants were trained and consented [21]. The video presented to participants were designed in such a way that they induce different types of emotions in the participants' brains and then can be captured by the EEG. These recorded signals after being downsampled to 128 Hz, are utilized for implementing ESNN. We have compared the results with LSTM, where LSTM achieved an accuracy of 57.81 with a training time of 295.81 s and ESNN has achieved an accuracy of 99.6 but has taken more time of EEG data of 1234.16 s due to large data samples and multiple features of EEG data during training. On analyzing the three datasets it was found that the proposed ESNN model with tuned parameters and top-5 fired neurons achieve better accuracy for all three datasets as compared to LSTM.

6 Conclusion

The paper presented a detailed experimental review of Evolving Spiking Neural Networks as a unique classifier. We tried to understand how ESNN architecture behaves with different kinds of inputs. The ESNN architecture was fed with three different datasets: one mobility data having locations and time details of various professionals, a 5G network slicing data, and EEG data recorded for emotion analysis for 32 users. We had also performed experimental analysis by tuning *c, mod,* and *SIM* parameters of the ESNN model. After using the tuned parameters, it was found that the ESNN as a classifier gives more accurate results. The ESNN results have also been compared with the most used recurrent neural network i.e., LSTM. We had also used the GRF encoding scheme

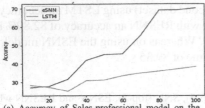
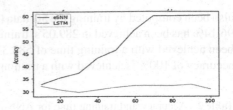

(a) Accuracy of Sales-professional model on the work-professional data by adding samples from the work-professional data. The Sales professional model (model trained on sales professional dataset) is tested on work-professional dataset by adding samples from work-professional dataset to check which model can learn in a smaller number of epochs.

(b) Accuracy of the random waypoint model on work-professional data by adding samples from work-professional data. The random way point professional model (model trained on random-waypoint dataset) is tested on random waypoint dataset by adding samples from random waypoint dataset to check which model can learn in a smaller number of epochs.

Fig. 4. (a) and (b) Accuracy of ESNN model on various combinations of the telecom datasets for the small number of samples and epochs.

which has significantly improved the time of the model. When compared to RESNN, ESNN takes very less time because of the simple spike generation process, whereas RESNN has a complex reservoir structure that requires more computational time. With these improvements and comparative analysis, we concluded ESNN, the third-generation neural network is going to bring the dawn in classification algorithms. With more experiments and improvements one can achieve better and efficient performance using ESNN architecture for different tasks.

References

1. Maass, W.: Networks of spiking neurons: the third generation of neural network models. Neural Netw. **10**(9), 1659–1671 (1997)
2. Kasabov, N.K.: Evolving spiking neural networks. In: Time-space, spiking neural networks and brain-inspired artificial intelligence, pp. 169–199. Springer, Heidelberg (2019)
3. Kasabov, N.K.: ECOS: Evolving Connectionist Systems and the ECO Learning Paradigm. Iconip, vol. 98 (1998). s
4. Schliebs, S., Nuntalid, N., Kasabov, N.: Towards spatio-temporal pattern recognition using evolving spiking neural networks. In: Wong, K.W., Mendis, B.S.U., Bouzerdoum, A. (eds.) ICONIP 2010. LNCS, vol. 6443, pp. 163–170. Springer, Heidelberg (2010). https://doi.org/10.1007/978-3-642-17537-4_21
5. Lobo, J.L., Laña, I., Del Ser, J., Bilbao, M.N., Kasabov, N.: Evolving spiking neural networks for online learning over drifting data streams. Neural Netw. **108**, 1–19 (2018)
6. Schliebs, S., Hamed, H.N.A., Kasabov, N.: Reservoir-based evolving spiking neural network for spatio-temporal pattern recognition. In the International Conference on Neural Information Processing, pp. 160–168. Springer, Heidelberg. https://doi.org/10.1007/978-3-642-24958-7_19
7. Schliebs, S., Defoin-Platel, M., Kasabov, N.: Integrated feature and parameter optimization for an evolving spiking neural network. In: Köppen, M., Kasabov, N., Coghill, G. (eds.) ICONIP 2008. LNCS, vol. 5506, pp. 1229–1236. Springer, Heidelberg (2009). https://doi.org/10.1007/978-3-642-02490-0_149

8. Hamed, H.N.A., Kasabov, N., Shamsuddin, S.M.: Integrated feature selection and parameter optimization for evolving spiking neural networks using quantum inspired particle swarm optimization. In: 2009 International Conference of Soft Computing and Pattern Recognition, pp. 695–698. IEEE, December 2009

9. Kasabov, N., Dhoble, K., Nuntalid, N., Indiveri, G.: On-line spatio-and spectro-temporal pattern recognition with evolving spiking neural networks utilizing integrated rank order-and spike-time learning. Neural Networks (2011)

10. Dhoble, K.: Spatio-/spectro-temporal pattern recognition using evolving probabilistic spiking neural networks (Doctoral dissertation, Auckland University of Technology) (2013)

11. Wysoski, S.G., Benuskova, L., Kasabov, N.: On-line learning with structural adaptation in a network of spiking neurons for visual pattern recognition. In: Kollias, S.D., Stafylopatis, A., Duch, W., Oja, E. (eds.) ICANN 2006. LNCS, vol. 4131, pp. 61–70. Springer, Heidelberg (2006). https://doi.org/10.1007/11840817_7

12. Wysoski, S.G., Benuskova, L., Kasabov, N.: Fast and adaptive network of spiking neurons for multi-view visual pattern recognition. Neurocomputing **71**(13–15), 2563–2575 (2008)

13. Wysoski, S.G., Benuskova, L., Kasabov, N.: Adaptive learning procedure for a network of spiking neurons and visual pattern recognition. In: Blanc-Talon, J., Philips, W., Popescu, D., Scheunders, P. (eds.) ACIVS 2006. LNCS, vol. 4179, pp. 1133–1142. Springer, Heidelberg (2006). https://doi.org/10.1007/11864349_103

14. Wysoski, S.G., Benuskova, L., Kasabov, N.: Text-independent speaker authentication with spiking neural networks. In: de Sá, J.M., Alexandre, L.A., Duch, W., Mandic, D. (eds.) ICANN 2007. LNCS, vol. 4669, pp. 758–767. Springer, Heidelberg (2007). https://doi.org/10.1007/978-3-540-74695-9_78

15. Wysoski, S.G., Benuskova, L., Kasabov, N.: Adaptive spiking neural networks for audiovisual pattern recognition. In: International Conference on Neural Information Processing, pp. 406–415. Springer, Heidelberg (2017). https://doi.org/10.1007/978-3-540-69162-4_42

16. Soltic, S., Wysoski, S.G., Kasabov, N.K.: Evolving spiking neural networks for taste recognition. In 2008 IEEE International Joint Conference on Neural Networks (IEEE World Congress on Computational Intelligence) (pp. 2091–2097). IEEE, June 2008

17. Soltic, S., Kasabov, N.: Knowledge extraction from evolving spiking neural networks with rank order population coding. Int. J. Neural Syst. **20**(06), 437–445 (2010)

18. Perepu, S.K., Dey, K.: CDDM: a method to detect and handle Concept Drift in Dynamic Mobility Model for seamless 5G Services. In: 2020 IEEE Globecom Workshops (GC Wkshps), pp. 1–6. IEEE, December 2020

19. Thantharate, A., Paropkari, R., Walunj, V., Beard, C.: DeepSlice: a deep learning approach towards an efficient and reliable network slicing in 5G networks. In: 2019 IEEE 10th Annual Ubiquitous Computing, Electronics & Mobile Communication Conference (UEMCON), New York City, NY, USA, 2019, pp. 0762–0767, doi: https://doi.org/10.1109/UEMCON47517.2019.8993066

20. Thantharate, A., Paropkari, R., Walunj, V., Beard, C., Kankariya, P.: Secure5G: a deep learning framework towards a secure network slicing in 5g and beyond. In: 2020 10th Annual Computing and Communication Workshop and Conference (CCWC), Las Vegas, NV, USA, 2020, pp. 0852-0857. doi: https://doi.org/10.1109/CCWC47524.2020.9031158

21. Koelstra, S., et al.: DEAP: a database for emotion analysis; using physiological signals. IEEE Trans. Affective Comput. **3**(1), 18–31 (2012). https://doi.org/10.1109/T-AFFC.2011.15

22. Hochreiter, S., Schmidhuber, J.: Long short-term memory. Neural Comput. **9**(8), 1735–1780 (1997)

Elements of TinyML on Constrained Resource Hardware

Vasileios Tsoukas, Anargyros Gkogkidis, and Athanasios Kakarountas(✉)

Intelligent Systems Laboratory, Department of Computer Science and Biomedical Informatics, University of Thessaly, Lamia, Greece
{vtsoukas,agkogkidis,kakarountas}@uth.gr

Abstract. The next phase of intelligent computing could be entirely reliant on the Internet of Things (IoT). The IoT is critical in changing industries into smarter entities capable of providing high-quality services and products. The widespread adoption of IoT devices raises numerous issues concerning the privacy and security of data gathered and retained by these services. This concern increases exponentially when such data is generated by healthcare applications. To develop genuinely intelligent devices, data must be transferred to the cloud for processing due to the computationally costly nature of current Neural Network implementations. Tiny Machine Learning (TinyML) is a new technology that has been presented by the scientific community as a means of developing autonomous and secure devices that can gather, process, and provide output without transferring data to remote third party organizations. This work presents three distinct TinyML applications to cope with the aforementioned issues and open the road for intelligent machines that provide tailored results to their users.

Keywords: Internet of Things · TinyML · Neural networks · Constrained hardware · Emerging technologies

1 Introduction

The Internet of Things, alternatively referred to as the Internet of Everything, is an emerging technology described as a network of interconnected machines and devices that collect data and provide valuable information for various sectors. The technology is widely recognized as a critical area of future innovation and is attracting significant attention from a diverse range of industries and researchers. Nowadays, competition forces businesses and individuals to experiment and evolve in order to overcome challenges, reduce cost, and produce high-quality products. Despite technology's high potential, the advantages may not be apparent initially, resulting in increased investment costs requiring careful evaluation from experts to ensure that business resources are used wisely [1].

Kevin Ashton originated the term "Internet of Things" in 1999 in reference to supply chain management [2,3]. Since then, the technology into consideration has

M. Singh et al. (Eds.): ICACDS 2022, CCIS 1614, pp. 316–331, 2022.
https://doi.org/10.1007/978-3-031-12641-3_26

been applied to various domains such as Automotive, Healthcare, Manufacturing, Environment Monitoring, and Agriculture [4]. According to the International Telecommunication Union (ITU), the IoT architecture is quite similar to the Open Systems Interconnection (OSI) reference model in network and data communication and consists of the following layers: The Sensing Layer, The Access Layer, The Network Layer, The Middleware Layer and finally, The Application Layer. On the contrary, the IoT Forum proposed that the architecture can be categorized into three different types, including Processors, Transpiration, and Applications [5]. A typical IoT system may consist of radio frequency identification (RFID), wireless sensor networks (WSN), middleware, cloud computing, and IoT application software [1]. The RFID is capable of identifying data and cooperating with WSN to monitor and track attributes of products such as the temperature [5]. Developers exploit the middleware to help them connect devices and data and achieve communication that will result in a valuable output. The Cloud works as data storage and also as a decision-maker that analyzes data, processes them, and extracts results. Finally, the IoT applications act as a bridge that enables device-to-device and human-to-device interactions [1].

Monitoring environmental data or machine data in a time series format that can be processed and then used to improve the quality of a product, predict machine failures or future outcomes is possibly the most known use case of an IoT system. Additionally, two of the most common environmental parameters measured by a system, such as the aforementioned, are temperature and humidity. An example of one or more IoT devices can be found in a Data Center (DC) room. A stable temperature management system is one of the most critical components of a DC. An air conditioning system serves as the foundation and ensures the DC's operation stability. This system is composed primarily of five components: an environmental parameter monitor, an air conditioning system, a ventilation system, a network communication system, and a multilevel intelligent temperature control system. Numerous high-precision devices work within a DC, allowing the generation of exceptionally concentrated heat loads. However, information technology devices are frequently susceptible to temperature and humidity shifts. A fast temperature change can have a detrimental effect on working devices and possibly result in economic loss. As a result, the most critical and fundamental requirement for a DC is for temperature and humidity to remain steady, which must be maintained by continuous environmental parameter monitoring through IoT devices [6]. Smart Agriculture is another sector trending the last decade and utilizes the technology of IoT. Devices could be used for environmental monitoring and provide real-time data such as the temperature, humidity and other qualitative characteristics, to inform the farmer or even control an entire greenhouse [7,8].

The first question that comes to mind is if that technology is flawless and capable of overcoming other known issues met in similar fields such as security. IoT comes with several challenges. The first one has to do with operational costs. Many devices must be connected to send data to each other and the Cloud. There is a high initial cost for buying those devices and an even higher operational cost for maintenance. Additionally, storing a vast amount of data and processing it on a cloud provider and using it as means for processing data to produce valuable information can highly increase the overall cost in a continuous manner. As mentioned above, data must be constantly transmitted between devices via the Cloud. This is achieved by utilizing wireless communications and protocols such as WIFI, Bluetooth, Zigbee, etc. The aforementioned wireless technologies are susceptible to various attacks such as Denial of service (DoS) attacks [9,10], Rogue Access Points threats [11,12], and Man in the middle (MITM) attacks [13–15].

Since IoT devices are used as smart devices able to provide intelligence and transform sectors to competent entities, e.g., smart agriculture and smart homes, it is worth exploring how smart those devices are and what intelligence can provide to conventional machines. A typical IoT device utilizes logical operations in the context of "IF this is TRUE or has the value x, THEN do that." The programmer is competent in choosing thresholds and hardcode the commands needed to be executed when the application is fed with a value that matches the thresholds described above. For example, if a machine produces three products per hour operates at $65°$ C and the accelerometer's reads for the machine's movement are between the initial set threshold then the machine operates as it is meant to do. If this isn't the case and a temperature read is above the required overall temperature, then the machine might be damaged or is about to fail. There are some cases, where the above statements might be factual and occasions that, although the readings are above the initial set threshold operates fine. This might happen due to the fact that the IoT device does not take into consideration the temperature and humidity of the building, other machines working next to it and producing heat, or other factors that were not forecasted in the initial configuration.

Deep Learning (DL) is the technology capable of making IoT devices intelligent able to provide tailored results by constantly learning from new data. Deep learning enables computational models built of numerous layers and nodes to understand multiple-level representations of data. These techniques have a significantly advanced state of the art in speech recognition, object recognition, object detection, and a variety of other disciplines, including machine monitoring, smart wearables, and security. DL uncovers intricate structures in big data sets by employing the backpropagation algorithm to advise how a machine's internal parameters should be changed in order to compute the representation in each layer from the previous layer's representation [16]. In the example mentioned above, with the help of DL, a neural network specifically designed and trained for each machine could be created and provide valuable insights by detecting outliers in time series data, utilizing anomaly detection. Due to the complexity of the DL models, DL algorithms, and the number of parameters required to utilize the technology of DL, IoT devices offload those heavy tasks to the Cloud. This results in several significant drawbacks that are critical in domains such as healthcare and the industry. Some of the most fundamental issues include information security and privacy, latency and bandwidth concerns, and the storage of massive amounts of raw data. Additionally, processing on Cloud premises is essential due to the high computational needs of Big Data processing and analysis. Even when utilizing expensive Cloud infrastructures, latency increases as a result of network traffic and data broadcasting. To overcome the aforementioned obstacles, we must move all processing jobs closer to the network's edge, which is the IoT device, by lowering the overall computational requirements of Neural Networks (NNs) and algorithms. This will result in an autonomous device capable of collecting data, processing it in real-time, and exhibiting the intended operational behavior, such as informing the industry of machine conditions or alerting employees in emergencies such as gas leaking, smoke detection, or machine failure.

This work presents the challenges of IoT devices as edge nodes, and introduces the TinyML perspective for creating autonomous and safe devices. The paper is organized as follows; Sect. 2 presents previous work regarding IoT devices. Section 3 introduces the technology of TinyML. Section 4 discusses how it is possible to embed Machine Learning into constrained hardware through optimization techniques. Section 5 introduces the TinyML applications developed. Finally, Sect. 6 concludes on the works findings and discusses future plans.

2   Previous Work

The authors of [17] examined the viability of a home refrigerator forecasting temperature using data from both internal and external sensors. The subject is discussed in terms of its potential applications and the use of various techniques, ranging from simple linear correlations to ARIMA models. They analyzed the precision and computing cost using real-world data from a refrigerator. The results indicate that modest average errors of up to 0.09° C are possible. The authors also discuss the possibility of running the models into constrained hardware such as an Arduino board and conclude that the results revealed that it is going to be feasible with the recent advances in hardware.

A green data center air conditioning system aided by cloud approaches is presented in paper [6]. The system is composed of two subsystems: a data center air conditioning system and a cloud management platform. The data center air conditioning system encompasses environmental monitoring, air conditioning, ventilation, and temperature control, whereas the cloud platform supports upper-layer applications with data storage and analysis. Additionally, the detailed design and implementation are described, including the temperature control dispatch algorithm, the sensor network's topological structure, and the framework for the environment monitoring node. Finally, a study is conducted to observe whether the proposed technology can considerably reduce data center energy usage without sacrificing cooling performance.

Another study [18] intends to use growing technology, such as IoT and smart agriculture, through the use of automation. Environmental monitoring is critical for increasing the production of efficient crops. The purpose of this paper is to demonstrate how to monitor temperature and humidity in an agricultural field using sensors based on the CC3200 single chip. The camera is interfaced with the CC3200 to shoot images and transfer them via MMS to the farmers' mobile devices via Wi-Fi.

A team [19] created an MQTT (Message Queue Telemetry Transportation) broker using Amazon Web Service (AWS). The MQTT broker has been used as a platform for the IoT applications such as monitoring and controlling room temperatures, as well as sensing, alarming, and suppressing fires. The IoT end device, based on Arduino, was utilized to connect sensors and actuators to the platform over a Wi-Fi channel. They also developed a scenario for smart homes and implemented IoT messages that met the scenario's requirements. Additionally, they constructed the intelligent system in hardware and software and validated its operation. They demonstrated that MQTT and AWS are both suitable technological platforms for modest IoT business applications.

The authors of [20] introduced a highly scalable intelligent system for managing and monitoring greenhouse temperature utilizing IoT technology. The system's primary purpose is to monitor the greenhouse environment and regulate the internal temperature in order to conserve energy while maintaining productive conditions. A Petri Nets (PN) model is utilized to monitor the greenhouse environment while also creating the appropriate reference temperature for later transmission to a temperature regulation block. The second purpose is to develop a scalable energy-efficient system design that can manage vast amounts of IoT big data collected from sensors via a dynamic graph data model for future analysis and prediction of production, crop growth rate, energy consumption, and other related issues.

Another study regarding the use of the IoT to monitor the temperature and humidity of a data center was conducted by Rahman et al. [21]. The authors utilized a simple monitoring system to identify the relationship and difference between temperature and humidity at various measurement points. The proposed framework was used to construct a temperature and humidity monitoring system, which was installed at Politeknik Muadzam Shah's data center, where values were captured and forwarded to an AT&T M2X IoT platform for storage. The data was then retrieved and analyzed, revealing a considerable difference in temperature and humidity detected at various sites. Additionally, the monitoring system was successful in identifying extreme variations in temperature and humidity and automatically notifying IT staff via e-mail, SMS, and mobile push notification for further action.

The authors of work [22] present a novel smart IoT-based agriculture stick that will aid farmers in obtaining real-time data (Temperature, Soil Moisture) for effective environment monitoring, enabling them to practice smart farming and boost their overall yield and product quality. The Agriculture stick described in this study is equipped with Arduino technology, a breadboard loaded with numerous sensors, and a live data feed retrieved via Thingsspeak.com. The suggested product has been validated on Live Agriculture Fields and has a high accuracy of over 98% in data flows.

In the field of healthcare, a research work [23] establishes a cost-effective solution for an embedded device capable of monitoring newborn infants in the incubator in real-time. It enables early detection of potentially life-threatening situations and ensures the infant's safety. Numerous existing medical technology companies (small and medium-sized) may be hesitant to adopt the most advanced technologies since their maintenance may be prohibitively expensive. On the other hand, the larger medical firms implementing them are cost-effective, although the average person cannot. Thus, the primary purpose of our research is to overcome these disadvantages and deliver environmentally friendly services to the general public.

In another work [24], a team described a system for monitoring and automatically correcting an aquaponics setup in a temperature-controlled greenhouse via an Android cellphone and the IoT. The system acquires real-time data from the light intensity sensor, as well as the air temperature and humidity sensors. Additionally, it comprises monitoring the pH and temperature of the system's recirculating water as well as the canopy area of the plant. If the acquired data is outside the threshold range, the system immediately triggers the correction devices, including grow lights, exhaust and inlet fans, evaporative cooler, aerator, and peristaltic buffer device, to correct and restore the system to regular operation. Internet remote access entails the successful wireless transmission and reception of data reports in real-time between the system and an Android unit via the Android application.

The authors of paper [25] assess and forecast temperature and humidity using IoT and machine learning's linear regression algorithm. They use the Message Queuing Telemetry Transport (MQTT) protocol to collect temperature and humidity data in several locations over a short period of time. As a result, they initialize the acquired data on the AWS cloud for five days. This data is stored in AWS and exported to a .csv file using Dynamo Database (DynamoDB). As a result, data is recorded, processed, and used to forecast temperature and humidity data using a linear regression approach in machine learning.

3   The Era of TinyML

As described before, sensors and logical operations are not enough to label an IoT device as intelligent, and those that employ machine learning models capture data and send it to the cloud for analysis. The motivation for transferring information to a different entity is process-dependent. Due to the increasing complexity of algorithms and machine learning models, they require significantly more computational resources and power than a compact IoT device can provide. This results in an enormous amount of data that the system is unable to store. Wireless technologies enable IoT devices to interact with other intelligent devices. Frequently, transmitted data is not secure, and the devices under consideration lack basic security safeguards.

TinyML is a comparatively recent developing technology that is gaining researchers' attention. The technology leverages meticulous hardware and software design to enable the deployment of Machine Learning (ML) models and DL algorithms on compact, low-cost, and energy-efficient devices. By using this new area, we may develop new services and solutions which do not require advanced hardware and address IoT device concerns including latency and bandwidth limitations. The Internet of Things devices will be used to gather, evaluate, and extract data. Due to the fact that this knowledge is not exchanged with other entities, the devices are more safe and private.

Additionally, the technology necessary to accomplish the tasks, microcontrollers, is believed to be incredibly energy efficient and low-power. It typically consumes a little over a milliwatt of energy and is capable of providing intelligence in a very short period of time. TinyML could be the field that significantly alters how developers approach developing innovative and secure home applications today. Notifying a user of a potential machine failure, crop disaster or as described above, monitoring newborn infants in the incubator, is critical, and there is no tolerance for latency or communication interruptions. These gadgets will perform real-time analysis and warn users in their homes or businesses without the need for data transfer, ushering in a new era of autonomous devices powered solely by a battery.

4   Optimization and Model Reduction

A typical TinyML operation is composed of three distinct procedures. The first step is to train the machine learning model on a machine with sufficient computational capability. The following step optimizes it using widely recognized optimization techniques or frameworks. Finally, the simplified TinyML design is ready to be implemented in the microcontroller in the final phase. The aforementioned operation could also be achieved by only utilizing a framework such as Tensorflow [26] or the Edge Impulse [27] framework, which is capable of collecting data, transforming it to an appropriate form, and providing the tools to train the neural network. Additionally, it creates the libraries and the firmware needed to embed the freshly trained model into the development board. In this section, the optimization methods and frameworks mentioned above are briefly discussed.

To train, analyze, and store NNs, powerful scientific workstations equipped with multiple high-end Graphics Processing Units (GPUs) are required. Additionally, as IoT devices evolve, a new requirement arises that encourages engineers to reduce and improve the networks described above in order to make them suitable with constrained hardware with limited resources. Likewise, when networks and models grow in size to achieve higher performance, even certain high-end machines are unable to accommodate them, resulting in a slew of challenges for the computer science community and also how we, as academics, may supply solutions through improved optimization strategies.

4.1   Optimization Techniques

Network pruning is one of the earliest and most widely used optimization techniques. This method eliminates connections and neurons with weights less than a predefined threshold in order to increase efficiency and minimize a network's computing requirements without sacrificing a substantial degree of accuracy [28–30].

The values of a network trained on powerful GPUs are stored in 32-bit Floating-Point (FP) single precision. To achieve faster inference and reduced processing requirements, researchers have concentrated on training networks using one or two-bit representations that can run on different types of hardware, such as microcontrollers. Quantization is a common approach for converting FP32 data to 8-bit representations [31,32]. This optimization method has an effect on overall accuracy and demands close scrutiny. The following works demonstrate the technique into consideration [33–35].

4.2   Frameworks

TensorFlow Lite [36] was built by Google as a comprehensive set of tools for developing TensorFlow NN models on mobile devices. It comprises two primary components: a converter and an interpreter. TensorFlow Lite Micro [37] is an adaptation of this framework designed to meet resource-constrained platforms' restrictions. The framework enables the transfer of deep learning models to embedded devices, considerably broadening the scope of machine learning, making it available to non-expensive devices, in the size of a coin, utilizing a few KB of memory and under 100 MHz of CPU clock speeds.

Edge Impulse (EI) is a platform that enables the integration of machine learning models on MCUs. One outstanding aspect of this framework is its ability to support a wide variety of devices, like the Arduino Nano 33 BLE Sense [38], the Raspberry Pi 4 [39], and the ST B-L475E-IOT01A [40]. It enables the user to gather data straight from the device, categorize it, and upload it to the cloud as a dataset. Edge Impulse includes prebuilt machine learning blocks that may be trained on datasets and deeply customized to meet the project's requirements. Additionally, users can obtain live categorization data while testing the model's performance on devices to ensure that the model operates as intended in real-time.

5   The System

Three distinct applications were developed to explore the feasibility of executing ML models and NNs on constrained hardware. The first one, which was based on the EI platform, is capable of monitoring and classifying temperatures in real-time. Similarly, the second application was constructed using EI and was designed to provide an anomaly detection system capable of notifying the user when temperature values change. The final application is capable of forecasting weather conditions from the device's ambient sensors. It is not flawless and cannot be described as a precise and accurate product. The purpose of designing this application was to determine the feasibility of running the aforementioned service on constrained hardware. A new column was added to the dataset to denote the weather condition at the time of the readings. Python 3.9 was used to write the code, and the TensorFlow Lite for microcontrollers framework was utilized to implement it. The development began on Google Collaborate. Following the model's construction and training, the TensorFlow converter was used to convert the model to a format compatible with the development board. The transformed model was downloaded and stored in a newly created folder for local development. This time, Visual Studio Code and the Platformio extension were used. After that, C++ code was built to activate the necessary sensors, initiate serial communication, and start the machine learning model. The application prints environmental values and a weather forecasting message.

Fig. 1. The Arduino Nano 33 BLE Sense in a 3d-printed case

5.1   Hardware

The Arduino Nano 33 BLE Sense is a well-known and widely used development board for TinyML applications. We chose the aforementioned board since it is entirely compatible with both the TensorFlow Lite for microcontrollers framework and the Edge Impulse platform, as well as having a temperature and humidity sensor. The board is based on the Nordic Semiconductor nRF52840, which contains a 32-bit ARM® CortexTM-M4 CPU running at 64 MHz, 256 KB SRAM, and 1 MB of flash memory. It operates at 3.3 V and has a 45 × 18 mm size factor, making it one of the smallest. It includes a motion, vibration, and orientation sensor, as well as sensors for gesture recognition, color, brightness, proximity identification, temperature, humidity, and pressure monitoring. Additionally, the board includes an integrated digital microphone, a communications chipset capable of Bluetooth and Bluetooth Low Energy data transmission, and 14 digital I/O pins. The board's official datasheet [41] contains additional information. Figure 1 depicts the Arduino Nano 33 BLE Sense in a 3d-printed case [46].

5.2 Dataset

The data was collected during a seven-day period. For eight hours every day, the device monitored a room. The lowest temperature was seven degrees and the highest was twenty-seven degrees. We desired a range of measurements in order to generate four distinct classes and hence complicate the anomaly detection system. Additionally, a second dataset comprising temperature, humidity, and pressure parameters was developed. Further, this information includes a weather description. There are five distinct weather classifications, including "Sunny," "Cloudy," and "Raining." It is far from a perfect dataset, as other data types are required to make a weather forecasting dataset. It was created to investigate the viability of incorporating Linear Regression into hardware with constraints. The Arduino Nano 33 BLE Sense and its integrated sensor for temperature, humidity, and pressure sensing were used to perform all of the monitoring. The dataset was built by extracting data into a.csv file and also by utilizing Edge Impulse's feed forwarder tool to transfer the data to the online platform.

5.3 Applications

Application 1: Temperature Classification. The goal of the initial application was to construct a simple device capable of monitoring and classifying the temperatures in a room in real time. Each class represents a distinct circumstance and warns the user appropriately by sending messages to the user via the LED light on the Arduino Nano 33 BLE Sense and the board's communication chipset. The application operates similarly to a standard IoT application, except that instead of alerting the user when the temperature hits a predefined threshold, it may be educated in the room to produce more personalized results. We used the EI platform for this application. The first step was to download and flash the board with the needed firmware from EI. The Arduino HTS221 library was then used to obtain the readings, along with the Arduino IDE. To collect data for various temperatures, the device was connected to EI using the feed forwarder tool. Following data cleansing, it was labeled to represent the four distinct dataset classifications. Next, an impulse was created for the time series data, which was processed and classified using the spectral analysis block in conjunction with Keras. Following that, the neural network setup and architecture were selected, and the model was evaluated using EI's model testing tab. Finally, from the EI the platform, the impulse model was converted into optimal source code for the Arduino in order to deploy the model. The final testing was carried out in the Arduino IDE while the model was running on the development platform. During the testing, the onboard BLE module was used, in order to notify the user on his/hers mobile phone.

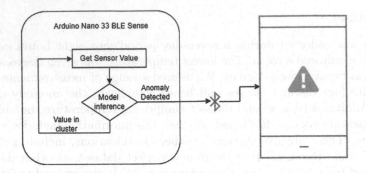

Fig. 2. Flowchart of the second application, notifying user when an anomaly is detected

Application 2: Anomaly Detection. The second application was created to provide an anomaly detection system capable of recognizing and notifying the user of outliers in temperature readings. Additionally, this application was created using the EI platform. The required temperatures were labeled as normal data following data gathering and cleaning, as depicted in Fig. 3. This time, the program was designed for time series data, with the processing block performing spectral analysis and the learning block performing anomaly detection with K-Means. After converting the impulse to optimized source code, the model was tested in EI's model testing tab and deployed to Nano 33 Sense. Utilizng the BLE technology the user receives an alert when an anomaly is detected, as depicted in Fig. 2. Finally, additional experiments were conducted to identify outliers in new data using the Arduino IDE.

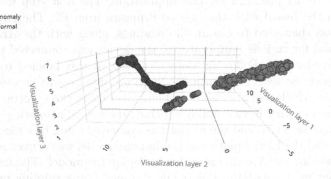

Fig. 3. Feature visualization of the training set, using the feature explorer before training for anomaly detection

Application 3: Weather Forecasting. The final application is capable of forecasting weather conditions from the device's ambient sensors. It is not flawless and cannot be described as a precise and accurate product. The purpose of designing this application was to determine the feasibility of running the afore-mentioned service on constrained hardware. A new column was added to the dataset to denote the weather condition at the time of the readings. Python 3.9 was used to write the code, and the TensorFlow Lite (TF Lite) for micro-controllers framework was utilized to implement it. The development started utilizing the Google Collaborate platform. Following the model's construction and training, the TensorFlow converter was used to convert the model to a for-mat compatible with the development board. The transformed model was down-loaded and stored in a newly created folder for local development. This time, Visual Studio Code and the Platformio [42] extension were used. After that, C++ code was built to activate the necessary sensors, initiate serial communication, and start the machine learning model. The application prints environmental values and a weather forecasting message.

Table 1. Applications' types, approaches and evaluation metrics

App. number	Type	Approach	Framework	Accuracy/F1-Score
App 1	Classification	Classification - Keras	EI	Acc. 92%
App 2	Anomaly Detection	K-Means	EI	F1. 0.82
App 3	Prediction	Linear Regression	TF Lite	Acc. 67%

Table 1 provides a summary of the applications types developed for the pur-pose of this work, the approaches used, the utilized frameworks, and their accu-racy or F1-Score as validation metrics. The Laboratory's cloud infrastructure was used to evaluate the results. The infrastructure stated above is funded by the project "ParICT CENG: Enhancing ICT research infrastructure in Central Greece" and enables the processing of Big data from sources such as sensor readings and simulations.

6   Conclusion

The rapid adoption of IoT devices presents a slew of privacy and security con-cerns regarding the data gathered and maintained by these services. Due to the computationally intensive nature of existing Neural Network implementations, this data must be transmitted to the cloud for processing to produce truly intelli-gent devices. TinyML is a recent approach proposed by the scientific community to construct autonomous and secure devices capable of gathering, processing, and warning without uploading data to remote agencies.

This work presented three examples of TinyML applications. The first appli-cation was developed to produce a basic device capable of monitoring and clas-sifying the temperatures in a room in real-time. Each class represents a distinct circumstance and warns the user appropriately by sending messages to the user via the LED light on the Arduino Nano 33 BLE Sense and the board's commu-nication chipset. The second application was developed to provide an anomaly detection system capable of detecting and alerting the user to temperature measurement outliers. The last application is capable of forecasting weather conditions from the device's ambient sensors.

The findings are encouraging and demonstrate the viability of developing intelligent gadgets. There have been numerous attempts and works throughout the previous decades to optimize and push machine learning to the edge. Recent trends with platforms such as Edge Impulse and NXP [43], as well as frameworks such as TensorFlow Lite for Microcontrollers and Microsoft's ELL [44], demonstrate the community's concerted effort to bring machine learning models and neural networks to constrained hardware via a variety of optimization methods and techniques.

With our future plans, we will investigate and seek to overcome two of the most significant obstacles encountered in the TinyML technology. The first one is about data transit and data security. The second difficulty arises while attempting to retrain a model. Numerous scientific publications have presented a range of methods, for example, Ravaglia et al. work [45] with on-device continuous learning with quantized latent replays.

Acknowledgment. We acknowledge support of this work by the project "Par-ICT_CENG: Enhancing ICT research infrastructure in Central Greece to enable processing of Big data from sensor stream, multimedia content, and complex mathematical modeling and simulations" (MIS 5047244) which is implemented under the Action "Reinforcement of the Research and Innovation Infrastructure", funded by the Operational Programme "Competitiveness, Entrepreneurship and Innovation" (NSRF 2014–2020) and co-financed by Greece and the European Union (European Regional Development Fund).

References

1. Lee, I., Lee, K.: The Internet of Things (IoT): applications, investments, and challenges for enterprises. Bus. Horizons **58**, 431–440 (2015). https://doi.org/10.1016/j.bushor.2015.03.008
2. Gubbi, J., Buyya, R., Marusic, S., Palaniswami, M.: Internet of Things (IoT): a vision, architectural elements, and future directions. Future Gener. Comput. Syst. **29**, 1645–1660 (2013). https://doi.org/10.1016/j.future.2013.01.010
3. Ashton, K.: That "Internet of Things" Thing. http://www.itrco.jp/libraries/RFIDjournal-ThatInternetofThingsThing.pdf. Accessed 12 Feb 2022
4. Sundmaeker, H., et al.: Vision and Challenges for Realising the Internet of Things. European Commision (2010)
5. Atzori, L., Iera, A., Morabito, G.: The Internet of Things: A survey. Comput. Networks **54**, 2787–2805 (2010). https://doi.org/10.1016/j.comnet.2010.05.010
6. Liu, Q., Ma, Y., Alhussein, M., Zhang, Y., Peng, L.: Green data center with IoT sensing and cloud-assisted smart temperature control system. Comput. Networks **101**, 104–112 (2016). https://doi.org/10.1016/j.comnet.2015.11.024
7. Gondchawar, N., Kawitkar, R.: IoT based Smart Agriculture
8. Mohd Kassim, M.R., Mat, I., Harun, A.N.: Wireless Sensor Network in precision agriculture application. In: 2014 International Conference on Computer, Information and Telecommunication Systems (CITS), pp. 1–5 (2014). https://doi.org/10.1109/CITS.2014.6878963

9. What is a denial of service attack (DoS)? https://www.paloaltonetworks.com/cyberpedia/what-is-a-denial-of-service-attack-dos. Accessed 13 Feb 2022
10. Arış, A., Oktuğ, S.F., Yalçın, S.B.Ö.: Internet-of-Things security: Denial of service attacks. In: 2015 23nd Signal Processing and Communications Applications Conference (SIU), pp. 903–906 (2015). https://doi.org/10.1109/SIU.2015.7129976
11. Forensic Analysis and Security. https://securitytoday.com/articles/2018/05/01/forensic-analysis-and-security.aspx. Accessed 13 Feb 2022
12. Rai, S., Chukwuma, P., Cozart, R.: Security and Auditing of Smart Devices: Managing Proliferation of Confidential Data on Corporate and BYOD Devices. Auerbach Publications (2017)
13. Melamed, T.: An active man-in-the-middle attack on Bluetooth smart devices. Int. J. Safety Secur. Eng. 8, 200–211 (2018). https://doi.org/10.2495/SAFE-V8-N2-200-211
14. Bluetooth Bug Opens Devices to Man-in-the-Middle Attacks. https://threatpost.com/bluetooth-bug-mitm-attacks/159124/. Accessed 13 Feb 2022
15. Hajian, R., ZakeriKia, S., Erfani, S.H., Mirabi, M.: SHAPARAK: scalable healthcare authentication protocol with attack-resilience and anonymous key-agreement. Comput. Networks 183, 107567 (2020). https://doi.org/10.1016/j.comnet.2020.107567
16. LeCun, Y., Bengio, Y., Hinton, G.: Deep learning. Nature 521, 436–444 (2015). https://doi.org/10.1038/nature14539
17. Monteiro, P.L., Zanin, M., Menasalvas Ruiz, E., Pimentão, J., Alexandre da Costa Sousa, P.: Indoor Temperature Prediction in an IoT Scenario. Sensors 18, 3610 (2018). https://doi.org/10.3390/s18113610
18. Prathibha, S.R., Hongal, A., Jyothi, M.P.: IOT based monitoring system in smart agriculture. In: 2017 International Conference on Recent Advances in Electronics and Communication Technology (ICRAECT), pp. 81–84 (2017). https://doi.org/10.1109/ICRAECT.2017.52
19. Kang, D.-H., et al.: Room temperature control and fire alarm/suppression IoT service using MQTT on AWS. In: 2017 International Conference on Platform Technology and Service (PlatCon), pp. 1–5 (2017). https://doi.org/10.1109/PlatCon.2017.7883724
20. Subahi, A.F., Bouazza, K.E.: An intelligent IoT-based system design for controlling and monitoring greenhouse temperature. IEEE Access 8, 125488–125500 (2020). https://doi.org/10.1109/ACCESS.2020.3007955
21. Rahman, R.A., Hashim, U.R., Ahmad, S.: IoT based temperature and humidity monitoring framework. Bull. Electr. Eng. Inform. 9, 229–237 (2020). https://doi.org/10.11591/eei.v9i1.1557
22. Nayyar, A., Puri, V.: Smart farming: IoT based smart sensors agriculture stick for live temperature and moisture monitoring using Arduino, cloud computing & solar technology. Presented at the November 9 (2016). https://doi.org/10.1201/9781315364094-121
23. Ashish, B.: Temperature monitored IoT based smart incubator. In: 2017 International Conference on I-SMAC (IoT in Social, Mobile, Analytics and Cloud) (I-SMAC), pp. 497–501 (2017). https://doi.org/10.1109/I-SMAC.2017.8058400
24. Tolentino, L.K.S., et al.: Development of an IoT-based aquaponics monitoring and correction system with temperature-controlled greenhouse. In: 2019 International SoC Design Conference (ISOCC), pp. 261–262 (2019). https://doi.org/10.1109/ISOCC47750.2019.9027722

25. Vamseekrishna, A., Nishitha, R., Kumar, T.A., Hanuman, K., Supriya, Ch.G.: Prediction of temperature and humidity using IoT and machine learning algorithm. In: Bhattacharyya, S., Nayak, J., Prakash, K.B., Naik, B., Abraham, A. (eds.) International Conference on Intelligent and Smart Computing in Data Analytics. pp. 271–279. Springer, Singapore (2021). https://doi.org/10.1007/978-981-33-6176-8_30

26. TensorFlow https://www.tensorflow.org/. Accessed 12 Feb 2022

27. Edge Impulse https://www.edgeimpulse.com/. Accessed 12 Feb 2022

28. Hagiwara, M.: Removal of hidden units and weights for back propagation networks. In: Proceedings of the International Joint Conference on Neural Networks, pp. 351–353. Publ by IEEE (1993)

29. Karnin, E.D.: A simple procedure for pruning back-propagation trained neural networks. IEEE Trans. Neural Netw. 1, 239–242 (1990). https://doi.org/10.1109/72.80236

30. Mozer, M.C., Smolensky, P.: Skeletonization: A Technique for Trimming the Fat from a Network via Relevance Assessment. In: Advances in Neural Information Processing Systems. Morgan-Kaufmann (1989)

31. Dettmers, T.: 8-Bit Approximations for Parallelism in Deep Learning. arXiv:1511.04561 [cs]. (2016)

32. Gysel, P., Pimentel, J., Motamedi, M., Ghiasi, S.: Ristretto: A Framework for Empirical Study of Resource-Efficient Inference in Convolutional Neural Networks. IEEE Transactions on Neural Networks and Learning Systems. 29, 5784–5789 (2018). https://doi.org/10.1109/TNNLS.2018.2808319

33. Settle, S.O., Bollavaram, M., D'Alberto, P., Delaye, E., Fernandez, O., Fraser, N., Ng, A., Sirasao, A., Wu, M.: Quantizing Convolutional Neural Networks for Low-Power High-Throughput Inference Engines. arXiv:1805.07941 [cs, stat]. (2018)

34. Park, E., Yoo, S., Vajda, P.: Value-aware Quantization for Training and Inference of Neural Networks. arXiv:1804.07802 [cs]. (2018)

35. Gholami, A., Kim, S., Dong, Z., Yao, Z., Mahoney, M.W., Keutzer, K.: A Survey of Quantization Methods for Efficient Neural Network Inference. arXiv:2103.13630 [cs]. (2021)

36. Edge Impulse https://www.edgeimpulse.com/. Last accessed 12 Feb 2022

37. David, R., Duke, J., Jain, A., Reddi, V.J., Jeffries, N., Li, J., Kreeger, N., Nappier, I., Natraj, M., Regev, S., Rhodes, R., Wang, T., Warden, P.: TensorFlow Lite Micro: Embedded Machine Learning on TinyML Systems. arXiv:2010.08678 [cs]. (2021)

38. Arduino Nano 33 BLE Sense https://store-usa.arduino.cc/products/arduino-nano-33-ble-sense. Accessed 12 Feb 2022

39. Raspberry Pi 4 https://www.raspberrypi.com/products/raspberry-pi-4-model-b/. Accessed 12 Feb 2022

40. ST B-L475E-IOT01A. https://www.st.com/en/evaluation-tools/b-l475e-iot01a.html. Accessed 12 Feb 2022

41. Arduino Nano 33 Datasheet. https://docs.arduino.cc/resources/datasheets/ABX00031-datasheet.pdf. Accessed 12 Feb 2022

42. Platformio. https://platformio.org/. Accessed 12 Feb 2022

43. NXP. https://www.nxp.com/. Accessed 12 Feb 2022

44. Microsoft's ELL. https://microsoft.github.io/ELL/. Accessed 12 Feb 2022

45. Ravaglia, L., Rusci, M., Nadalini, D., Capotondi, A., Conti, F., Benini, L.: A TinyML Platform for On-Device Continual Learning with Quantized Latent Replays. arXiv:2110.10486 [cs]. (2021)
46. Arduino Nano case. https://www.thingiverse.com/thing:4412970/files. Accessed 12 Feb 2022

A Meta-heuristic Based Clustering Mechanism for Wireless Sensor Networks

M. P. Nidhish Krishna and K. Abirami(✉)

Department of Computer Science and Engineering,
Amrita School of Engineering, Amrita Vishwa Vidyapeetham, Coimbatore, India
cb.en.p2cse20032@cb.students.amrita.edu, k_abirami@cb.amrita.edu

Abstract. Wireless Sensor Networks are usually the application specific networks that consists of nodes which gathers some data from its surrounding environment continuously or intermittently based on user specifications. In practical use case scenarios, WSNs will consist of densely populated sensor nodes where each node will have a greater number of neighboring nodes. Establishing direct communication with more neighboring nodes results high energy consumption because large transmission power is required. To overcome the most critical issue of poor network lifetime, efficient use of energy resources is required. The goal of the proposed study is to create an efficient cluster head selection model based on Ant Colony Optimization technique that helps to save energy and extend network lifetime by selecting cluster heads wisely.

Keywords: Clustering · Wireless Sensor Networks · ACO

1 Introduction

As a well-developed technology, Wireless Sensor Networks is recognized as one of the integral parts of the fourth industrial revolution (Industry 4.0) and also it is the developing direction of next-generation computer networks. For the first time, the notion of "Industrial Internet" or "Industrial IoT" was defined, i.e., connecting devices and equipments, people, and data analysis over an open, worldwide network [12, 13]. Because of the significant growth and advancement of sensor technology, Wireless Sensor Networks (WSNs) have recently become widely used for both military and commercial applications and are mostly deployed to perform periodic monitoring activities in hostile environments. Sensors in such networks usually have limited resources, such as energy, storage capacity, and processing capabilities. Furthermore, resources, particularly the energy of sensors, may not be replaceable due to the hazardous working environment. As a result, the network's lifetime is determined by its total energy usage. Therefore, high-performance routing protocol and decreasing energy consumption are the critical requirements for WSN applications. Some researchers proposed the concept of cluster network topology as the first clustering routing protocol in WSN to reduce energy consumption. The essential idea here is to select cluster nodes equiprobably in a random cycle, assigning total network resource energy to each sensor node averagely. This decreases the energy consumption and prolongs the network survival lifetime. But the way of randomly selecting a cluster node makes each node qualified to become the cluster node, and it also makes it possible for the low-energy node to qualify as the cluster node. If a cluster node has lost all its energy from its energy source, the node becomes a blind, ending up in premature death. Clustering is one of the energy management strategies used in wireless sensor networks, in which the network is divided into a number of clusters, each with its own cluster head. Rather than each node transmitting its data to the base station directly, nodes in a cluster have the freedom of sending their data to the cluster head sensor node, which again, aggregates the data and sends it to the respective base station. Because fewer nodes or only appropriate nodes provide data to the base station and this mechanism introduces an approach for conserving energy in resource-constrained WSNs. Computationally, selecting m cluster heads among n nodes of network has a total of nC_m possibilities [11]. So, Cluster Head selection in WSNs is an NP hard problem. Hence, the proposed work uses Ant Colony Optimization based approach for clustering instead of brute force approaches. The rest of the paper is organized as follows. Section 2 presents the review of related works. Section 3 explains about the Ant Colony Optimization algorithm. Section 4 explains about the proposed work. The simulation results and analysis are discussed in Sect. 5 followed by conclusion in Sect. 6.

2 Related Work

Halil Yetgin et al. [1], in their paper, describe about recent developments in network lifetime maximization approaches that were in the technical literature. Halil Yetgin et al. [1] also gave a clear definition of Network Lifetime design objective along with practical design constraints of Wireless Sensor Networks for maximizing the Network Lifetime. Halil Yetgin et al. also has tabulated and suggested various network lifetime optimization techniques in the context of various methods like resource allocation techniques, opportunistic transmission techniques, sleep-wake scheduling techniques, routing techniques, mobile relay/sink techniques, coverage and connectivity improvement techniques, optimal deployment techniques. Halil Yetgin et al. [1] also suggest some design guidelines for maximizing the network lifetime of wireless sensor networks like QoS requirements, NL Design objective based on application, computational complexity. Similar to [1, 16] explains and surveys various implementations of the k-means clustering combined evolutionary based optimization approaches for clustering of WSNs. Work in [17] elaborates about the use of low power communication protocols to improve battery life. Paper [18] proposes a strategy that uses clustering and resource scheduling approach that brings energy efficiency. L. Xu et al. in [2] present an overview of several clustering algorithms in Wireless Sensor Networks, as well as obstacles in using such strategies in 5G IoT applications. L. Xu et al. examine existing protocols from the perspective of Quality of Service (QoS), with three goals in mind: energy efficiency, reliable communication, and latency awareness. Rather than providing a limited perspective on how to improve network lifetime, [2] suggests Quality of Service is a crucial factor when it comes to developing intelligent Wireless Sensor Networks. In paper [3], Heinzelman et al. proposed the famous Low Energy Adaptive Clustering Hierarchy Protocol (LEACH), which was one of the oldest works proposed in researches for saving energy in applications of wireless sensor networks. In the research [3], Heinzelman et al. demonstrated that direct approaches, such as traditional direct transmission protocols, multi-hop routing, and static clustering, are not ideal for practical use-case scenarios of wireless sensor networks. The suggested LEACH protocol [3] is primarily a clustering-based technique that employs a randomized station of cluster heads to distribute the network's energy load equitably among other available sensor nodes. The simulation work by Heinzelman et al. from [3] claims to have reduced the energy consumption by as much as a factor of 8 compared to that of conventional protocols, which had the potential to almost double the lifetime of the network. In [4], A. Verma et al. had proposed a clustering scheme called FLEC. The algorithm proposed in [4] uses a fuzzy logic model with average energy of cluster head sensor nodes as one of the extra input membership functions for the model and battery power, Base Station mobility, Base station centrality as the other input membership functions. The proposed method in [4] was compared with DEEC, LEACH-Fuzzy, LEACH clustering methods and quite surprisingly FLEC had a better performance in terms of stability, throughput and network lifetime when compared with all of them. In [5], T. M. Behera et al. suggested a cluster head selection technique that took into account the network's challenging circumstances and increased energy dynamics. The Fuzzy model in [5] attempts to select the cluster head by taking into account the optimal number of clusters in the network, which is determined by the network's residual energy. Also, the simulation results from [5] proves that the solution is one of the effective ones in terms of robustness and efficacy as it had improved performance in throughput by 46% and reduced energy consumption by 66% in the considered simulation scenarios. In [6], H. Ali et al. had proposed a novel cluster head selection algorithm called ARSH-FATI. The main novel idea employed in proposed work from [6] is the use of a separate meta-heuristic called novel rank-based clustering (NRC) scheme with main aim of reducing the communication energy consumption of the sensor nodes. During the run-time of cluster head selection, the methodology from [6] employs a metaheuristic that dynamically alternates exploration and exploitation search nodes. When ARSH FATI is combined with the NRC technique, [6] claims that the suggested proposed algorithm achieves an overall improvement of 60%, 40% over LEACH and PSO-C respectively. H. El Alami and A. Najid present a strategy called ECH in their study [7], which uses a waking mechanism for overlapping and neighboring nodes to achieve efficiency in terms of energy usage in Wireless Sensor Networks. The work from [7] also assumes that all sensor nodes and Base stations are motionless after a network distribution is made, and every node in the network, irrespective of the type, performs a set of periodic tasks (Sensing, computing, communication). The main idea of the algorithm proposed in [7] is that the energy consumption of Wireless networks is optimized by reducing data redundancy in the network. Using the sleeping-waking mode, the ECH approach from [7] optimizes energy consumption in WSNs while simultaneously lowering node failure probability. The key idea here is that only waking nodes are used to detect data from the environment of interest, and takes care of aggregating these data and communicating it to BS via the CHs. [7] compared the performance of the ECH method along with other approaches like LEACH, TEEN SEP, DEEC, and the results also were in favour of the method in terms of the overall energy consumed. In paper [8], Gupta P and Sharma

A.K. had proposed a clustering-based optimized Hybrid Energy-Efficient Distributed Protocol (HEED), which is claimed to be a modified variant of normal HEED protocol. Paper [8] proposes six different variants of Optimized HEED protocols (OHEED). The key contribution of [8] is the use of intra-cluster and inter-cluster communication in the conventional HEED protocol to provide better load balancing across sensor nodes, hence avoiding overburdening of respective cluster heads. The Bacterial Foraging Optimization technique is also used in the approach proposed in [8] for selecting optimal cluster heads, with residual energy as the major parameter. The simulation of both 1 tier and 2 tier protocols from [8] showed that there is a good increase in terms of network lifetime, especially the ICFLOH1TC and ICFLOH2TC showed a 350% and 275% increase in network lifetime, respectively in the considered simulation scenario. In paper [9], a novel improved energy-efficient clustering mechanism for wireless sensor networks is proposed by A. A. H. Hassan et al. In [9], the fuzzy C-means method is paired with a method to reduce and regulate energy consumption of nodes in the network. Secondly, Cluster heads are formed by a new algorithm called the CH Selection-rotation algorithm, which has been integrated with a back-off timer mechanism. Mohamed Elshrkawey et al. suggested a new cluster head selection technique as an addition to the standard LEACH protocol for lowering energy consumption in wireless sensor networks in their study [10]. The work from [10] addresses the energy-hole problem by employing a 2D elliptical Gaussian distribution function to define the position of sensor nodes and base stations. The suggested work in [10] is based on a cluster head election method, which is a variant or modified form of the LEACH algorithm where the threshold function is modified by a factor that represents the sensor nodes' current energy level. The aim is to close the energy gap between all sensors in each cluster, and the algorithm is effective at doing so since a modified TDMA schedule has been applied in this proposed algorithm, which differs from the original one used in the LEACH protocol. In paper [11], an evolutionary computing algorithm is employed for clustering in Wireless Sensor Networks, fairly similar to the work proposed in [8]. In [11], a new energy-efficient clustering technique is proposed, which is based on variable population-based swarm intelligence meta-heuristics inspired by the chemical reaction process. For the Chemical reaction optimization process, the potential energy function of the molecules are chosen based on parameters like intra-cluster distance, sink distance, energy ratio. The results from the proposed work in [11] were compared with various existing algorithms like LDC (Least Distance Clustering), PSO-C (Particle Swarm Optimization), DECA (Differential Evolution based Clustering Algorithm) etc. and it was very promising in terms of features like network lifetime, Packet Delivery Success, Convergence rate.

3   Ant Colony Optimization

Ant Colony Optimization is a meta-heuristic technique in which a set of artificial ants collaborates to solve complex discrete optimization problems with clean solutions. Ant Colony Optimization allocates resources to a group of relatively simple artificial ants based on indirect interaction indirectly via stigmergy. Best solutions are an emergent property of cooperative interaction between agents. The ants release a chemical called pheromones in the path through which they move. As more number of ants move in a direction, the pheromone concentration in that path becomes more. So, the idea here is that the higher the concentration of pheromones, the higher is the possibility for any new ant to take up that respective path to reach the food from the anthills. The path that gets chosen by the ants will be the least distance from anthills to food. The mechanism is very similar to a distributed optimization mechanism. Every single ant has a major contribution to the obtaining the solution. The ants use an incremental methodology to obtain a viable optimal solution. The given problem is solved in Ant Colony Optimization by simulating a number of artificial ants moving across a graph that encapsulates the problem itself. Each graph vertex has a node, and presence of a graph edge that indicates the existence of connectivity between two nodes. A pheromone value is always present for each edge. The quantity of pheromone can be read and modified as the iteration proceeds further. An ant chooses the next vertex to visit at each phase of the solution construction process based on a stochastic mechanism influenced by the pheromone: when in vertex i, the following vertex is selected among the previously unselected ones. In particular, if a site j has not been visited before, it can be selected with a probability value depending to the pheromone linked with edge (i, j). After each iteration, based on the quality of the solutions constructed by the ants, the pheromone values are modified to develop solutions that are similar to the finest ones that have already been built.

In terms of optimization problem, to apply ACO, a model is required as follows:

A model $P = (S, \Omega, f)$ of a combinatorial optimization problem consists of:

- A search space S - defined over a finite collection of discrete variables X_i, $i = 1, 2 \ldots n$
- Ω - a set of constraints that exist between variables
- An objective function $f : S \rightarrow R_0^+$ to be minimized

The generic variable X_i takes values in $D_i = \{v_i^1, \ldots, v_i^{|D_i|}\}$. By moving over a fully connected construction graph $G_C(V, E)$, where V is a set of vertices and E is a set of edges, an artificial ant creates a solution using the ACO technique. Also, ants deposit a certain quantity of pheromone on the components i.e. either on vertices or on the edges that they traverse. The volume of $\Delta \tau$ of pheromone deposited on a path depends on the solution that is found and its quality. The ants that follow utilise the pheromone information to find the most promising areas of the search space.

Construct Ant Solution: A group of m artificial ants aims to find solutions using elements from a finite set of solution components $C = \{C_{ij}\}, i = 1, 2, \ldots, n, j = 1, 2, \ldots |D_i|$. Starting with an empty partial solution $S^P = \varnothing$, a solution is built. At each solution step, the partial solution S^P is enhanced by the addition of a viable solution item from the set $N(S^P) \subseteq C$, which is defined as a set of components that can be included to a partial solution that has already been acquired S^P without disregarding any of the given requirements in Ω. The choosing of a component from $N(S^P)$ is controlled by a technique that is directly influenced by the volume of pheromones linked with each element of the set $N(S^P)$. The rule for selection of solution component varies between Ant Colony Optimization algorithms, although it is all based on real-world ant behaviour.

Apply Local Search: It is prevalent practice to improve the quality of solutions acquired by the all of the ants via a local search after they have been formed and before updating the pheromone. This step is optional, but it is prevalent in most use cases of ACO.

Update Pheromones: The pheromone update process increases pheromone values linked all promising solutions while lowering those associated with irrelevant ones. It is achieved (i) by using the pheromone evaporation process to minimize all pheromone values, and (ii) by raising the levels of pheromones linked with a chosen set of good solutions.

The basic version on ACO i.e. Ant System algorithm is being implemented initially for the current problem statement chosen. Its main feature is that the pheromone matrix values are modified at each iteration by all the m ants which created a solution during that iteration. The pheromone quantity τ_{ij}, linked with the edge between nodes i and j, is modified according to the following equation:

$$\tau_{ij} \leftarrow (1 - \rho).\tau_{ij} + \sum_{k=1}^{m} \Delta\tau_{ij}^{k} \tag{1}$$

where,

ρ - evaporation rate m - Number of ants, and
$\Delta\tau_{ij}^{k}$ - the quantity of pheromone deposited on the edge (i, j) by ant k, given by:

$$\Delta\tau_{ij}^{k} = \begin{cases} \frac{Q}{L_k}, \text{ if ant } k \text{ used edge } (i, j) \text{ in its tour} \\ 0, \text{ otherwise} \end{cases} \tag{2}$$

In the construction of a solution, ants select the node to be visited next via a probabilistic or stochastic mechanism. If an ant k is in node i and has so far has come up with the solution S^P, the probability of proceeding to node j is:

$$p_{ij}^{k} = \begin{cases} \frac{\tau_{ij}^{\alpha} \cdot \eta_{ij}^{\beta}}{\sum_{c_{i,l} \in N(S^P)} \tau_{il}^{\alpha} \cdot \eta_{il}^{\beta}}, \text{ if } C_{ij} \in N(S^P) \\ 0, \text{ otherwise} \end{cases} \tag{3}$$

where $N(S^P)$ - the a group of components that are feasible; that is, edges (i, l) where l is a city that is unvisited by the ant k, α and β – weightage factor of the pheromone and the heuristic information η_{ij} is given by $\eta_{ij} = 1/d_{ij}$; where d_{ij} - distance value between nodes i and j.

4 Proposed Work

For the Ant Colony Optimization to be used for chosen clustering problem, on a weighted graph, the network lifetime maximisation problem is viewed as a challenge of finding the best optimum path. As the ants walk over the graph, they will incrementally build up solutions. The solution construction in Ant Colony Optimization is highly influenced by a pheromone model [14] i.e. set of parameters associated with graph components whose values are modified dynamically at runtime by the controlled ants. WSNs are made up of a number of distributed set of sensor nodes that are spread out over a small geographical area and have limited power. It is considered that renewing the sensor node's battery is a challenging operation. The primary objective is to achieve efficiency in terms of total energy consumption in WSNs. The node which has the highest probability value and the maximum energy is selected to become the dominant cluster. The Energy function includes Alive and dead nodes as one of the parameters, so the overall performance of the network is aggraded. In this scheme, the Energy value of each and every sensor node present in the network is calculated. The Energy value will get changed dynamically. Each node's Energy value is calculated based on its neighbour nodes. The ratio of the value collected from a neighbour sensor is assigned as the Energy value of the node. For the particular period, the setup server should update the Energy value based on its neighbour's vote. The below given Fig. 1. depicts the overall working of the algorithm. The sensors are considered to be placed randomly in the area of interest by an uncontrolled method and to create and form an ad-hoc network. Then the network is split into a set of interconnected substructures i.e. clusters. Each cluster present, has a particular node which is acting as Cluster Head (CH) that is chosen based on a specific metric or a combination of multiple metrics (id, degree of node, weight). Within its network substructure, the Cluster Head serves as a coordinator. CH has the information including a contained list of nodes present in the cluster and the direct path to every node. The main role of the CH is to communicate with all the nodes of its cluster in the communication network. A Cluster Head, on the other hand, should be able to have a communication going on with nodes in other clusters, either directly or via the respective CH or gateways. Three steps are involved in communication. The CH firstly, receives the data sent by its sub-ordinate nodes. It then compresses the data before sending it to the BS or another CH, allowing the relevant CH to decrease energy consumption and relatively maximize the network lifetime. None of the algorithms make use of a set of key metrics combined together like: power, lower id, maximum degree, mobility, and node connectivity etc. CH is chosen in the proposed work, based on the total number of neighbours, residual energy, and distance of a node from the cluster's centre. The Cluster Head selection model is applied only to nodes near the cluster's centre. Hence, in WSNs, a method for selecting a CH is presented based on the following iteration parameters: residual energy (E_{res}), connection capability of (C_n), and node degree (D). During the setup phase, the Base Station sends a message to all of the sensor nodes in the network. The sensor nodes will respond with their respective position, node ID, and distance between them in a reply message. All sensor nodes will keep their receivers turned when the CH selection phase is happening.

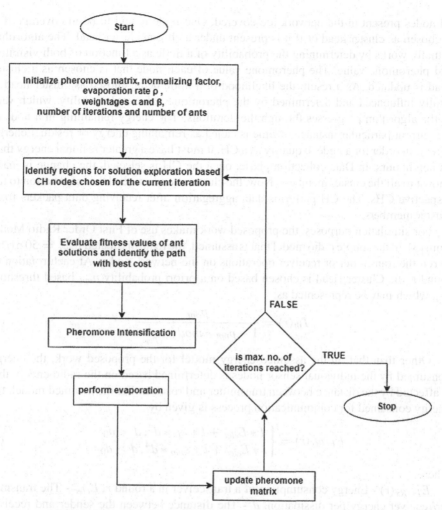

Fig. 1. Flowchart for ACO based algorithm for WSN CH selection

The peer-to-peer sensor network is considered to be a graph, and the location of each uniquely numbered sensor node is represented using Cartesian coordinates. In the chosen case, one node is within the range of the other node if the Euclidean distance value between the two nodes is within a specific limit. The Ant Colony Optimization algorithm identifies the set of optimal cluster heads using an iterative process that obtains a local solution. The cluster head selection process is performed based on two aspects, i.e. pheromone value of each node's edge and its visibility. The topology structure of the network affects visibility. Also, pheromone value associated with each node and it reduces or increases during each iteration of the ACO algorithm. The idea is, for each iteration, one node is selected as cluster head, and further, the next cluster head and so on, which is selected and influenced based on visibility and pheromone of the neighbouring nodes. The iteration process in the ACO algorithm does not terminate until all nodes present in the network are covered. One node is said to be in coverage if it is chosen as cluster head or if it is present under a chosen cluster head. The algorithm actually works by determining the probability of a node as a function of both visibility and pheromone value. The pheromone value of each node that is chosen as a cluster head is updated. As a result, the likelihood of a node being chosen as cluster head is totally influenced and determined by the pheromone value and visibility, which vary as the algorithm progresses through the iterations. The energy remaining in a node at the current particular instance of time is called as remaining energy or residual energy (Eres). In order for a node to qualify as a CH, it must have a greater residual energy than its neighbours. In Data collection phase, once the CH is selected, the change is made known to all the cluster members. Now, the cluster members send the information to the respective CHs. The CH performs data aggregation after receiving data packets from cluster members.

For simulation purposes, the proposed work makes use of First Order Radio Model from [3]. In the simple radio model that is assumed, the radio dissipated $E_{elec} = 50\,\text{nJ/bit}$ to run the transmitter or receiver operations on the network nodes. At each rotation or round r, the Cluster Head is chosen based on a priori probability p_{opt} based threshold T_{th}, which may be represented as:

$$T_{th}(r) = \frac{p_{opt}}{1 - p_{opt} * mod\left(r, \frac{1}{p_{opt}}\right)} \tag{4}$$

Other than that, in the assumed energy model for the proposed work, the energy consumed by the individual sensor nodes is determined based on the radio energy that is affected by the distance between transmitter and receiver. For the assumed model, the energy consumed for communication process is given by,

$$E_{Tx/Rx}(r) = \begin{cases} l * E_{elec} + l * \varepsilon_{fs} * d^2, d < d_0 \\ l * E_{elec} + l * \varepsilon_{mp} * d^4, d \geq d_0 \end{cases} \tag{5}$$

where,

$E_{Tx/Rx}(r)$ - Energy consumption of a transceiver in a round r, E_{elec} - The transmitter/receiver energy per dissipation, d - The distance between the sender and receiver nodes ε_{fs} and ε_{mp} - the amplifier energy dissipation in free space and multipath respectively, and the threshold distance $d_0 = \sqrt{\varepsilon_{fs}/\varepsilon_{mp}}$ which is the boundary between free space and multi-path.

The presented work assumes that there are n number of sensor nodes and one main base station (BS) in the model of network. The considered scenario for WSN operation has the following properties/assumptions:

1. The sensor nodes are scattered randomly over a field (2-D Cartesian plane) such that no any exact overlap of nodes occur for one particular point.
2. Replenishing the battery energy source is impossible.
3. In terms of processing and communication capabilities, all of the sensors in the network are identical.
4. The Base Station is considered to be present at a fixed position inside of the WSN's area of operation and also no any constraints are given for the position of Base Station inside this particular sensing field.

5. Each Cluster Head assigned in the network, aggregates the data received. It then transmits the received aggregated data to the BS. The nodes may use a varying power level for transmission power according to the distance of transmission for the data to be sent.

6. Every nodes in the network are equiprobable in terms of getting qualified as Cluster Head and having the capability to operate as Cluster Head Node.

7. The communication links between any of the sensor nodes in the network are assumed to be wireless and coverage among nodes exist only if their fall under the specified range of distance within each other.

5 Results and Discussion

The network is simulated and tested for performance comparison in MATLAB R2021b with 100 network nodes. The chosen 100 nodes are distributed and scattered over an area of 100×100 sq. units to evaluate the performance of the suggested protocol. The network's nodes are assumed to be homogeneous in nature, with similar processing and communication capabilities. All nodes in the network are initially assigned a base energy value of 0.5 J. The below given Table 1 shows the parameters that are chosen for the simulation of Efficient Cluster Head selection algorithm using Ant Colony Optimization for Wireless Sensor Networks.

Table 1. Simulation characteristics for Wireless Sensor Networks

Parameter name	Value
Network Field size	100×100 sq. units
Number of Sensors (n)	100
Base Station position value	(50, 0)
Initial energy assigned to each node	0.5 J
Cluster Radius (R)	10 m
Energy consumption for transceiver operation (E_{elec})	50 nJ/bit
Energy dissipation by amplifier in free path propagation	10 pJ/bit/m^2
Energy dissipation by amplifier in multi-path propagation	0.0013 pJ/bit/m^4
Message size (l)	4000 bits
Threshold distance (d_0)	87.7 m

The existing models that are considered for the performce comparison of the proposed model are LEACH protocol and TEEN Protocol. The same network analysis is used to assess the performance of all current and new mechanisms. To assess the different simulation cases, the source and destination are given as inputs one by one. During the simulations, routes are chosen based on the following next hop option. The below given Fig. 2 shows an example iteration in the simulation. The red coloured nodes indicate the node selected as Cluster Head for the current iteration of the simulation. The star shaped node indicated the Base Station node at the co-ordinate (50, 0). Since, the initial energy of the nodes was 0.5 J, the simulation was tested for 3500 rounds and appropriate for the testing of current scenario because almost all the algorithms may run out of alive nodes. If a different value of Initial energy is chosen, then number of rounds can be increased or decreased based on the value chosen.

The following metrics are chosen for the comparison of the performance of the algorithm with standard algorithms:

Alive and dead nodes count: total number of alive and dead nodes respectively at each ith round of simulation.

Throughput: Total number of packets of data sent from respective Cluster head to the base station.

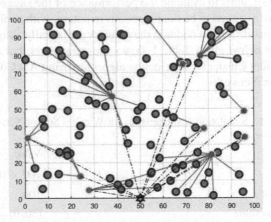

Fig. 2. One iteration of network simulation (Color figure online)

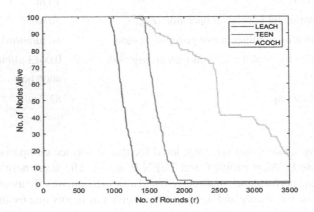

Fig. 3. Graph for alive node count

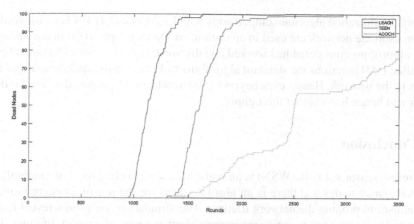

Fig. 4. Graph for dead node count

The above given Figs. 3 and 4 depicts the number of alive nodes and dead nodes respectively, which are present at each instance of the test iteration. Clearly, the above graphs indicate that the proposed CH selection scheme using Ant Colony Optimization is better as number of alive nodes are higher and hence the network lifetime is relatively higher in the proposed scheme. The key thing to be noted here is that, after 2000 iterations the standard LEACH and TEEN algorithms have more number of dead nodes (almost 90) and the proposed scheme outperforms these both algorithms as number of dead nodes are very less. One another key thing to be noted here is that, in standard algorithms considered, if one or more nodes begin to die, the decrease of alive nodes happens vary rapidly and exponentially.

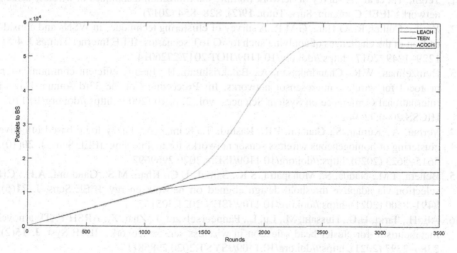

Fig. 5. Graph for throughput comparison

The above given Fig. 5 depicts the throughput for the proposed scheme in comparison with the standard algorithms (LEACH and TEEN). Clearly, the proposed Ant

Colony Optimization algorithm outperforms both LEACH and TEEN because available resources in the network are used in an optimized manner. Upto 1000 iterations all the three algorithms considered had worked equally well, but, the point to be noted here is that after 1500 iterations the standard algorithms fail because of rapid decrease of alive nodes in the network. Hence, even beyond 1500 iterations the proposed scheme is doing better and hence has a higher throughput.

6　Conclusion

A wireless sensor network (WSN) is an embedded system comprised of randomly distributed sensor nodes and there is an absolute need for the use of an energy-efficient mechanism to prolong the network lifetime. The simulation and results reveal that the suggested technique has a significant improvement in terms of network lifetime. Thus, the use of proposed method can deliver a more extended stability period with more effective clustering as its better ability in exploring and exploiting the solution space. The suggested method, which utilizes the ant colony optimization algorithm, to determine the least number of cluster heads in a WSN. The method used is analogous to determining the topology graph's least dominating set, in which each CH is a component of the overall set. For future enhancements, bringing in features of secure routing or hybridization of the algorithm along with clustering can be taken forward for further feature enhancement of the proposed algorithm.

References

1. Yetgin, H., et al.: A survey of network lifetime maximization techniques in wireless sensor networks. IEEE Commun. Surv. Tutor. **19**(2), 828–854 (2017)
2. Xu, L., Collier, R., O'Hare, G.M.P.: A survey of clustering techniques in WSNs and consideration of the challenges of applying such to 5G IoT scenarios. IEEE Internet Things J. **4**(5), 1229–1249 (2017). https://doi.org/10.1109/JIOT.2017.2726014
3. Heinzelman, W.R., Chandrakasan, A., Balakrishnan, H.: Energy-efficient communication protocol for wireless microsensor networks. In: Proceedings of the 33rd Annual Hawaii International Conference on System Sciences, vol. 2, p. 10 (2000). https://doi.org/10.1109/HICSS.2000.926982
4. Verma, A., Kumar, S., Gautam, P.R., Rashid, T., Kumar, A.: Fuzzy logic based effective clustering of homogeneous wireless sensor networks for mobile sink. IEEE Sens. J. **20**(10), 5615–5623 (2020). https://doi.org/10.1109/JSEN.2020.2969697
5. Behera, T.M., Nanda, S., Mohapatra, S.K., Samal, U.C., Khan, M.S., Gandomi, A.H.: CH selection via adaptive threshold design aligned on network energy. IEEE Sens. J. **21**(6), 8491–8500 (2021). https://doi.org/10.1109/JSEN.2021.3051451
6. Ali, H., Tariq, U.U., Hussain, M., Lu, L., Panneerselvam, J., Zhai, X.: ARSH-FATI: a novel metaheuristic for cluster head selection in wireless sensor networks. IEEE Syst. J. **15**(2), 2386–2397 (2021). https://doi.org/10.1109/JSYST.2020.2986811
7. El Alami, H., Najid, A.: ECH: an enhanced clustering hierarchy approach to maximize lifetime of wireless sensor networks. IEEE Access **7**, 107142–107153 (2019). https://doi.org/10.1109/ACCESS.2019.2933052

8. Gupta, P., Sharma, A.K.: Clustering-based optimized HEED protocols for WSNs using bacterial foraging optimization and fuzzy logic system. Soft. Comput. **23**(2), 507–526 (2017). https://doi.org/10.1007/s00500-017-2837-7

9. Hassan, A.A.H., Shah, W.M., Habeb, A.-H.H., Othman, M.F.I., Al-Mhiqani, M.N.: An improved energy-efficient clustering protocol to prolong the lifetime of the WSN-based IoT. IEEE Access **8**, 200500–200517 (2020). https://doi.org/10.1109/ACCESS.2020.3035624

10. Elshrkawey, M., Elsherif, S.M., Wahed, M.E., An enhancement approach for reducing the energy consumption in wireless sensor networks. J. King Saud Univ.-Comput. Inf. Sci. **30**(2), 259 (2018). ISSN 1319-1578, https://doi.org/10.1016/j.jksuci.2017.04.002

11. Srinivasa Rao, P.C., Banka, H.: Energy efficient clustering algorithms for wireless sensor networks: novel chemical reaction optimization approach. Wirel. Netw. **23**(2), 433–452 (2015). https://doi.org/10.1007/s11276-015-1156-0

12. Qin, W., Chen, S., Peng, M.: Recent advances in industrial internet: insights and challenges. Digit. Commun. Netw. **6**(1), 1–13 (2020)

13. Joseph, J., et al.: A survey on wireless networks: classifications, applications and research challenges. Perspect. Commun. Embed. Syst. Sig. Process. PiCES2(9), 200–209 (2018)

14. Dorigo, M., Stützle, T., The ant colony optimization metaheuristic. In: Ant Colony Optimization. MIT Press, Cambridge, pp. 25–64 (2004)

15. Venkateswarao, T., Sreevidya, B.: An energy-efficient wireless sensor deployment for lifetime maximization by optimizing through improved particle swarm optimization. In: Kaiser, M.S., Xie, J., Rathore, V.S. (eds.) Information and Communication Technology for Competitive Strategies (ICTCS 2020). LNNS, vol. 190, pp. 49–63. Springer, Singapore (2021). https://doi.org/10.1007/978-981-16-0882-7_3

16. Bhavani, K.D., Radhika, N.: K-means clustering using nature-inspired optimization algorithms - a comparative survey. Int. J. Adv. Sci. Technol. **29**(Special Issue 6), 2466–2472 (2020)

17. Vidhya, S.S., Mathi, S.: Investigations on Power-aware solutions in low power sensor networks. In: Ranganathan, G., Fernando, X., Shi, F. (eds) Inventive Communication and Computational Technologies. LNNS, vol. 311, pp. 911–925. Springer, Singapore (2022). https://doi.org/10.1007/978-981-16-5529-6_69

18. Gokuldev, S., Jathin, R.: Range smart cluster monitor based guesstimate approach for resource scheduling in small size clusters. Int. J. Eng. Technol. **7**(2), 837–841 (2018)

Univariate Time Series Forecasting of Indian Agriculture Emissions

Abhay Deshpande[1(✉)], Tanmay Belsare[1], Neha Sharma[2], and Prithwis De[3]

[1] Symbiosis Statistical Institute, Symbiosis International University, Pune, India
abhaypdeshpande18@gmail.com
[2] Analytics and Insights, Tata Consultancy Services, Pune, India
[3] MBU, Tata Consultancy Services, London, UK

Abstract. The first ever increased contribution to the GHG from human activities was due to emissions from agriculture. Because of the large area covered under agriculture and comprehensive practices, the agriculture sector has a significant impact on the earth's GHG. Therefore, agriculture emissions being one of the main emissions and GHG being the cause of climate change, it is necessary to determine the pattern of emission and forecast of the future. The aim of this paper is to determine the emission pattern from manure management, agriculture soils, enteric fermentation and rice cultivation, also forecasting the emissions. The study object was India. Different univariate time series models were built and best model was selected on the basis of evaluation criteria for forecasting. The performed modelling was based on the FAOSTAT data of India.

Keywords: Climate change · Agriculture emissions · GHG · Forecasting · Time series

1 Introduction

Climate change, according to the United Nations, refers to long-term changes in temperature and weather patterns [1]. The Greenhouse effect arises when the accumulation of the Greenhouse Gases (GHG) is altered in the atmosphere and these gases capture heat in the atmosphere. Carbon dioxide (CO_2), Nitrous oxide (N_2O), Methane (CH_4), are some of the examples of GHGs. Emissions from CH_4 are contributed by livestock, land-use, other agricultural techniques, breakdown of organic material in solid waste of municipal landfills. Activities concerning agriculture, combustion of solid waste and fossil fuels, industrial, land-use and treatment of waste water leads to the release of N_2O [2].

The "Intergovernmental Panel on Climate Change" (IPCC) is a body created for evaluating the science behind climate change and providing consistent reports to the policymakers, so that they can make effective and better decisions [3]. Each GHG has a varied global warming potential (GWP) and stays in the atmosphere for a varying

M. Singh et al. (Eds.): ICACDS 2022, CCIS 1614, pp. 346–358, 2022.
https://doi.org/10.1007/978-3-031-12641-3_28

amount of time. "The number of metric tonnes of CO_2 emissions with the same global warming potential as one metric tonne of another greenhouse gas" is referred to as CO_2e [4].

Food and agriculture organization (FAO) is one of the agencies in the United Nations with a count of approximately 130 member countries. For over 245 nations and territories FAOSTAT provides free data related to food and agriculture statistics [5]. IPCC provides the GHG inventory guidelines for a sector named "Agriculture, Forestry and other land use" (AFLOU). Emissions from sources namely "Enteric fermentation", "Rice Cultivation", "Manure placed on pasture", "Crop leftovers", "Synthetic Fertilizers", "Manure Management", "Manure applied to soils" and "Biomass Burning" are all included in FAOSTAT. Under IPCC inventory rules, enteric fermentation, rice cultivation, and biomass burning are recorded individually, while remaining categories are grouped together as 'agricultural soils' [4].

The emissions from the agriculture are contributed through the bacterial process associated with breaking down of organic and inorganic material which are produced or feed in to the management of agricultural systems [6]. The AFLOU sector can be divided into two major parts: one being the CO_2 emissions from removals of management and change of land-use, and other being the emissions from agriculture [7]. One of the major sources of CH_4 and N_2O is the agriculture sector and 30% of the global GHG are being contributed by agriculture and deforestation [3]. Livestock farming is one of the most crucial subsectors under agriculture which accounts for almost 70% of the total emissions from agriculture [3]. Indirect emissions from agriculture are mostly contributed by irrigation activities, usage of pesticides, farm machinery and production of fertilizers [3].

The aim of this paper is to:

1. To determine the pattern of the Indian agriculture emissions from sources like Manure management, agriculture soils, enteric fermentation and rice cultivation.
2. Building different univariate time series model on the agriculture emissions and comparing the model based on the evaluation metric.
3. Using the best time series model forecasting the emissions for the years 2020, 2021, 2022, 2023, 2024, 2025.

This research paper consists of 6 sections. After brief introduction on climate change and GHGs, the literature review covers Agriculture sector as one of the major contributors towards climate change and factors affecting agriculture emissions followed by steps taken in agriculture sector to mitigate the climate change. Methodology section consists of data preparation and approach taken to obtain solution of problem at hand. Section 4 contains the insights obtained while performing Exploratory Data Analysis (EDA). Obtained results are reported in Sect. 5 with discussion on the performance of methods used. Findings are discussed in Sect. 6 along with future scope of the study.

2 Literature Review

William et al. (2021) reviewed the sector-wise GHG emissions from 1990–2018 globally, the results showed that since 2000 there was an upward trend in the emissions of AFLOU at the rate of 0.8% per year, also emissions were higher in the developing countries than in the developed countries. The authors also found that, particularly in Africa, Middle East, southern East and southern Asia trends in the AFLOU sector emissions were driven because of increase in population [7].

a) **Factors affecting the agriculture emissions**

Bin et al. (2017) explored the driving factors behind China's agriculture CO_2 emissions with the help of geographically weighted Regression Model across the eastern, western and central regions, revealing that economic growth, urbanization and energy intensity being positively related to the emissions. The authors also found out that all the three factors are significantly different across provinces and regions of China, also suggesting that emissions reductions policies should be devised such that considering provincial and regional difference [8]. Awais et al. (2021) carried out a global metal analysis of GHG emissions due to agriculture soils, finding out that acidic soils, neutral soils and alkaline soils resulted in highest contribution to the N_2O, CO_2 and CH_4 respectively. Further, factors like texture of soil, crop type and climate zone were responsible for increase in the emissions from agriculture, also Global Warming Potential (GWP) of soil's emission was increased due to the application of poultry manure [9].

b) **Mitigation in Agriculture**

With the Paris Agreement coming into existence in 2015, pledge was taken by more than 100 countries for reducing the emissions but information about the quantity of mitigation required in each sector was missing [10]. Eva et al. (2016) estimated that about ~1 $GtCO_2¬$ yr-1 of mitigation by 2023 is needed to reach the goals of Paris Agreement, moreover the authors discussed about the achievability and the following suggestions for reduction in the emissions from agriculture sector.

1. Finance and Funding for the low emission technologies like methane inhibitors, varieties of wheat and maize that restrain N_2O production and identification of low producing methane Cattle breeds [10].

2. Need in giving technical assistance to farmers such as web-based information sites, innovation hubs for farmers and two-way cell phones support [10].

   One of the techniques namely Climate Smart Agriculture (CSA) endorsed by the FAOSTAT who's one of the main objectives is to reduce or remove the GHG from agriculture can be adopted as a mitigation technique. Not only the technique meets the goal of reducing the emissions, but also it meets Sustainable Development Goals (SDGs) and achieve food security. By FAOSTAT, Several Case studies of CSA has been documented to prove its effectiveness [11]. The International Atomic Energy Agency (IAEA) helps the member states to tackle climate change by reducing GHG in agriculture, achieving sustainable increase in the agriculture productivity and building food security systems through the nuclear techniques. By using Global Warming Potential (GWP) 300 times larger than that of CO_2, carbon sequestration, low-cost inhibitors regulating nitrogen processes in soils and isotopic techniques like nitrogen-15, carbon-13 etc. can accomplish the goals [12].

c) **Time Series Forecasting in Agriculture Emissions**
Dennis et al. (2016) aims to examines the pattern of emissions through time series data and, as a result, anticipate the linear trend in GHG emissions from the Kenyan Savanna. The study inspects and forecasts the time series data from 1993 to 2012 using Autoregressive (AR) modelling. The study's main finding is that due to the continuous burning of Savanna grassland emissions would keep on rising unless major mitigation measures are implemented to change the status quo [13]. Akcan et al. (2018) investigated GHG emissions with respect to sectors like energy, waste, industrial processes, and agriculture in Turkey, furthermore, used time series modelling methods such as moving average, exponential smoothing and exponential smoothing with trend for forecasting the emissions which can be contributed for the national GHG emissions inventory of Turkey [14]. Abderrachid et al. (2020) forecasted soil CO_2 and N_2O emissions from an agricultural field in Quebec using machine learning categories like classical regression, shallow learning and deep learning, resulting in the LSTM model's performance as the best one and evaluated through lowest RMSE among others [15].

Despite the fact that multiple studies have analysed various sources of agriculture emissions, associated impacts, influencing factors, and mitigation techniques, there is no empirical information available for analysing and forecasting diverse sources of Indian agriculture emissions. This study aims to determine the pattern of emissions through time and, as a result, anticipate the trend in GHG emissions from Indian agricultural sources.

3   Methodology

1. Data Preparation
Data related to agriculture and food for the whole world is been provided by the Food and Agriculture Organization Corporate Statistical Database (FAOSTAT). The data for this paper was taken from climate change domain in the FAOSTAT database.

In FAOSTAT, the country-wise aggregated emissions data is available. The bulk data for all the countries was downloaded from the emissions total section and was used. For the paper, the emissions from the years 1961 to 2019 were considered, as these were the available years in the data for India. Before using the data for the exploratory data analysis purpose and time series purpose, following preprocessing was performed in order to subset the required data as per the objective

   i)   Deleting unrequired columns
   ii) Filtering the FAOSTAT source and CO_2e emissions from the data
   iii) Subsetting the relevant emissions sources for the study
   iv) Structuring the data from the wide format to the long format in order to perform EDA and build models
   v) Filtering the area as "India" and building India specific time series model
   vi) For modelling purpose, the emissions were converted into mega tonnes (Mt)

The analysis was carried out in Python. Data preparation is performed manually as per above mentioned steps. After importing the required libraries, the dataset is then imported into Python. Agricultural Soils, Enteric Fermentation, Manure Management, and Rice Cultivation are extracted for India to form a data-frame that spans the years 1961 to 2019.

2. Univariate Time Series Models

2.1 Linear Trend Model

If the time series variable is under study, it's reasonable to want to find and fit any systematic time patterns that could exist. It's a type of basic regression model in which the dependent variable is simply a time index variable, such as 1, 2, 3, … or another evenly spaced sequence of numbers. The linear trend model aims to discover the slope and intercept that best suit all of the historic information.

2.2 Holt's Linear Trend

Simple exponential smoothing was modified by Holt for allowing trend in the data to be forecasted. Three equations including forecast equation, smoothing equations for the level and the trend are used in this method:

Equation for Forecast: $\hat{y}_{t+s|t} = lv_t + sm_t$

Equation for Level: $lv_t = \varphi y_t + (1 - \varphi)(l_{t-1} + m_{t-1})$

Equation for Trend: $m_t = \omega(lv_t - lv_{t-1}) + (1 - \omega)m_{t-1}$

lv_t is an estimate level of time series; m_t is slope estimate of time series at time t; φ is level's smoothing parameter; ω is trend's smoothing parameter; $0 \leq \varphi \leq 1; 0 \leq \omega \leq 1$.

2.3 Holt's Linear Damped Trend

The linear approach of Holt's shows a trend which either continuously increases or decreases towards the future. These methods overly predict for forecast periods that are longer. Gardner and McKenzie (2011) presented a parameter that "dampens" the trend such that it becomes flat line in the future at some instance [16].

This approach contains a damping parameter $0 < \tau < 1$:

Equation for Forecast: $\hat{y}_{t+s|t} = lv_t + (\tau + \tau^2 + ... + \tau^s)m_t$

Equation for Level: $lv_t = \varphi y_t + (1 - \varphi)(l_{t-1} + \tau m_{t-1})$

Equation for Trend: $m_t = \omega(lv_t - lv_{t-1}) + (1 - \omega)\tau m_{t-1}$

In addition, the method includes a normalizing criterion and (magnitude in between 0 and 1). Put $\tau = 1$, we get Holt's linear method. In case $0 < \tau < 1$, τ reduces the trend till it reaches a steady state at some point in the future.

2.4 ARIMA

ARIMA means AutoRegressive Integrated Moving Average. An ARIMA model with no seasonal component model is created by combining differencing with autoregression and a moving average model. Both lagged y_t values and lagged errors are included in the "predictors" on the right-hand side.

ARIMA (a,b,c) model is:

$$\left(1 - \vartheta_1\delta - \cdots - \vartheta_a\delta^a\right)(1 - \delta)^b y_t = \omega + (1 + \varnothing_1\delta - \cdots - \varnothing_c\delta^c)e_t$$

where a is the autoregressive part order, b is first differencing degree involved, c is the moving average order. And $\omega = \cup(1 - \vartheta_1 - \cdots - \vartheta_a)$; $\cup$ is mean of $(1 - \delta)^b y_t$. The auto_arima() function makes it easy to include a constant in your code. A constant is included by default for b = 0 or b = 1 if it improves the AIC value; for b > 1, the constant is always excluded.

3 Exploratory Data Analysis

See Figs. 1 and 2.

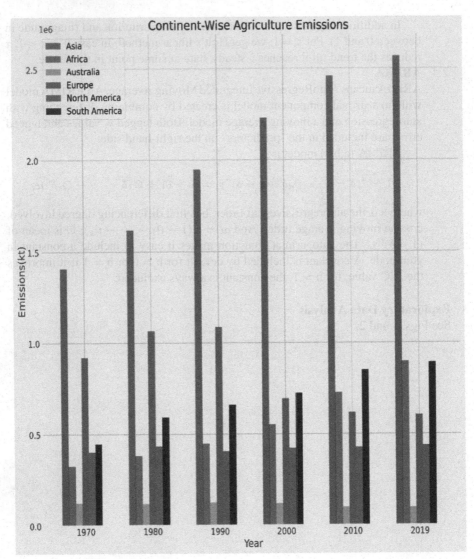

Fig. 1. Multiple bar graph gives a general idea of continent-wise Agriculture GHG emissions of past 5 decades. Emissions from *Asia* are subsequently higher than *Europe, South America, North America, Africa* and *Australia*, respectively. This trend can be seen till year 2000. In year 2000 a significant decrease in Europe's Agriculture GHG emissions. In year 2019 order of emitters from highest to lowest was *Asia, Africa, South America, Europe, North America and Australia*, respectively.

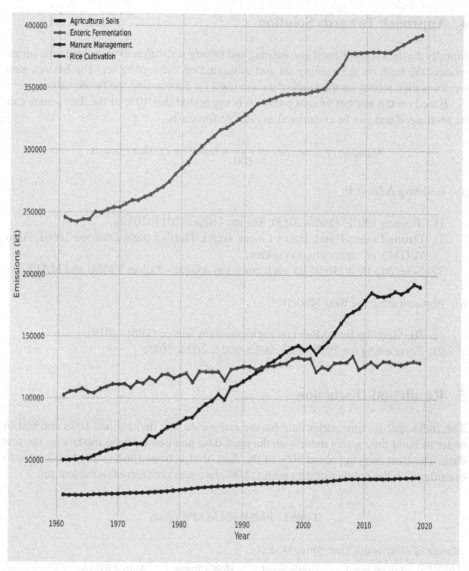

Fig. 2. Line chart shows year-on-year increase in GHG emissions of subsectors in Agriculture sector of India over the course of past 50 years. In 2019, GHG emissions from *Enteric Fermentation* has increased to 3.93 lakh kilotonnes from 2.4 lakh kilotonnes from 1961, emissions from *Agricultural Soil* have increased to 1.90 lakh kilotonnes from 0.50 lakh kilotonnes. Similarly, from *Rice Cultivation*, GHG emissions increased from 1.02 lakh kilotonnes to 1.29 lakh kiltonnes and from *Manure Management* the emissions stood at 0.37 lakh kilotonnes.

4   Approach Towards Solution

Initially the data is partitioned into training and testing set, different univariate time series models are built on the training set and evaluated on the testing set. The best-chosen model is then rebuilt on the whole data and used for forecasting the future values.

Based on the number of data points, it is suggested that 10% of the data points can be forecasted and can be considered as reliable forecasts.

$$Number \ of \ forecasts = \frac{10}{100} * Number \ of \ Data \ points$$

a) Building A Model:

1) Training (90%) (1961–2013), Testing (10%) (2014–2019).
2) Fitting Linear Trend, Holt's Linear Trend, Holt's Linear Damped Trend, Auto ARIMA on each emission source.
3) Selecting Best Model for each emission source – Lower RMSE and MAPE.

b) Implementing the Best Model:

1) Building the Best Model on each emission source (1961–2019).
2) Forecasting for 2020, 2021, 2022, 2023, 2024, 2025.

5   Result and Discussion

The India specific time series data for the emissions was divided into train and test in order to build time series models on the train data and compare the models on the test data. The train carrying about 90% of the data, that is from 1961 to 2013. Taking into consideration the number of data points, 10% data was considered in the test set.

Table 1. RMSE and MAPE values

Source of emission	Univariate Time Series Models								
	Linear trend model		Holt's linear trend		Holt's linear damped trend		Auto ARIMA		
	RMSE	MAPE	RMSE	MAPE	RMSE	MAPE	Parameters (a, b, c)	RMSE	MAPE
1	12.89	3.329	10.41	2.274	10.09	2.207	**0,1,3**	**5.724**	**1.633**
2	**2.602**	**1.220**	4.628	2.124	3.113	1.330	0,1,0	4.435	2.838
3	6.583	5.007	**1.024**	**0.647**	2.204	1.513	0,1,1	2.478	1.698
4	1.676	4.543	0.762	1.778	0.761	1.778	**2,1,0**	**0.336**	**1.194**

1: Enteric Fermentation, 2: Agriculture Soils, 3: Rice Cultivation, 4: Manure Management

In Table 1, RMSE and MAPE values of each model for Enteric Fermentation, Agriculture Soils, Rice Cultivation and Manure Management are given. The best model is chosen such that model has the lowest RMSE and MAPE values (Fig. 3).

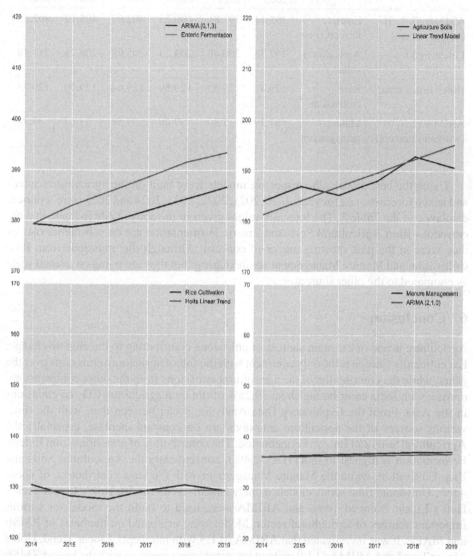

Fig. 3. Describes the best model forecast and the test set. It can be seen that the difference between the forecasts and test set is less. *ARIMA (0,1,3), Linear Trend model, Holt's Linear trend model* and *ARIMA (2,1,0)* are the best suitable models for *Enteric Fermentation, Agriculture Soils, Rice cultivation, Manure Management* respectively.

As there are 58 data points, i.e., 1961 to 2019, next 6 years data points (10% of data points) are considered for the forecasts.

Table 2. Best time series model's forecasts

Time series model	Source	Emissions (Mt)					
		2020	2021	2022	2023	2024	2025
ARIMA (0, 1, 3) (with intercept)	Enteric fermentation	395.44	397.81	400.25	402.69	405.12	407.55
Linear trend	Agricultural soils	197.73	200.48	203.23	205.98	208.73	211.48
Holt's linear trend	Rice cultivation	129.47	129.53	129.59	129.64	129.70	129.75
ARIMA (2, 1, 0) (without intercept)	Manure management	37.29	37.39	37.49	37.58	37.67	37.74

Using the best models, the respective models were built on the agriculture sources and next 6 forecasts, i.e., for years 2020, 2021, 2022, 2023, 2024 and 2025 were reported, displayed in the Table 2. The forecasts values given in the above table indicate that the emissions from Agricultural Soils and Enteric Fermentation are on substantial rise, as they were in the past showing matter of concern. Although the emissions from Rice Cultivation and Manure Management are increasing, but they are rising on a small scale as compared to the other sources.

6 Conclusion

Agriculture is one of the main sources of emissions contributing to the climate change. Exceptionally Europe is the only continent with the fall of agriculture emissions over the years, while the contribution of the agricultural emissions from the other continents are on rise, with India contributing about 29.25% of the total agriculture CO_2 eq emissions in the Asia. From the Exploratory Data Analysis, it can be seen that, with the time, various sources of the agriculture emissions are on constant increase, especially the Agricultural Soils and Enteric Fermentation. The contribution of emissions from Enteric Fermentation is highest in India (i.e. 52.4%), consequently the Agricultural Soils and Rice Cultivation, while the Manure Management with the least contribution of about 4.9%. Univariate time series models namely Linear Trend model, Holt's Linear Trend, Holt's Linear Damped Trend and ARIMA were used to build the model for various important sources of agricultural sector. Model were evaluated on the basis of RMSE and MAPE. It was investigated that ARIMA (0,1,3 with intercept), Linear Trend, Holt's Linear Trend and ARIMA (2,1,0 without intercept) were reported with the lowest RMSE and MAPE value for the Enteric Fermentation, Agricultural Soils, Rice Cultivation and Manure Management sector respectively. In FAOSTAT, as the emission data for 2020 and 2021 were not available, same were forecasted, along with 2022, 2023, 2024 and 2025 using the best-chosen models. Historically, as the emissions were on rise from all the sectors, the forecasted values also move with the same pattern. This study shows the rising emissions from the Indian agriculture emissions, which can be matter of concern in context to the climate change.

The best univariate models found out in the study can be used for:

1. Forecasting the future values of the emissions.
2. Filling gaps in the data.

Proposed models only use past emissions to predict future emission values, whereas other factors that may cause emissions are not considered. Investigating the multi-variate relationship between the sources can be the future scope, also a study can be carried out to find and examine various important factors causing the rise in agricultural emissions.

References

1. What is Climate Change?, United Nations, Climate Action. https://www.un.org/en/climatech ange/what-is-climate-change
2. Overview of Greenhouse Gases, United States Environmental Protection Agency (US EPA). https://www.epa.gov/ghgemissions/overview-greenhouse-gases#:~:text=GHG% 20emissions%20are%20often%20measured,CO2%2C%20per%20unit%20mass
3. About the IPCC, The Intergovernmental Panel on Climate Change. https://www.ipcc.ch/
4. Smith, P., et al.: IPCC report on agriculture, agriculture, forestry and other land use (AFOLU). In: Edenhofer, O., et al. (eds.) Climate Change 2014: Mitigation of Climate Change. Contri- bution of Working Group III to the Fifth Assessment Report of the Intergovernmental Panel on Climate Change. Cambridge University Press, Cambridge (2014)
5. Home, Food and Agriculture Organization of the United Nations. https://www.fao.org/hom e/en/
6. Samrat, N.H., Islam, N.: Greenhouse gas emissions trends and mitigation measures in aus- tralian agriculture sector—a review. Agriculture 11, 85 (2021). https://doi.org/10.3390/agricu lture11020085
7. Lamb, W., et al.: A review of trends and drivers of greenhouse gas emissions by sector from 1990 to 2018. Environ. Res. Lett. 16(7), 073005 (2021). https://iopscience.iop.org/article/10. 1088/1748-9326/abee4e
8. Xu, B., Lin, B.: Factors affecting CO2 emissions in China's agriculture sector: evidence from geographically weighted regression model. Energy Policy, Elsevier 104(C), 404–414 (2017). https://ideas.repec.org/a/eee/enepol/v104y2017icp404-414.html
9. Shakoor, A., et al.: Effect of animal manure, crop type, climate zone, and soil attributes on greenhouse gas emissions from agricultural soils—a global meta-analysis. J. Clean. Prod. 278, 124019 (2021). https://doi.org/10.1016/j.jclepro.2020.124019
10. Wollenberg, E., et al.: Reducing emissions from agriculture to meet the 2 C target. Glob. Change Biol. 22(12), 3859–3864 (2016). https://onlinelibrary.wiley.com/doi/full/10.1111/ gcb.13340
11. FAO: Climate-smart agriculture case studies 2021 – projects from around the world. Rome (2021). https://doi.org/10.4060/cb5359en
12. IAEA: Greenhouse gas reduction. IAEA, 13 April 2016. https://www.iaea.org/topics/greenh ouse-gas-reduction. Accessed 19 Jan 2022
13. Wasonga, O.V., Olila, D.O.: Climate change, savanna grassland, autoregressive model, time series data (2016). https://www.semanticscholar.org/paper/Climate-change%2C-savanna-gra ssland%2Cautoregressive-Wasonga-Olila/8fbf81bf357e3b68ed7639d7acf1e4277de49441
14. Akcan, S., Kuvvetli, Y., Kocyigit, H.: Time series analysis models for estimation of greenhouse gas emitted by different sectors in Turkey. Hum. Ecol. Risk Assess. Int. J. 24(2), 522–533 (2018). https://doi.org/10.1080/10807039.2017.1392233

15. Hamrani, A., Akbarzadeh, A., Madramootoo, C.A.: Machine learning for predicting green-house gas emissions from agricultural soils. Sci. Total Environ. **741**, 140338 (2020). https://doi.org/10.1016/j.scitotenv.2020.140338
16. Gardner, E., McKenzie, E.: Why the damped trend works. J. Oper. Res. Soc. **62**, 1177–1180 (2011). https://doi.org/10.1057/jors.2010.37

A Novel Multimodal Fusion Technique for Text Based Hate Speech Classification

Pranav Shah[✉] and Ankit Patel

Department of Information Technology, Lalbhai Dalpatbhai College
of Engineering (LDCE), Ahmedabad 380015, Gujarat, India
200280723020@ldceahm.gujgov.edu.in, acpatel@ldce.ac.in

Abstract. Hate speech is when someone or a group of people is insulted
or stigmatized based on their background, colour, gender, religion, or
other traits. Additionally, social media generates massive amounts of
data every day. However, due to data peculiarities, a single classifier cannot deliver the heterogeneous feature for text classification. As
a result, a novel fusion RNN (BiLSTM-BiGRU)-Multichannel CNN-
Capsule Network-Attention (RMCCA) is presented in this research. The
proposed approach improves in classification improvement. By eliminating ambiguity and text granularities, the suggested method facilitates in
strengthening classification accuracy and ground truth evidence. Separate data sets are used to validate the suggested models. The empirical
results show that the offered methods produce sufficient hate speech classification results.

Keywords: Hateful-speech · Deep learning techniques · Fusion
multi-model techniques · Attention with capsules

1 Introduction

Many of people who use the internet, which becomes incredibly common as each
day. Every user can share their feelings which are easily available and accessible
[7,16]. There are various advantages of social media, such as the ability to connect
individuals, bring people together, and build businesses more quickly but modern
systems promotes vile content quickly published. It has increased recently, and
so have the incidence of cyberhate incidents. As a corollary, it has gained the
attention of hate speech researchers. According to the literature, those who have
negative attitudes are prejudiced against certain communities, minority groups,
females, or themselves [13]. As a result, these hostile words have a direct influence
on the victim's and their family's respect or loss of revenue [11,13]. When certain
incidents happen, cyberhate campaign, discriminatory speech, and remarks on
social media sites are usually triggered.

In latest days, intelligent techniques have been used widely to handle detection and classification difficulties, resulting in a potent approach. Machine learning techniques have been actively used [10]. For predicting and improving accuracy, this study proposes deep learning models Bi-directional LSTM-GRU and

M. Singh et al. (Eds.): ICACDS 2022, CCIS 1614, pp. 359–369, 2022.
https://doi.org/10.1007/978-3-031-12641-3_29

MCNN and capsule network with Attention mechanism [6]. The presented research approach is CNN-RNN with Capsule Network with Attention that elicits the attributes from MCNN-Bi-directional LSTM-GRU layers [2]. Dynamic routing can be thought of as a parallel attention mechanism that permits every capsule on one layer to pay attention to some active capsules underneath it while dismissing others. For a better understanding of features, the next Capsule network layer performs dynamic feature routing. It detects the word's real ground truth, which benefits the model in making better accurate predictions [14].

2   Related Research Work

A purpose of hateful content investigation means to discover, evaluate people's reactions toward others. This increased on popularity in latest decade, and as a result, it is becoming a significant research topic in the internet domain [17]. In 2019, they introduced a proposed feature algorithm for efficiently detecting this content. Then, to protect and detect online public humiliation. The spammers could be silenced and blocked by the application, that classifies humiliation into six categories: abusive, comparison, casting judgement on a person, sarcasm, and false equivalence. The MANDOLA have various advantages to identify hateful speech on social media. It employs a stacking ensemble master-slave classifier with three layers [17]. It built with combination of two classification models (i.e. logistic regression, Linear SVM). It produces excellent performance but takes a lot of time to calculate. Word embedding with various deep learning approaches have been used to classify drug addiction tweets. The suggested paradigm locates the paragraph's uncertainty for more accurate classification. As an outcome, identification was used to contend with ambiguous insights and evaluate performance. Research gap found for this related work is related to Inter sectionality of text which can be check further to get the more text diversity.

2.1   Motivation

Various nations have laws against hateful speech, and India, the nation's biggest democracy, has its own series of rules and regulations. In India Hate speech, whether said or written, is criminal under Indian law under sections 153(A) and 295(A). It results in a three years, or a fine [3,18]. As a consequence, it encourages us to look for and suppress hateful speech on social media. Furthermore, hate speech can grow to uncontrollable protests and riots, as well as antisocial behaviour [13].

3   Datasets Information

Datasets describes data gathering as well as needed information to identify and measure hate speech in a variety of contexts. In addition, for the creation of the dataset, the most recent trending topics in the world are taken into account.

3.1 Data Source

It is critical to realise that the most important component of hate speech analysis is gathering the dataset based on an events. As a consequence, we evaluated three datasets one is the current world level pandemic situation based i.e. COVID-19 Variant dataset, Second Vaccination dataset and final Indian Farms law Bill 2021 dataset using the hashtags protest of farmers, It has gathered information from Twitter. The timeline of this research datasets has taken from Nov 2021 to Jan 2022. All these datasets are collected from Twitter[1]. Initially data prepossessing steps applied on these datasets to remove unwanted noise, outliers and duplicates as a part of cleaning datasets (Table 1).

Table 1. Datasets description

Dataset	COVID-19 Variant	Farm law bill	Vaccination
	DS1	DS2	DS3
Total tweets	19000	57327	21118
Hate	6000	3049	5689
Non-hate	21000	57327	17804
% of hate	31.57	5.31	26.93

3.2 Datasets Processing and Preparation

As a part of data pre processing step, various Datasets prepossessing steps has been first. Thereafter, Stemming used in order to reduce noise and fasten task and then distribute it into hate and non-hate-related tweets. PyPI[2] is then used to gather hate-related vocabulary and offensive words. Subsequently, we added the identified high-intensity words that were determined to be hatred terms (Table 2).

4 Proposed Fusion Multi Modal Approach

It represents comprehensive description of hatred content classification approach. As illustrated in Fig. 1, this research presents a novel Fusion RNN (BiLSTM-BiGRU)-Multichannel CNN-CapsNet-Attention (RMCCA) method. The presented approach has been compared with existing methodologies. When it comes to distinguishing hate speech, the current research and development this novel fusion model outperform well then existing machine learning models and fuzzy logic Fuzzy-Minimum-normalization (FuzzyM1), Fuzzy-Product-Normalization (FuzzyM2), Fuzzy -lukasiewicz-Normalization (FuzzyM3), Fuzzy-Yager-Normalization (FuzzyM4), Fuzzy Hybrid with KNN (FuzzyM5 and M6).

---

[1] https://developer.twitter.com/en.
[2] https://pypi.org/.

Table 2. Hyperparameter setting for RMCCA fusion multi-modal

Hyperparameter	Values
Vector size	100
Convolution filters	100
Multi-CNN-dropout percentage	40%
Numbers of capsules	10
Routing	5
Dimensions of capsules	16
Capsule net-dropout percentage	25%
Convolution kernel size	3,4,5
Bidirectional GRU units	100
Batch size	64
Epochs	50

The basis for fusion model is to achieve more accuracy by using both features of RNN and Multi-channel CNN in addition to attention layer.

In the domain of text classification, CNN along with RNN has indeed been extensively explored that have shown remarkable outcomes. Given the fact that CNN could elicits local features across successive words in a phrase [20], it misses the contextual semantic features among sentences, there is an anterior-posterior relationship among texts. Bidirectional GRU could compensate for CNN's inability to extract contextual semantic features from large texts, As a consequence, we present a RNN (BiLSTM-BiGRU)-Multichannel CNN-CapsulNetwork-Attention (RMCCA) (Fig. 2).

By using attention mechanism after capsule network layer, the model could indeed concentrate attention towards the parts of the sentence (words) that really are essential for sentiment polarity categorization, combining the benefits of CNN for extracting local textual features and bi-directional LSTM and bi-directional GRU for extracting ground truth [15].

Since each distinct word is represented a feature, the text classification creates a vast dimensional corpus. As an outcome, the Fusion model is made up of several parallel architectures. Multi-channel CNN, Bi-LSTM-Bi-GRU, Capsule network with an attention mechanism are all included. Elicits unique features extracted using a single-channel CNN for text data categorization [3,19]. We have implemented multi-channel CNN to gain different attributes. Localized word features are obtained by multi-channel CNN algorithms. In addition, different hyperparameters for 3 separate CNN kernel in order to fetch different features [8,19]. The vector size of each feature was set to 30. In multi channel CNN, The filter of Three, four, five have applied to these word vector outputs [5]. In fundamental CNN architecture, the ReLU activation function is used to evaluate only active neurons for higher prediction.

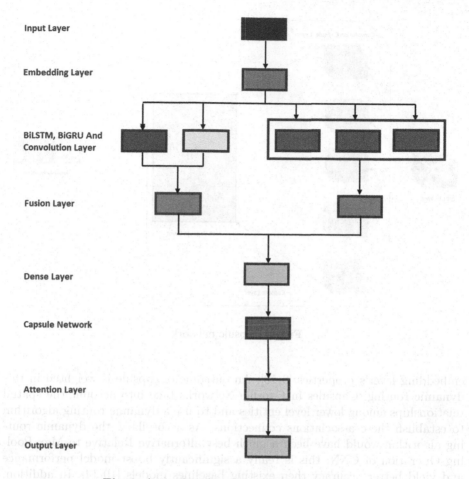

Fig. 1. A novel multi model fusion architecture

A presented neural network models have obtained significant application performance in specific sectors, demonstrating the Bi-vitality LSTM's in the discipline of sequential data processing. Bi-directional Gate Recurrent Units (Bi-GRU) by Cho in [4]. Unlike Bi-directional LSTM, Bi-directional Gate Recurrent Units only importantly requires two gates: an update and a reset gate. These two gates collect data and discover the hidden root truth of the words at the same time [1]. The update gate has been used to pass relevant data at a specific point in time. The reset gate removes all non-essential data and provides only the information needed [8]. This presented mechanism comprehend the true semantic meaning of a sentence in text categorization. We have employed word attention technique to choose ground truth. [19]. Moreover, This developed RNN-Multichannel CNN-CapsNet-Attention (RMCCA) Multi model CNN and Bi-directional GRU-Bidirectional LSTM-Capsule Network with Attention technique. Capsule Network for text categorization employs a deep learning based

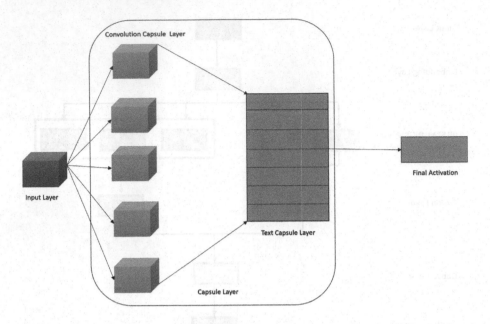

Fig. 2. Capsule network

embedding layer's properties [9,10]. An outcome of capsule is weighted in this dynamic routing. Capsules in Capsule Networks take into account the spatial relationships among lower level entities and to use a dynamic routing algorithm to establish these associations connections. As a corollary, the dynamic routing algorithm would have been a much best alternative Relative to Max Pooling Operation of CNN, this is really a significantly boost model performance and yield better accuracy then existing baselines models [10,14]. In addition, the inner product of the source capsule's outcomes is consumed. The output is then transferred to the higher who has the same capsule as the lower level text features. Then, CapsNet can readily processed and performed to solve problem, when other approaches fails to identify the relationships in whole sentence. For each of the Three datasets, this technique has been developed. The word embedding method that used Twitter for example transforms hundred matrix [19] (Fig. 4).

Three channels having filter sizes of three, four, and five [19]. A CapsNet detects the features as well as the contextual importance of every feature and how they interact [12]. We have implemented with total of ten capsules, fixed-sized at sixteen.

Confusion Matrix		Actual Classes	
		In fact is a positive (Positive)	In fact negative (Negative)
Predicted Classes	Positive prediction Class	TP (True Positive)	FP (False Positive) Type I Error
	Negative prediction Class	FN (False Negative) Type II Error	TN (True Negative)

Fig. 3. Confusion matrix for binary classification

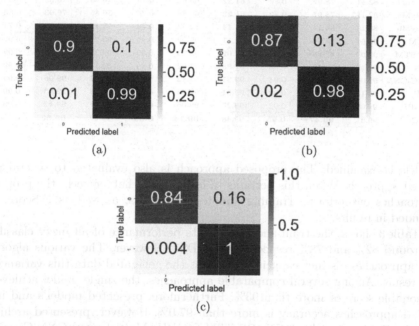

(a)

(b)

(c)

Fig. 4. Multichannel CNN-BiLSTM-BiGRU-CapsNet with attention models confusion matrix for (a) DS1, (b) DS2, (c) DS3.

5   Results and Discussion

Two distinct groups to assess this fusion model established in the class of unlabeled text i.e. hate speech and not hate speech the various sets of data are used assess algorithm implementation tp all four numbers constitute a confusion matrices as in Fig. 3.

The projected Fusion RNN (BiLSTM-BiGRU)-Multichannel CNN-CapsNet-Attention (RMCCA) findings are discussed in this section. It is essential to evaluate the research system's capacity on a variety of datasets under various conditions. With three datasets for hate speech analysis, the performance of RNN (BiLSTM-BiGRU)-Multichannel CNN-CapsNet-Attention (RMCCA) described

Table 3. Performance analysis of various methods.

Methods	Training Acc (%)	Testing Acc (%)	F-Score	Training Acc (%)	Testing Acc (%)	F-Score	Training Acc (%)	Testing Acc (%)	F-Score
	DS1			DS2			DS3		
GBT	82.45	72.26	0.85	94.78	93.63	0.95	97.76	97.16	0.97
DT	100	68.15	0.70	100	92.05	0.94	100	94.14	0.95
NB	71.65	71.19	0.77	84.98	84.63	0.89	84.95	82.78	0.92
SVM	72.54	71.31	0.83	94.42	94.06	0.95	96.55	95.15	0.97
DNN	82.16	76.25	0.82	94.56	93.04	0.98	95.26	95.97	0.97
FuzzyM1	82.11	78.97	0.82	61.22	60.82	0.74	76.85	76.03	0.83
FuzzyM2	82.11	78.97	0.82	61.22	60.82	0.74	76.85	76.03	0.83
FuzzyM3	82.11	78.97	0.82	61.22	60.82	0.74	76.85	76.03	0.83
FuzzyM4	82.11	78.97	0.82	61.22	60.82	0.74	76.85	76.03	0.83
FuzzyM5	82.11	78.97	0.82	61.22	60.82	0.74	76.85	76.03	0.83
FuzzyM6	88.15	84.14	0.87	76.14	72.23	0.76	75.12	73.37	0.74
LSTM	98.04	88.66	0.91	99.11	99.02	0.99	99.28	99.16	0.99
Bi-LSTM	99.03	87.68	0.90	99.02	98.73	0.99	99.35	99.14	0.99
CNN	99.12	94.16	0.95	97.83	97.02	0.99	99.11	99.04	0.99
MulC-CNN	99.06	93.04	0.94	99.37	97.12	0.98	99.17	99.06	0.99
GRU	99.11	91.05	0.93	99.31	99.04	0.99	99.28	99.03	0.99
Bi-GRU	99.08	92.08	0.92	99.27	99.05	0.99	99.31	99.12	0.99
RMCCA	99.31	**96.47**	**0.95**	99.48	**99.14**	**0.99**	99.42	**99.53**	**0.99**

models is examined. The proposed approach is also evaluated to several presented approach. When this pertains to categorising hate speech, the proposed approaches outperform. Training and testing results, as well as F-Score, are included in results.

Table 3 shows, the train and test results performance of all fuzzy classifiers is around 82% and 78% respectively for DS1. However, The various algorithmic approaches distinct capacity, represent the generated data this variance in the result. Among several comparable approaches, the single model achieves a reasonable score of more than 93%. Furthermore, presented model's and individual approaches accuracy is more than 97.0%. However, presented architecture for Dataset 1 yields RNN (BiLSTM-BiGRU)-Multichannel CNN-CapsNet-Attention (RMCCA) has the best result, with 96.47%. In covid-19 variant dataset, Several deep learning techniques obtain near-perfect training and testing performance of 97.0% and 99.0%. RNN (BiLSTM-BiGRU)-Multichannel CNN-CapsNet-Attention (RMCCA) has a performance of around 99.00%. The training accuracy for the suggested architectures is greater than 99.4%, while the testing accuracy is greater than 99.1%, as illustrated in the confusion matrix for the proposed RNN (BiLSTM-BiGRU)-Multichannel CNN-CapsNet-Attention (RMCCA). The impact of this empirical result is this fusion model saves the time as well as outperforms well in terms of accuracy than existing base-line models.

The proposed technique outperformed existing baseline models with a performance of about 99.0%. On the DS2 and DS3 datasets, the proposed model classifier has accuracy of 99.14% and 99.53%, respectively. Table 3 illustrated, on the other hand, indicates that various existing algorithms do not perform well

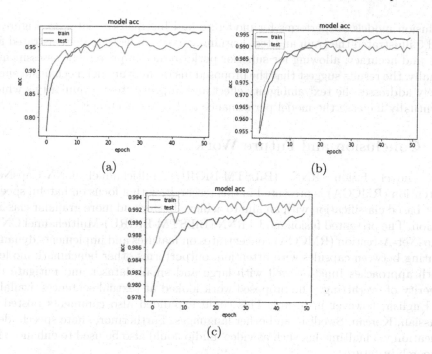

(a)

(b)

(c)

Fig. 5. Multichannel CNN-BiLSTM-BiGRU-CapsNet with attention model accuracy (a) DS1, (b) DS2, (c) DS3.

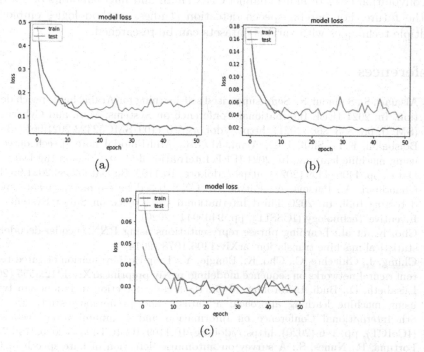

(a)

(b)

(c)

Fig. 6. Multichannel CNN-BiLSTM-BiGRU-CapsNet with attention model loss for (a) DS1, (b) DS2, (c) DS3.

as fusion models outperforms for moderate and big massive datasets. Figures 6 and 5 illustrated the train and test learning curve graphs were demonstrated for loss and accuracy, allowing for superior performance improvement assessments. Finally, the results suggest that the proposed fusion multimodel mechanism effectively addresses the text ambiguity effect and improves text granularity's which eventually increase the model performance and hence accuracy.

6 Conclusion and Future Work

The novel fusion RNN (BiLSTM-BiGRU)-Multichannel CNN-CapsNet-Attention (RMCCA) is presented in this research, with a focus on hateful speech text based classification with an aim to achieve better and more granular classification. The presented fusion model RNN (BiLSTM-BiGRU)-Multichannel CNN-CapsNet-Attention (RMCCA) concentrates on features and implements dynamic routing between capsules with attention, outperforms other benchmark models. Both approaches function well with large and small dataset and mitigate the severity of overfitting. The proposed work looked at textual sentences available in English, however in future, This could expanded also comments posted in Russian, Korean, Swedish, and other languages. Furthermore, hate speech identification via multimedia, such as video, audio could also be used to enhance this research in future.

To explore the feature diversity more, It can analyse with higher kernel size at convolution layer of multi channel CNN small and huge amounts of datasets in the future. For hateful speech detection studies, methodologies employing multiple techniques with various datasets can be researched.

References

1. Alsafari, S., Sadaoui, S.: Semi-supervised self-learning for Arabic hate speech detection. In: 2021 IEEE International Conference on Systems, Man, and Cybernetics (SMC), pp. 863–868 (2021). https://doi.org/10.1109/SMC52423.2021.9659134
2. Boishakhi, F.T., Shill, P.C., Alam, M.G.R.: Multi-modal hate speech detection using machine learning. In: 2021 IEEE International Conference on Big Data (Big Data), pp. 4496–4499 (2021). https://doi.org/10.1109/BigData52589.2021.9671955
3. Chaudhari, A., Parseja, A., Patyal, A.: CNN based hate-o-meter: a hate speech detecting tool. In: 2020 Third International Conference on Smart Systems and Inventive Technology (ICSSIT), pp. 940–944 (2020)
4. Cho, K., et al.: Learning phrase representations using RNN encoder-decoder for statistical machine translation. arXiv:1406.1078 (2014)
5. Chung, J., Gulcehre, C., Cho, K., Bengio, Y.: Empirical evaluation of gated recurrent neural networks on sequence modeling. arXiv preprint arXiv:1412.3555 (2014)
6. Elisabeth, D., Budi, I., Ibrohim, M.O.: Hate code detection in Indonesian tweets using machine learning approach: a dataset and preliminary study. In: 2020 8th International Conference on Information and Communication Technology (ICoICT), pp. 1–6 (2020). https://doi.org/10.1109/ICoICT49345.2020.9166251
7. Fortuna, P., Nunes, S.: A survey on automatic detection of hate speech in text. ACM Comput. Surv. 51(4), 1–30 (2018)

8. Goodfellow, I., Bengio, Y., Courville, A., Bengio, Y.: Deep Learning, vol. 1. MIT Press, Cambridge (2016)
9. Greff, K., Srivastava, R.K., Koutník, J., Steunebrink, B.R., Schmidhuber, J.: LSTM: a search space odyssey. IEEE Trans. Neural Netw. Learn. Syst. **28**(10), 2222–2232 (2016)
10. Huang, X., Xu, M.: An inter and intra transformer for hate speech detection. In: 2021 3rd International Academic Exchange Conference on Science and Technology Innovation (IAECST), pp. 346–349 (2021). https://doi.org/10.1109/IAECST54258.2021.9695652
11. Khan, H., Yu, F., Sinha, A., Gokhale, S.S.: A parsimonious and practical approach to detecting offensive speech. In: 2021 International Conference on Computing, Communication, and Intelligent Systems (ICCCIS), pp. 688–695 (2021). https://doi.org/10.1109/ICCCIS51004.2021.9397140
12. Kim, J., Jang, S., Park, E., Choi, S.: Text classification using capsules. Neurocomputing **376**, 214–221 (2020)
13. Liu, H., Burnap, P., Alorainy, W., Williams, M.L.: A fuzzy approach to text classification with two-stage training for ambiguous instances. IEEE Trans. Comput. Soc. Syst. **6**(2), 227–240 (2019)
14. Mayda,, Demir, Y.E., Dalyan, T., Diri, B.: Hate speech dataset from Turkish tweets. In: 2021 Innovations in Intelligent Systems and Applications Conference (ASYU), pp. 1–6 (2021). https://doi.org/10.1109/ASYU52992.2021.9599042
15. Naidu, T.A., Kumar, S.: Hate speech detection using multi-channel convolutional neural network. In: 2021 3rd International Conference on Advances in Computing, Communication Control and Networking (ICAC3N), pp. 908–912 (2021). https://doi.org/10.1109/ICAC3N53548.2021.9725696
16. Naseem, U., Razzak, I., Eklund, P.W.: A survey of pre-processing techniques to improve short-text quality: a case study on hate speech detection on twitter. Multimedia Tools Appl. 1–28 (2020)
17. Paschalides, D., et al.: Mandola: a big-data processing and visualization platform for monitoring and detecting online hate speech. ACM Trans. Internet Technol. **20**(2), 1–21 (2020)
18. Sachdeva, J., Chaudhary, K.K., Madaan, H., Meel, P.: Text based hate-speech analysis. In: 2021 International Conference on Artificial Intelligence and Smart Systems (ICAIS), pp. 661–668 (2021). https://doi.org/10.1109/ICAIS50930.2021.9396013
19. Wang, J., Yu, L., Lai, K.R., Zhang, X.: Tree-structured regional CNN-LSTM model for dimensional sentiment analysis. IEEE/ACM Trans. Audio Speech Lang. Process. **28**, 581–591 (2020)
20. Watanabe, H., Bouazizi, M., Ohtsuki, T.: Hate speech on twitter: a pragmatic approach to collect hateful and offensive expressions and perform hate speech detection. IEEE Access **6**, 13825–13835 (2018). https://doi.org/10.1109/ACCESS.2018.2806394

Detection of Breast Tumor in Mammograms Using Single Shot Detector Algorithm

S. Ruban[1]([✉]), M. M. Jabeer[1], and Ram Shenoy Besti[2]

[1] St Aloysius College (Autonomous), Mangalore, India
rub2kin@gmail.com
[2] Father Muller Medical College, Mangalore, India

Abstract. The influence of AI over Healthcare is increasing day by day. Lot of efforts are being made to increase the efficiency of Cancer diagnosis at the early stages, where medical images play a vital role. Among the women, Breast cancer is on a steep rise with an alarming statistics of, Indian woman being diagnosed with Breast cancer, for every four minutes. Though Breast cancer is a curable disease, identifying the problem early is an important metric for the successful outcome. Earlier days, Mammograms were the only way of detecting Breast cancer. However, Mammogram is not effective for women belonging to every age, and hence yields to many false positive and false negative cases. This leads to disastrous results. To overcome this limitation, this experimental study tells, about using one of the Deep Learning algorithm called Single Shot Detector algorithm to detect the cancerous portion in the Mammograms. This experiment was done in the Real Time Data set of mammograms collected from a 1250 bed hospital, after the scientific and Ethical committee approval. With a total of 80 mammography images belonging to either category of being, either Malignant or Benign. The Single Shot Detector algorithm gave an accuracy of 74%, which was better compared with CNN (64%) and VGG (60%) deep learning algorithms.

Keywords: Deep learning · Convolutional neural network · Single shot detector · X-ray · Mammography · Breast cancer detection · Visual geometry group

1 Introduction

Breast cancer is becoming more common in our country as a result of a lack of awareness and other causes. Though this cancer can be diagnosed very easily, it is considered to be the most prevalent cancer [1] that leads to deaths. This Cancer comes second among the U.S. women [2]. Lifestyle changes [3], and also the lack of awareness among the women in the community [4] are few reasons for the increase. Breast cancer cells usually creates a lump that can be felt or seen on an x-ray. It's important to keep in mind that most breast lumps aren't cancerous (malignant). Breast tumors that aren't cancerous and don't spread outside the breast are known as noncancerous. Although benign breast sizes may not provide a life-threatening threat, they do increase a woman's risk of developing cancer. Any lump or alteration in the breast should be evaluated by a healthcare professional

M. Singh et al. (Eds.): ICACDS 2022, CCIS 1614, pp. 370–380, 2022.
https://doi.org/10.1007/978-3-031-12641-3_30

to find any risk associated with that. Many studies are done globally to understand the awareness among the women [5]. Early detection and progressive cancer treatment are the most essential strategies for avoiding breast cancer deaths [6].

It is possible to treat a carcinoma if it is identified early. Cancer that is diagnosed early, when it is little and has not spread, is easier to treat. Frequent screening examinations are the most accurate way to detect breast cancer early [7]. Mammograms, or low-dose chest x-rays, are the most common method of screening. Mammograms can detect cancerous changes in the breast, years before physical symptoms appear, and decades of research shows that, women who have regular mammograms get identified early [8], and hence require fewer treatments such as breast removal surgery (mastectomy), and are more likely to be cured.

Mammograms do have several problems. Some malignancies cannot be traced. And, in rare occasions, a woman will want additional evidence to establish whether something spotted on a mammogram is malignant. A radiologist is responsible for writing a report after evaluating the mammography, and in some cases could be false positive or negative depending upon many factors. However, Radiologists have been able to give an improved suggestion using Artificial Intelligence [9]. AI promises more accurate diagnosis [10]. This study looks at how the Single Shot Detector algorithm, can assist the radiologists make better decisions when it comes to detecting breast cancer from mammograms. The model generated using the above technique can assist the radiologist by providing additional input from the machine, indicating whether the mammography image contains malignant tissue or not, and thereby reducing false positive and negative cases. and improvises the accuracy of detecting breast cancer. The next section deals with the existing work in this area, Sect. 3 describes the methodology that is adopted for this research study, Sect. 4 describes the results and discussion and finally the conclusion.

2 Literature Survey

Few of the earlier works in detecting Breast Cancer, by applying Deep Learning Algorithms, points to the usage of CNN algorithm [11] to detect lesions in ultrasound pictures. The researchers automated the entire procedure and, it improves the accuracy of lesion diagnosis and, as a result, lowers breast cancer mortality. Similarly, the K-medoid clustering approach is used to acquire and assess breast cancer data [12]. Association Rule Mining and Neural Networks [13] was employed to diagnose breast cancer in their experimental investigation.

Another novel hybrid technique for classifying cancer that integrates decision trees with many other techniques [14] was proposed. Another experimental investigation was performed to classify cancer by developing a classifier [15] based on collaborative representation. Machine Learning was utilized [16] to locate and classify cancer main sites. The authors developed a scoring system to detect the presence of malignant cells in mammography images in the experimental trials stated in [17, 18].

Few of the other researchers, of the experimental investigation [19] describes the medical imaging modalities, that are used to evaluate breast images and shows that cancer symptom detection accuracy can be improved. The authors [20] employed machine learning algorithms for lung cancer diagnosis. Another study cited for this work employed patch-level categorization and a deep learning technique to detect lung cancer [21]. If cancer is suspected, the patient goes through a series of tests before being screened for malignant tumor cells. To diagnose lung cancer, doctors offer medical imaging methods. Imaging techniques help with the investigation of suspicious areas, the diagnosis of cancer stage, the confirmation of treatments, and the detection of signs that cancer cells are returning. Another team of researchers [22] developed an end-to-end breast cancer detection system for detecting breast cancer in mammography pictures. The first and most crucial task in the planned diagnosis of cancer is to identify cells.

In another research work [23], the researchers adopted a hybrid methodology and combined the CNN algorithm with other strategies, to make it considerably more efficient. The research of the authors reported in the work [24] focuses on the use of machine learning algorithms for early detection. In the experimental investigation [25], another similar effort on detection is given.

This study is aimed at improvising the detection of breast cancer, for patients who undergo Mammography tests in the Department of Radio Diagnostics and Imaging in the Father Muller Hospital. Hence the data was captured from the Father Muller Hospital and the performance evaluation was done over the Real time data.

3 Methodology

The best approach to prevent the disease from spreading is to detect cancer cells at an early stage. Cancer cells can be discovered at an early stage by utilizing deep learning algorithms, to detect the location of the cancer cell in the mammogram, which may not be visible by the naked eye. Hence to model such a method, the Mobinet SSD model is used. It is used to detect and highlight cancer cells in the mammographic images. The architecture is elaborated below. The proposed framework explains how the Mammogram, which is a DICOM image, is annotated by the Radiologist and the system is trained to identify the Cancerous cells in the Mammogram. Once annotated and trained, the algorithm can detect the Cancerous cells in the Mammogram, which helps the radiologist to fine tune his diagnosis. The workflow and the Methodology followed in this experimental study is elaborated, and is represented in Fig. 1.

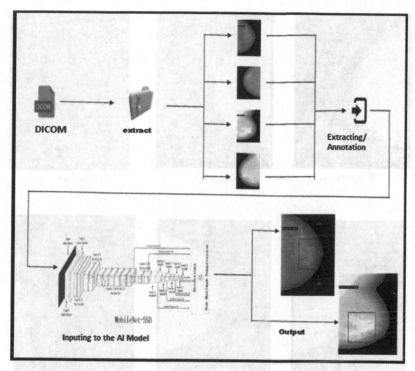

Fig. 1. Methodology breast tumour detection using SSD algorithm

3.1 Data Collection

The Mammography Images were taken from the Department of Radio Diagnostics and Imaging of the Father Muller Hospital in Mangalore. The data set, comprises of mammography images of patients, who underwent Breast cancer screening in the previous years. The Data set was blinded, considering the confidentiality of the patients as recommended by the Ethical committee. It has a total of 80 Mammogram Images in it. The data collection includes both benign and malignant mammography images. There were more benign mammography images and fewer malignant mammography images. Few of the images are presented below in Fig. 2.

3.2 Data Pre-processing

After receiving the mammogram Dicom format Images, they were annotated and labelled. The tool that was utilized for this was LabelImg. It's a graphical picture annotation tool which is available as open source. XML files were used to save the annotations. After that, the XML files were transformed to CSV files. The tfrecord format was used to transform the CSV files. Finally, the SSD models were trained using the tfrecord files as input data. The image was normalized, and the label indicates whether or not the image is malignant. The dataset is then split into two sections: training and testing.

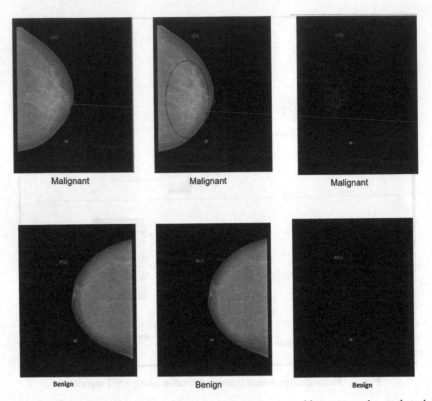

Fig. 2. A snapshot of the various Mammography Images used in our experimental study

3.3 Data Processing Using Single Shot Detection Algorithm

Deep Learning approaches for working with Images will be extremely valuable for data scientists in charge of obtaining, analyzing, and interpreting large volumes of data. Though there are various algorithms such as CNN, ANN etc., The Single Shot Detector (SSD) algorithm was used in this experimental study, owing to its strength and usage in object detection. The architecture is presented in Fig. 3.

The SSD algorithm is a popular object detection method. In the vast majority of circumstances, it outperforms Faster RCNN. The SSD design is made up of a single neural network that learns to predict and classify bounding box positions in one step. As a result, the SSD can be built from start to finish. A core architecture (in this example, Mobile Net) with multiple levels of convolution makes up the SSD network. SSD just needs one shot to distinguish many items in an image, whereas RPN-based systems like the RCNN series require two shots of view, one to generate regional proposals and the other to determine the topic of each proposal. As a result, SSD outperforms RPN-based methods.

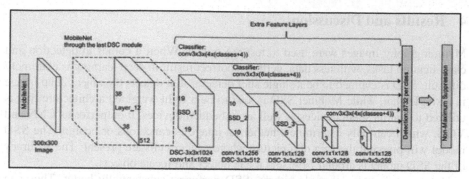

Fig. 3. A snapshot of the Single Shot Detection (SSD) architecture

3.4 Working Principle of the Single Shot Detector Algorithm

The Single shot multibox detector is a method that is accomplished with the help of a single deep neural network. This detector can detect items of various sizes and scales in the image since it works at a range of scales. For feature extraction, SSD often employs an auxiliary network. This is also known as the foundation network. With the intermediate tensors kept, more convolution layers are added, resulting in a stack of feature maps of varying sizes for detection. Suppose, there are feature layer with an x, b, and c channel dimension. Then, convolution is applied to an x, b, and c feature layer (usually 3 x 3). So there are k bounding boxes conceivable for each location of the objects recognized, each with a probability score. Finally, Non-max suppression is employed to ensure that an object is surrounded by only one bounded box. As a result, there is just one bounded box around a single identified object. The method is named Non-maxima Supression because it suppresses any boxes with non-maximum values. The working principle of SSD algorithm is presented in Fig. 4.

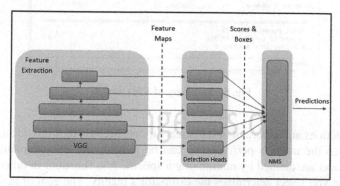

Fig. 4. A snapshot of the Single Shot Detection (SSD) working principle

4   Results and Discussion

Mammography Images were used to test the model. When it comes to detection and classification tasks, with less data, getting a correct result in deep learning is extremely difficult. SSD is claimed to be a single shot detector that detects the object in an image in a single shot, while Mobinet SSD is said to be a light weighted architecture that is utilized to run on small devices as well as embedded devices. It outperforms CNN and VGG, which can only determine whether an image is cancerous or benign. The SSD model will pinpoint the area of a picture where cancer cells are present. The accuracy of the SSD model is around 74%, and it can detect and locate objects.

In comparison to other algorithms, SSD performs substantially better. They can detect whether a picture is benign or malignant, as well as find cancerous cells in the image, with a 74% accuracy. As a result, the SSD approach performs better for this real-time data collection. Few of the insights related to the Accuracy and Loss of the model, confusion matrix that were generated are presented below from Figs. 5, 6 and 7.

```
Model: "sequential"

Layer (type)                 Output Shape              Param #
=================================================================
conv2d (Conv2D)              (None, 126, 126, 32)      896

max_pooling2d (MaxPooling2D  (None, 63, 63, 32)        0
)

conv2d_1 (Conv2D)            (None, 60, 60, 64)        32832

max_pooling2d_1 (MaxPooling  (None, 15, 15, 64)        0
2D)

conv2d_2 (Conv2D)            (None, 13, 13, 32)        18464

max_pooling2d_2 (MaxPooling  (None, 6, 6, 32)          0
2D)

flatten (Flatten)           (None, 1152)              0

dense (Dense)               (None, 128)               147584

dense_1 (Dense)             (None, 1)                 129

=================================================================
Total params: 199,905
Trainable params: 199,905
Non-trainable params: 0
_____
None
```

Fig. 5. Summary of SSD model.

Loss functions are important in statistical models because they provide a benchmark against which the model's performance can be judged, and the parameters obtained from the model are defined by minimizing a specific loss function. In other words, the loss function you select determines the estimator's quality. The goal of this study is to examine loss functions, their usefulness in validating predictions, and the many loss functions that are used. The Classification Loss of the model is presented in the Fig. 8.

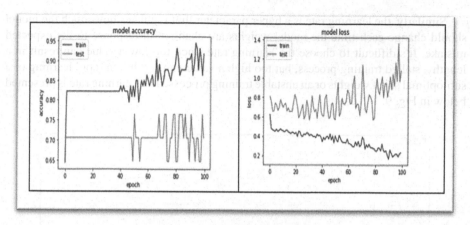

Fig. 6. Accuracy and loss of the SSD model.

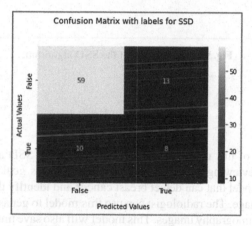

Fig. 7. Confusion matrix of SSD model.

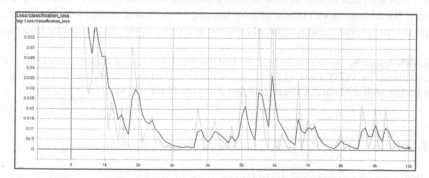

Fig. 8. Classification loss of SSD model.

Similarly, the learning rate is a hyper parameter that regulates how much the model should change each time the model weights are changed in response to the expected mistake. It's difficult to choose the learning rate since too low a value can result in a lengthy, stalled training process, but too high a value can result in a rapid learning of a sub-optimal set of weights or an unstable training process. The Learning rate is presented below in Fig. 9.

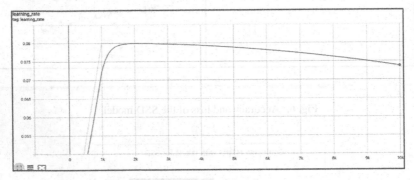

Fig. 9. Learning rate of the SSD algorithm.

5   Conclusion

The major purpose of this research work is to see how the SSD algorithm can assist the radiologist in reviewing the mammography image. This goal was proven by the development of a model that can detect breast cancer and identify the malignant cell in a mammography image. The radiologist can use this model to get assistance, during the examination of mammography images. This model will also save time for the radiologist while interpreting the mammography image and drafting the report. As a result, this model will aid radiologists in their decision-making when it comes to detecting Breast Cancer. By adding more mammography images to the training data set, from different hospital settings and various other places, the accuracy of the model can be improved.

Acknowledgments. Authors acknowledge, that this work was carried out in the Big Data Analytics Lab funded by VGST, Govt. of Karnataka, under K-FIST(L2)-545, and the data was collected from Father Muller Medical College, protocol no: (FMMCIEC/CCM/2165/2021).

References

1. World Health Organization. Breast Cancer? https://www.who.int/news-room/fact-sheets/det ail/breast-canceruse. Accessed 28 Oct 2021
2. American Cancer Society. How common is Breast Cancer? https://www.cancer/breast-can cer/about/howcommon-is-breast-cancer.html. Accessed 28 Jan 2022

3. Antony, M.P., Surakutty, B., Vasu, T.A., Chisthi, M.: Risk factors for breast cancer among Indian women: a case-control study. Niger J. Clin. Pract. **21**(4), 436–442 (2018)
4. Prusty, R.K., Begum, S., Patil, A., et al.: Knowledge of symptoms and risk factors of breast cancer among women: a community based study in a low socio-economic area of Mumbai, India. BMC Women's Health **20**, 106 (2020)
5. Abeje, S., Seme, A., Tibelt, A.: Factors associated with breast cancer screening awareness and practices of women in Addis Ababa, Ethiopia. BMC Womens Health **19**(1), 4 (2019)
6. Agide, F.D., Sadeghi, R., Garmaroudi, G., Tigabu, B.M.: A systematic review of health promotion interventions to increase breast cancer screening uptake: from the last 12 years. Eur. J. Pub. Health **28**(6), 1149–1155 (2018)
7. Sathwara, J., Balasubramaniam, G., Bobdey, S., Jain, A., Saoba, S.: Sociodemographic factors and late-stage diagnosis of breast cancer in India: a hospital-based study. Indian J. Med. Paediatr. Oncol. **38**(3), 277–281 (2017)
8. Vishwakarma, G., Ndetan, H., Das, D.N., Gupta, G., Suryavanshi, M., Mehta, A., et al.: Reproductive factors and breast cancer risk: a meta-analysis of case-control studies in Indian women. South Asian J Cancer. **8**(2), 80–84 (2019)
9. Rodríguez-Ruiz, A., et al.: Detection of breast cancer with mammography: effect of an artificial intelligence support system. Radiology **00**, 1–10 (2019)
10. Rodriguez-Ruiz, A., et al.: Stand-alone artificial intelligence for breast cancer detection in mammography: comparison with 101 radiologists. J. Natl. Cancer Inst. **111**(9), 916–922 (2019)
11. Yap, M.H., et al.: Automated breast ultrasound lesions detection using convolutional neural networks. IEEE J. Biomed. Health Inform. **22**(4), 1218–1226 (2018)
12. Ping, Q., Yang, C.C., Marshall, S.A., Avis, N.F., Ip, E.H.: Breast cancer symptom clusters derived from social media and research study data using improved K-Medoid clustering. IEEE Trans. Comput. Soc. Syst. **3**(2), 63–74 (2016)
13. Both, A., Guessoum, A.: Classification of SNPs for breast cancer diagnosis using neural-network-based association rules. In: 12th IEEE International Symposium on Programming and Systems (2015)
14. Vosooghifard, M., Ebrahimpour, H.: Applying grey wolf optimizer-based decision tree classifier for cancer classification on gene expression data. In: 5th International Conference on Computer and Knowledge Engineering (ICKE), Mashhad, pp. 147–151 (2015)
15. Wang, S., Chen, F., Gu, J., Fang, J.: Cancer classification using collaborative representation classifier based on non-convex lp-norm and novel decision rule. In: Seventh International Conference on Advanced Computational Intelligence (ICACI), Wuyi, pp. 189–194 (2015)
16. Chen, Y., et al.: Classification of cancer primary sites using machine learning and somatic mutations. BioMed. Res. Int. (2015)
17. Wahab, N., Khan, A.: Multifaceted fused-CNN based scoring of breast cancer whole-slide histopathology images. Appl. Soft Comput. (97), 106808 (2020)
18. Gravina, M., Marrone, S., Sansone, M., Sansone, C.: DAECNN: exploiting and disentangling contrast agent effects for breast lesions classification in DCE-MRI. Pattern Recogn. Lett. **145**, 67–73 (2021)
19. Murtaza, G., et al.: Deep learning-based breast cancer classification through medical imaging modalities: state of the art and research challenges. Artif. Intell. Rev. **53**(3), 1655–1720 (2019). https://doi.org/10.1007/s10462-019-09716-5
20. Tripathi, P., Tyagi, S., Nath, M.: A comparative analysis of segmentation techniques for lung cancer detection. Pattern Recogn. Image Anal. **29**(1), 167–173 (2019). https://doi.org/10.1134/S105466181901019X
21. Vu, Q.D., et al.: Methods for segmentation and classification of digital microscopy tissue images. Front. Bioeng. Biotechnol. **7**(53) (2019)

22. Kumar, P., Srivastava, S., Mishra, R.K., Sai, Y.P.: End-to-end improved convolutional neural network model for breast cancer detection using mammographic data. J. Defense Model. Simul. Appl. Methodol. Technol. (2020)
23. Ramadan, S.Z.: Using convolutional neural network with cheat sheet and data augmentation to detect breast cancer in mammograms. Comput. Math. Methods Med. (2020)
24. Mehmood, M., Ayub, E., Ahmad, F., et al.: Machine learning enabled early detection of breast cancer by structural analysis of mammograms. Comput. Mater. Continua **67**(1), 641–657 (2021)
25. Zhang, Y., Chan, S., Park, V.Y., et al.: Automatic detection and segmentation of breast cancer on MRI using mask R-CNN trained on non-fat-sat images and tested on fat-sat images. Acad. Radiol. 1–10 (2020)

Monthly Runoff Prediction by Hybrid CNN-LSTM Model: A Case Study

Dillip Kumar Ghose, Vinay Mahakur, and Abinash Sahoo[(✉)]

Department of Civil Engineering, National Institute of Technology Silchar, Silchar, Assam 788010, India
bablusahoo1992@gmail.com

Abstract. Prediction of runoff plays a vital part in planning and managing water resources. To model hydrological processes, advances in artificial intelligence methods can act as powerful tools. A novel hybrid model named CNN-LSTM is employed by incorporating a long short-term memory and convolutional neural network to improve prediction accuracy. Application of applied hybrid model is exemplified utilising hydrologic data from Barak River Basin in Cachar district of Assam, India, to test model accuracy. The best performing model was chosen on basis of different performance assessment standards, i.e., R^2, IA and RMSE. The outcomes revealed that CNN-LSTM model performs superiorly with $R^2 = 0.9875$, RMSE = 1.364 and IA = 0.9897 during training phase compared to standalone CNN model. The present study's result suggests that proposed CNN-LSTM network is a vital tool for runoff prediction in catchments and assists in providing viable measures for a catchment.

Keywords: Runoff · Cachar · CNN · CNN-LSTM

1 Introduction

Predicting runoff generated by a watershed is a key area of hydrologic research. Because of the large number of variables involved in the modelling process and the high geographical and temporal variability of watershed parameters, it is most difficult to understand the hydrological processes for measuring runoff. Forecasting runoff and rainfall variables are important for reservoir management, risk assessment, disaster management, flood prevention, and irrigation, among other things [1, 2]. And so, rainfall-runoff modelling is very significant in both environmental and hydrological studies. Forecasting models can be classified into data-driven models and process-based models [3, 4]. A physical-based model is based on physical codes, providing intuitions into physical procedures [4], depending on various simplifying presumptions, and requiring a huge amount of data [5].

© The Author(s), under exclusive license to Springer Nature Switzerland AG 2022
M. Singh et al. (Eds.): ICACDS 2022, CCIS 1614, pp. 381–392, 2022.
https://doi.org/10.1007/978-3-031-12641-3_31

In contrast, a data-driven model is an empirical model based on historical recorded data, straightforward and simple to implement, and do not need details about the underlying physical process [1]. To construct data-driven models, machine learning approaches are frequently used. In reality, the implementation of conventional quality models is reduced due to model complexity and a lack of understanding of stormwater runoff quality. Rainfall–runoff modelling [6–8], precipitation [9, 10] and flow forecast [11–13], and drought forecasting [14, 15] have successfully used ANNs. ANN model is widely used in prediction of runoff, but there is some limitation, as lots of datasets are required to develop a runoff model. Still, ANN is not able to compute huge data sets accurately and extract invariant structures and hidden features in datasets. Therefore, to eliminate these disadvantages of ANN, some researchers focused on a new model based on CNNs.

1.1 Literature Review

Convolutional neural networks (CNNs) use separate convolution operations on basis of a filter bank to detect local conjunctions of features [16, 17]. CNNs are useful for extracting hidden features and invariant structures in data, which is why they have been utilised in various applications [18, 19]. Gao et al. [20] employed two popular alternates of RNN (Recurrent Neural Networks), namely LSTM and GRU (Gated Recurrent Unit) networks and ANN for runoff simulation in Yutan gauge station, Fujian Region, China. Results showed that prediction accurateness of GRU and LSTM models increases with an increase in time step, and both models performed better than ANN models. Van et al. [21] established a rainfall-runoff model applying novel 1D-CNN with a ReLU activation function. Results of developed model are compared with that of LSTM and other conventional models. Their findings revealed that LSTM and CNN performed better than traditional models, with CNN performing slightly superior to LSTM. Ouma et al. [22] compared WNN (wavelet neural network) and LSTM to predict spatio-temporal variation of trends in rainfall-runoff time-series in hydrologic basins with scarce gauge stations. Han and Morrison [23] presented an application of ANN, SVM (Support vector machine), and LSTM approaches to predict runoff in Russian River basin, California, USA. They found that LSTM and SVM models outperformed ANN model in forecasting hourly runoff.

On the other hand, traditional approaches frequently encounter difficulties while solving optimization problems in the real world, needing a huge amount of memory, a long computing time, generating low-quality solutions, and perhaps becoming stuck in local optima. Ding et al. [24] developed a flood forecasting model on basis of STA-LSTM (Spatio-Temporal Attention). Wu et al. [25] proposed a CNN-LSTM model for merging TRMM 3B42 V7 satellite data, thermal infrared images, and rain gauge data by simultaneously developing their temporal and spatial correlations in China. Baek et al. [26] combined CNN-LSTM with deep learning to predict the Nakdong river basin's water level and quality. Barzegar et al. [27] projected a hybrid CNN-LSTM deep learning (DL) model to estimate accurate lake water levels. Ghimire et al. [28] developed a combined CNN-LSTM model to calculate hourly streamflow (short-term) at Teewah Creek and Brisbane River, Australia.

The main purpose of this study is to discover the hybrid CNN-LSTM model to predict monthly runoff in Cachar district, Assam. This is the first time CNN-LSTM has been used for monthly runoff prediction in this specific region, which is most remarkable.

2   Study Area

The district Cachar is located in southern region of Assam between 92° 24'E, and 93° 15'E longitudes, 24° 22'N, and 25° 8'N latitudes. It is bounded by Jayantia and Barail hills in north, Mizoram state in south, and Hailakandi and Karimganj districts in west with a total geographical area of 3,786 km^2. Major river of this district is Barak, and it receives a mean annual precipitation of around 3,000 mm. In Cachar, the climate is significantly humid, and because of this, the summer is intolerable. Proposed study area is showing the specified gauge station is presented in Fig. 1.

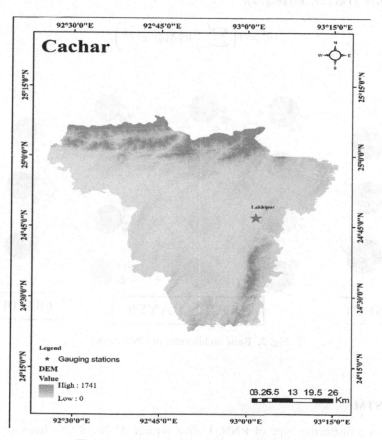

Fig. 1. Location of selected rain-gauge station

3 Methodology

3.1 CNN

CNN is a deep learning network structure learning from input directly, eliminating the necessity of manual feature extraction. A CNN can include tens or hundreds of layers, learning to recognize various picture aspects. Filters are employed at different resolutions for each trained image, and outcome of each convolved image is utilized as input to subsequent layer. Filters can begin with essential qualities like edges and brightness and progress to more complicated characteristics that uniquely describe the object. The convolutional layer, which comprises many convolution kernels, seeks to study feature input representations (or filter banks). Before being fed into the next layer, the convolutional layer's outputs are agreed via a nonlinear activation function. A convolutional layer h, is built using a series of $k = 1, \ldots, N_K$ when the input is a 1-D signal, small filters ($L \times 1$) as [29, 30] (Fig. 2):

$$h_i^k = f\left(\sum_{l=1}^{L} w_l^k X_{i+l} + b^k\right) \tag{1}$$

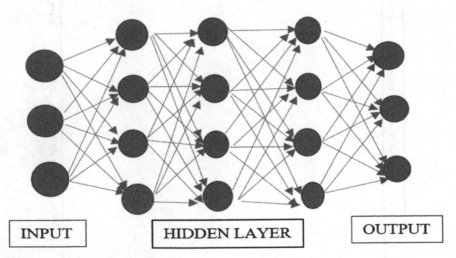

INPUT   HIDDEN LAYER   OUTPUT

Fig. 2. Basic architecture of CNN model

3.2 LSTM

LSTM is a distinctive type of RNN. Unlike typical ANNs, RNNs have a recurrent hidden component forming a self-coiled cycle to keep track of the history of all previous components in the arrangement [16, 31]. Each time step in LSTM comprises a distinct status termed cell state C, which holds information about long-term memory.

In CNN-LSTM, a heap of convolutional layers is employed for capturing sequential structures of variables. Components obtained by CNN are served into a two-layer LSTM network having a compact layer on the top. After that, the entire CNN-LSTM model is trained for computing p-stage ahead forecasts utilising q lag records. Different steps involved in a CNN-LSTM forecast model are:

1. Let $Y, t = 1, 2, 3, \ldots, n$, denoting objective variable time-series; and $X_1, X_2, X_3 \ldots, X_N, t = 1, 2, 3, \ldots, n$, denoting descriptive variables time-series.

2. Deciding lagged-time (q, number of preceding time steps for using as inputs).

3. Selecting size and number of convolutional screens.

4. Dividing time-series into testing and training datasets (X_{train}, Y_{train}) and (X_{test}, Y_{test}) respectively.

5. Using the train set for training CNN-LSTM for p-stage ahead forecasts.

6. Using the test set for obtaining p-stage ahead forecasts.

The specified flowchart of CNN-LSTM is given in Fig. 3.

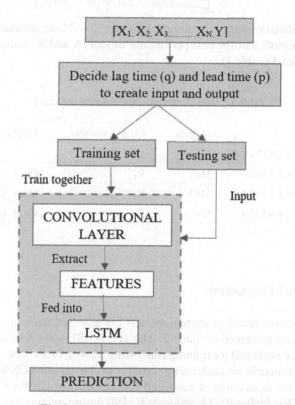

Fig. 3. Flowchart of CNN-LSTM algorithm

3.3 Evaluating Constraint

For estimating forecasting accurateness of applied hybrid and conventional models, performance metrics, namely coefficient of determination (R^2) [32, 33], index of agreement (IA) [34, 35], and root-mean-square error (RMSE) [36] were used. The three performance measures are expressed using following equations:

$$RMSE = \sqrt{\frac{1}{n} \sum_{i=1}^{n} (y_e - y_o)^2} \tag{2}$$

$$R^2 = \left(\frac{\sum_{i=1}^{n} y_e - \overline{y_e})(y_o - \overline{y_o})}{\sqrt{\sum_{i=1}^{n} y_e - \overline{y_e})^2 (y_o - \overline{y_o})^2}} \right)^2 \tag{3}$$

$$IA = 1 - \left[\frac{\sum_{k=1}^{N} (y_o - y_e)^2}{\sum_{k=1}^{N} (|y_e - \overline{y_o}| + |y_e - \overline{y_e}|)^2} \right] \tag{4}$$

where y_e = simulated runoff; y_o = observed runoff; $\overline{y_e}$ = Mean simulated runoff; $\overline{y_o}$ = Mean observed runoff. For the best performing model IA and R^2 must be highest and RMSE must be least (Table 1).

Table 1. Modelling input combination structure

Model description		Models	Output variable	Input combinations
CNN	CNN-LSTM			
CNN1	CNN-LSTM1	M#1	R_t	Q_{t-1}
CNN 2	CNN-LSTM 2	M#2	R_t	Q_{t-1}, Q_{t-2}
CNN 3	CNN-LSTM 3	M#3	R_t	Q_{t-1}, Q_{t-2}, P_t

4 Results and Discussion

The outcomes of two runoff prediction models are assessed utilising quantitative statistical metrics and presented in Table 2. The R^2, RMSE and IA of CNN and CNN-LSTM models are evaluated for training and testing phases. From Table 2, it is observed that CNN-LSTM models are performing superior to conventional CNN models. For all input scenarios, the assessment of estimated runoff values of CNN-LSTM models and observed runoff has higher R^2, IA and least RMSE during training and testing than the CNN models.

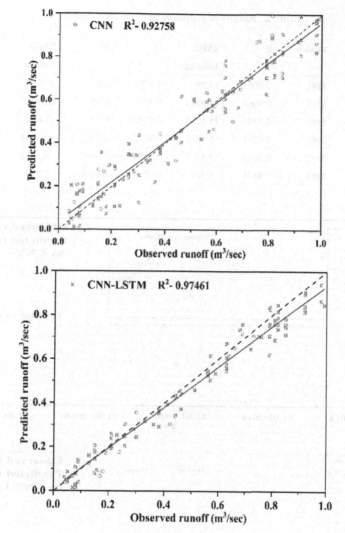

Fig. 4. Scatter plots of testing phase with CNN and CNN-LSTM models

Table 2. Results of statistical performance metrics

Station name	Models	R^2	RMSE	IA	R^2	RMSE	IA
			Training			Testing	
Lakhipur	M#1	0.8904	14.35	0.9704	0.8765	25.367	0.9619
	M#2	0.9063	13.9381	0.9738	0.8916	22.002	0.9634
	M#3	0.9387	11.5218	0.9786	0.9157	20.654	0.9718
	M#1	0.9541	5.4965	0.9822	0.9463	7.8534	0.9805
	M#2	0.9613	3.63	0.9858	0.9502	6.3962	0.9829
	M#3	0.9875	1.364	0.9897	0.9746	5.983	0.9876

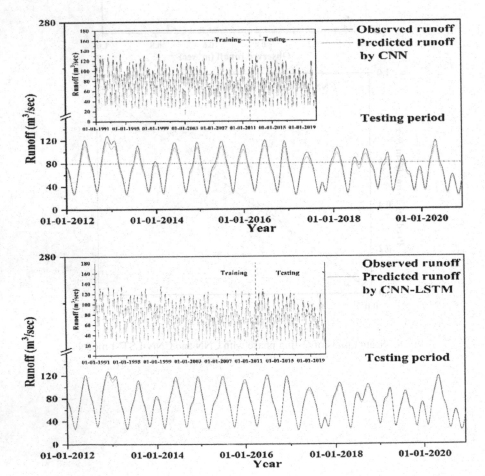

Fig. 5. Predicted outcomes of entire period (1/1/1991 to 1/12/2020) for Lakhipur station

Figure 4 demonstrates scatter plots of predicted vs observed runoff of Lakhipur station utilising CNN and CNN-LSTM models. Predicted runoff values of the CNN-LSTM model are very close to observed values. They can be fitted well by linear regression equation ($y = 0.955 * x + 0.135$) that can be used for modifying predicted peak values. Contrast to the CNN-LSTM model, predicted runoff values of CNN models are ill-organised and much scattered, signifying much lower prediction accurateness. Figure 5 shows final forecast results in testing phase utilising CNN-LSTM and CNN models. Clearly, the CNN-LSTM model has considerable forecasting precision compared to CNN model, and the CNN-LSTM model can well forecast peak scale. In addition to the statistical metrics, graphical assessments of predicted runoff with CNN-LSTM and CNN in Figs. 4, 5 and 6 indicate that CNN-LSTM outcomes accorded closely with

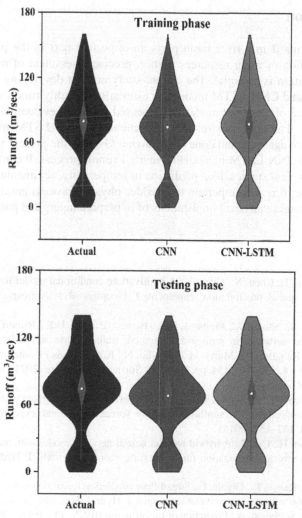

Fig. 6. Violin Plot of observed and predicted runoff of CNN, and CNN-LSTM models

observed dataset compared to CNN outcomes. Moreover, regression lines for runoff in Fig. 4 show that CNN-LSTM predicted the runoff values nearer to line of fit (45° line) than CNN.

The violin plots of the four models are illustrated in Fig. 6. In this figure, average of estimations by CNN model is lower than that CNN-LSTM model. The lower bound of the CNN-LSTM model indicates that it has outperformed others in estimating minimum values. A decrease in deviations proved effectiveness of hybrid model amid predicted and observed monthly runoff values, especially during rainy season and peak flow occurrences. The study of hybrid model's structure shows that the model effectively deals with variations in runoff values. This shows one of the significant features of monthly runoff time series in arid and semi-arid regions because of rainfall fluctuations in those areas.

5 Conclusion

Estimation of runoff in a river basin plays an important part in the proper planning and administration of water resources. Hence, precise assessment of runoff from historical precipitation is essential. The current study aims at developing and evaluating different CNN and CNN-LSTM models for estimating monthly runoff of Barak River. The performance of proposed model was assessed with two evaluation measures: R^2, RMSE, and IA. The outcomes revealed that enhanced CNN-LSTM model presented best performance against standalone CNN model. Overall, the present work illustrated effectiveness of CNN-LSTM in modeling rainfall-runoff process. It can also be applied to other hydrological studies, like, prediction of temperature, sedimentation, flood, etc. In the meantime, it is still important to consider physical process parameters in an AI model and explaining physical implications of hyperparameters and parameters.

References

1. Liu, Z., Zhou, P., Chen, X., Guan, Y.: A multivariate conditional model for streamflow prediction and spatial precipitation refinement. J. Geophys. Res. Atmospheres **120**, 10–116 (2015)
2. Samantaray, S., Sahoo, A., Mohanta, N.R., Biswal, P., Das, U.K.: Runoff prediction using hybrid neural networks in semi-arid watershed, India: a case study. In: Satapathy, S.C., Bhateja, V., Ramakrishna Murty, M., Gia Nhu, N., Kotti, J. (eds.) Communication Software and Networks. LNNS, vol. 134, pp. 729–736. Springer, Singapore (2021). https://doi.org/10.1007/978-981-15-5397-4_74
3. Zhang, X., Peng, Y., Zhang, C., Wang, B.: Are hybrid models integrated with data preprocessing techniques suitable for monthly streamflow forecasting? Some experiment evidences. J. Hydrol. **530**, 137–152 (2015)
4. He, X., Guan, H., Qin, J.: A hybrid wavelet neural network model with mutual information and particle swarm optimization for forecasting monthly rainfall. J. Hydrol. **527**, 88–100 (2015)
5. Mehr, A.D., Kahya, E., Olyaie, E.: Streamflow prediction using linear genetic programming in comparison with a neuro-wavelet technique. J. Hydrol. **505**, 240–249 (2013)
6. Samantaray, S., Sahoo, A.: Prediction of runoff using BPNN, FFBPNN, CFBPNN algorithm in arid watershed: a case study. Int. J. Knowl.-Based Intell. Eng. Syst. **24**, 243–251 (2020)

7. Samantaray, S., Sahoo, A.: Estimation of runoff through BPNN and SVM in Agalpur watershed. In: Satapathy, S.C., Bhateja, V., Nguyen, B.L., Nguyen, N.G., Le, D.-N. (eds.) Frontiers in Intelligent Computing: Theory and Applications. AISC, vol. 1014, pp. 268–275. Springer, Singapore (2020). https://doi.org/10.1007/978-981-13-9920-6_27

8. Vidyarthi, V.K., Jain, A., Chourasiya, S.: Modeling rainfall-runoff process using artificial neural network with emphasis on parameter sensitivity. Model. Earth Syst. Environ. 6(4), 2177–2188 (2020). https://doi.org/10.1007/s40808-020-00833-7

9. Chaudhury, S., Samantaray, S., Sahoo, A., Bhagat, B., Biswakalyani, C., Satapathy, D.P.: Hybrid ANFIS-PSO model for monthly precipitation forecasting. In: Bhateja, V., Tang, J., Satapathy, S.C., Peer, P., Das, R. (eds.) Evolution in Computational Intelligence. Smart Innovation, Systems and Technologies, Proceedings of the 9th International Conference on Frontiers in Intelligent Computing: Theory and Applications (FICTA 2021), vol. 267. Springer, Cham (2022). https://doi.org/10.1007/978-981-16-6616-2_33

10. Liu, Q., Zou, Y., Liu, X., Linge, N.: A survey on rainfall forecasting using artificial neural network. Int. J. Embedded Syst. 11, 240–249 (2019)

11. Singh, U.K., Kumar, B., Gantayet, N.K., Sahoo, A., Samantaray, S., Mohanta, N.R.: A hybrid SVM–ABC model for monthly stream flow forecasting. In: Advances in Micro-Electronics, Embedded Systems and IoT, vol. 838, pp. 315–324. Springer, Singapore (2022). https://doi.org/10.1007/978-981-16-8550-7_30

12. Samanataray, S., Sahoo, A.: A comparative study on prediction of monthly streamflow using hybrid ANFIS-PSO approaches. KSCE J. Civ. Eng. 25(10), 4032–4043 (2021). https://doi.org/10.1007/s12205-021-2223-y

13. Zhang, R., Chen, Z.Y., Xu, L.J., Ou, C.Q.: Meteorological drought forecasting based on a statistical model with machine learning techniques in Shaanxi province, China. Sci. Total Environ. 665, 338–346 (2019)

14. Khan, M.M.H., Muhammad, N.S., El-Shafie, A.: Wavelet-ANN versus ANN-based model for hydrometeorological drought forecasting. Water 10, 998 (2018)

15. Jabbari, A., Bae, D.H.: Application of Artificial Neural Networks for accuracy enhancements of real-time flood forecasting in the Imjin basin. Water 10(2018), 1626 (2018)

16. LeCun, Y., Bengio, Y., Hinton, G.: Deep learning. Nature 521(7553), 436–444 (2015)

17. Liu, W., Wang, Z., Liu, X., Zeng, N., Liu, Y., Alsaadi, F.E.: A survey of deep neural network architectures and their applications. Neurocomputing 234, 11–26 (2017)

18. Gu, J., et al.: Recent advances in convolutional neural networks. Pattern Recogn. 77, 354–377 (2018)

19. Krizhevsky, A., Sutskever, I., Hinton, G.E.: ImageNet classification with deep convolutional neural networks. Adv. Neural. Inf. Process. Syst. 25, 1097–1105 (2012)

20. Gao, S., et al.: Short-term runoff prediction with GRU and LSTM networks without requiring time step optimization during sample generation. J. Hydrol. 589, 125188 (2020)

21. Van, S.P., Le, H.M., Thanh, D.V., Dang, T.D., Loc, H.H., Anh, D.T.: Deep learning convolutional neural network in rainfall–runoff modelling. J. Hydroinf. 22, 541–561 (2020)

22. Ouma, Y.O., Cheruyot, R., Wachera, A.N.: Rainfall and runoff time-series trend analysis using LSTM recurrent neural network and wavelet neural network with satellite-based meteorological data: case study of Nzoia hydrologic basin. Complex Intel. Syst. 8, 1–24 (2021). https://doi.org/10.1007/s40747-021-00365-2

23. Han, H., Morrison, R.R.: Data-driven approaches for runoff prediction using distributed data. In: Stochastic Environmental Research and Risk Assessment, pp. 1–19 (2021)

24. Ding, Y., Zhu, Y., Feng, J., Zhang, P., Cheng, Z.: Interpretable spatio-temporal attention LSTM model for flood forecasting. Neurocomputing 403, 348–359 (2020)

25. Wu, H., Yang, Q., Liu, J., Wang, G.: A spatiotemporal deep fusion model for merging satellite and gauge precipitation in China. J. Hydrol. 584, 124664 (2020)

26. Baek, S.S., Pyo, J., Chun, J.A.: Prediction of water level and water quality using a CNN-LSTM combined deep learning approach. Water **12**, 3399 (2020)

27. Barzegar, R., Aalami, M.T., Adamowski, J.: Coupling a hybrid CNN-LSTM deep learning model with a boundary corrected maximal overlap discrete wavelet transform for multiscale lake water level forecasting. J. Hydrol. **598**, 126196 (2021)

28. Ghimire, S., Yaseen, Z.M., Farooque, A.A., Deo, R.C., Zhang, J., Tao, X.: Streamflow prediction using an integrated methodology based on convolutional neural network and long short-term memory networks. Sci. Rep. **11**, 1–26 (2021)

29. Laloy, E., Hérault, R., Jacques, D., Linde, N.: Training-image based geostatistical inversion using a spatial generative adversarial neural network. Water Resour. Res. **54**, 381–406 (2018)

30. Ince, T., Kiranyaz, S., Eren, L., Askar, M., Gabbouj, M.: Real-time motor fault detection by 1-D convolutional neural networks. IEEE Trans. Industr. Electron. **63**, 7067–7075 (2016)

31. Lipton, Z.C., Berkowitz, J., Elkan, C.: A critical review of recurrent neural networks for sequence learning. arXiv preprint arXiv:1506.00019 (2015)

32. Agnihotri, A., Sahoo, A., Diwakar, M.K.: Flood prediction using hybrid ANFIS-ACO model: a case study. In: Smys, S., Balas, V.E., Palanisamy, R. (eds.) Inventive Computation and Information Technologies, vol. 336, pp. 169–180. Springer, Singapore (2022). https://doi.org/10.1007/978-981-16-6723-7_13

33. Samantaray, S., Sahoo, A.: Prediction of suspended sediment concentration using hybrid SVM-WOA approaches. Geocarto International, pp. 1–27 (2021)

34. Sahoo, A., Ghose, D.K.: Application of hybrid MLP-GWO for monthly rainfall forecasting in Cachar, Assam: a case study. In: Bhateja, V., Satapathy, S.C., Travieso-Gonzalez, C.M., Adilakshmi, T. (eds.) Smart Intelligent Computing and Applications, Volume 1: Proceedings of Fifth International Conference on Smart Computing and Informatics (SCI 2021), vol. 282, p. 307. Springer, Cham (2022a). https://doi.org/10.1007/978-981-16-9669-5_28

35. Sahoo, A., Samantaray, S., Ghose, D.K.: Multilayer perceptron and support vector machine trained with grey wolf optimiser for predicting floods in Barak river, India. J. Earth Syst. Sci. **131**, 1–23 (2022)

36. Sahoo, A., Ghose, D.K.: Imputation of missing precipitation data using KNN, SOM, RF, and FNN. Soft Computing **26**, 5919–5936 (2022b)

Analysis of Malaria Incident Prediction for India

Poonkuntran Shanmugam[✉] and Ankit Shrivastava

VIT Bhopal University, Madhya Pradesh, India
s_poonkuntran@yahoo.co.in

Abstract. This paper presents a detailed analysis of the malaria incident prediction for India. It uses the government of India's open dataset of malaria incidents in all the states and union territories of India. The five classes of methods are experimented on the data to understand the best fit model. The robust linear, fine tree, linear SVM, ensemble boosted tree and neural network regression models are used for the training. The experimental results showed that the neural network model produces 85% of accuracy and it outperforms all other methods.

Keywords: Regression · Machine learning · Neural networks · Malaria · Data science

1 Introduction

Malaria is one of the important diseases to be taken care of by the government as per World Health Organization (WHO) reports [1]. It requires constant efforts and policy implementation for reducing the number of malaria incidents. As per the WHO malaria report [2], the malaria case incident rate per 1000 population decreased from 80% in 2000 to 56.8% in 2019. The death rate is also reduced from 24.7% in 2000 to 10.1% in 2019. India recorded the largest malaria deaths of 86% in the WHO Southeast Asia region [2]. India is still a constant contributor to malaria and almost all the states and union territories (UT) of India contribute to the malaria incidents. India is investing funds and initiating various initiatives to control malaria. The government of India recently launched the Web-based Malaria Information System (MMIS) under National Vector Bone Disease Control Programme (NVBDCP) to monitor malaria observation through an integrated system with dashboard and visualization supports [3]. The endemic countries are conducting research for forecasting the trends of malaria in different methods. They use incident data, climate information, and clinical information [4].

Data-driven prediction becomes popular after the use of ML in healthcare and much research has been proposed in this direction. It opened a new way of data-centric analysis rather than epidemiological analysis. It could improve the health care performance and helps early diagnosis of the diseases [21]. The work reported in [22] uses ensemble classifiers and random forest for detecting diabetics and it could provide 93% of accuracy. Another work in [23] uses gradient boosting algorithms to forecast the risk of sepsis in patients. It provides an accuracy of around 70%. The genetic algorithm is used in heart disease prediction in multiple stages through its mutations. It helps in predicting heart

M. Singh et al. (Eds.): ICACDS 2022, CCIS 1614, pp. 393–403, 2022.
https://doi.org/10.1007/978-3-031-12641-3_32

failures in advance [24]. The neural networks are used to detect diabetic retinopathy in earlier stages. It is supported by the principal component analysis [25].

The support vector machine and random forest are used to locate the brain tumors from Magnetic Resonance Imaging (MRI) [26]. The ML is also employed in post-treatment analysis in healthcare that includes the estimation of risk of readmission, cost of treatments, and treatment period required for complete cure [27]. The recent COVID-19 pandemic is also witnessed with ML techniques on the estimation of spread, treatment facilities required, and next trends [28].

In a similar direction, the data-driven approach for malaria prediction is rare and supported by a very less number of data sets and models. The work in [29], proposes the malaria prediction for the district of Visakhapatnam, India using 6 years of data collected from various health centers at Visakhapatnam. The results showed that gradient booster methods are good in predicting malaria prediction.

Machine Learning (ML) provides support to extract the knowledge from the data to group similar data into classes and forecast the future data by identifying the relationship among existing data. Forecasting requires intensive analysis to find the best relation to map the existing data to predict the future [5]. The ML techniques are used to identify the outliers in the data and K-means clustering is effective for outlier detection [6]. The feature engineering and classification are facilitated by Extreme Gradient Boosting (XGBoost) technique [7]. The XGBoost has accelerated the classification and improved the accuracy [8].

The ML provides the best platform to find the relation among the data and it poses a risk of bias due to the poor training samples. The training samples should be chosen with care for better results without bias [24]. The biased decisions from ML will not be recommended and proper selection of training data will be important for improving the performance of healthcare analysis [30, 31].

This paper presents the analysis of regression models to predict the malaria incident in India. We use the Indian State /UT-wise malaria incident data from 2016 to 2019. The dataset is taken from the Open Government Data Platform of India [9].

2 Related Works

Malaria prediction using machine learning attracted many researchers in recent times. The African countries are in the endemic category of malaria incidents. The literature showed significant contributions for those countries on malaria prediction. The prediction uses different parameters such as clinical information and climate information [10, 11]. The climate variables such as temperature, humidity, and radiation were used to model malaria in [11]. In [12], a stochastic time series model auto-regressive integrated moving average (ARIMA) was introduced to predict malaria using climate variables. The time-series data of malaria incidents have been collected and analyzed from various environmental parameters for South Africa [13]. In [14], climate-based malaria prediction has been discussed for Andhra Pradesh in India. Many other models have been developed for malaria predictions using clinical values, climate values, environmental parameters, and early signs of malaria [15–20]. It is shown from the survey that model fitting varies from data to data and requires attention in choosing the right model by analyzing the relationship of the data.

In [32], an image processing solution is proposed for detecting malaria from RBC images. It takes the RBC image of the subject and preprocesses the same for reducing noise. Then, the image is segmented to extract the RBC structures. The extracted RBC structures are then applied for feature extraction and mapping. Finally, it applies classification techniques to classify the feature as malaria-infected or not. It is a simple and data-driven approach where prior training samples are fed to the classifier on both the labels infected and not infected.

The convolution neural networks (CNN) are used in [33] for detecting malaria parasites in the blood smear images of the subjects. This CNN replaces the requirements of segmentation of RBC images into structures and a handcrafted feature mapping process.

The CNN is used at different levels for obtaining better results. In [34], the CNN is applied for Cell Segmentation; the Region-Based CNN for an end-to-end framework is presented in [35]. The most commonly used classification techniques for malaria predictions are KNN, Support Vector Machines, ANN, and Naïve Bayes.

3 Methods and Dataset

3.1 Malaria Incident Dataset

The malaria incident data set is a time-series dataset and it contains the number of malaria incidents that happened in respective States/UTs of India. The dataset contains 36 records and each record represents the State/UTs of India. There are four attributes for each record that represents the malaria incidents in year wise from 2016 to 2019. The statistics of the dataset are shown in Table 1.

Table 1. Malaria incident dataset for India – statistics

Year	2016	2017	2018	2019
Max	444843	347860	86486	86591
Min	2	1	5	3
Average	30202	23460	11942	7947
Median	3926.5	3354	1664	1074

The malaria incidents for 2019 are considered as a response to which the models are analyzed for better prediction. Figure 1 shows the distribution of the malaria incidents across different states of India.

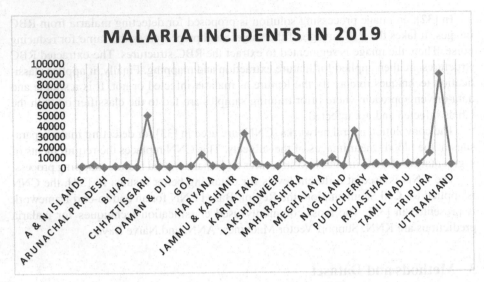

Fig. 1. Malaria incident in India for 2019

3.2 Methods and Parameters for Prediction

To forecast the malaria incidents, linear regression models, regression trees, support vector machines, ensemble trees, and neural networks are employed to analyze the performance. The classes of methods are chosen to understand the linearity and non-linearity of the data to choose the best model for the prediction.

The linear regression analysis is simple and well-suited for linearly dependent data. It comes with parametric and non-parametric models. The non-parametric model does not require any optimization parameters. However, the residuals are playing a vital role in regression. Any outlier in the residual will require attention. The malaria incident data sets will have outlier residuals by nature since they record the incident of malaria and it will be a random value across the locations. Hence, robust linear regression is used in the experiment to measure the linear relationship among the data. The linear regression uses robust linear methods without any optimization parameters.

The decision tree is a supervised model in which classification and regression are carried out. It is well suited for non-linear data. It has two parts decision nodes and leaves. The leaves will have decision values or outcomes of the model. The decision node has conditions and it is associated with entropy and information gain. The entropy is given by Eq. 1 and it is denoted by H(S) for a finite set S, which is the measure of the amount of uncertainty or randomness in data.

$$H(s) = \sum_{x \in X} p(x) \log 2(1/p(x)) \tag{1}$$

The information gain is denoted by IG(S, A) for a set S is the effective change in entropy after deciding on a particular attribute A. The IG(S, A) is given by Eq. 2.

$$IG(S, A) = H(S) - H(S, A) \tag{2}$$

The IG(S, A) is the information gained after applying feature A. H(S) is the Entropy of the entire set, while the second term calculates the Entropy after applying the feature A, where P(x) is the probability of event x. The decision tree will have a discrete set of values when it is being used for classification and have continuous values if it is used for regression. The fine tree model of the decision tree is chosen for the experiment to analyze the non-linear relationship of the malaria incident data. The minimum leaf size is fixed as 4.

The support vector machine (SVM) uses the simple linear model without optimization to locate the linearity of the data. The SVM supports linearly separable data. It uses a line or hyperplane to separate the data.

The ensemble model uses the boosted tree method with a minimum number of leaf nodes of 8 and learners of 30. The simple medium neural network model is employed to locate the non-linearity over layered optimization. The first layer size was increased to 25 nodes as the maximum for the training. The ReLU activation is used in the neuron activation. The neural network is fully connected.

The ReLU is a rectifier linear activation function and it activates the neuron by outputting the input as such if the input is positive. If the input is negative, it outputs zero. Such ReLU behavior is well recommended in the many neural networks that target faster and better training outputs. The ReLU eliminates the vanishing point issues in the activation function such as sigmoid and hyperbolic tangent.

For the quantitative evaluation and comparison, the following parameters were used in the analysis.

Root Mean Square Error (RMSE) is the variance between the actual values and predicted values for all the samples. The RMSE is given by Eq. 3.

$$RMSE = \frac{\sum\limits_{i=1}^{n} (Actual(i) - \text{Pr}edcited(i)}{n} \tag{3}$$

The R-Squared value is used to analyze the extent to which all the independent variables of the model explain the dependent variable. It takes the proportion of the unexplained variance of the model against the total available variance of the data. The R^2 value is given by the Eq. 4 and it usually ranges from 0 to 1. The R^2 value of 0.7& above is moderate and 0.9 & above will be good.

$$R^2 = 1 - \left[\frac{UnExplianedVarianceoftheModel}{TotalVaraince} \right] \tag{4}$$

4 Experiments and Analysis

The experiment of this analysis is done on Matlb2021using regression models discussed in Sect. 2. Table 2 shows the results of the experiments. It is observed from the results that the linear SVM model failed to train the data and it could train only 30% of the malaria incidents for the predicted year. At the same time, the neural network model could map 84% of the malaria incidents to the predicted year and it improves the R-Square of the model by 65% at approximately. The robust linear regression can provide

64% of accuracy whereas the fine tree produces 65% accuracy. The neural network model improves the accuracy by 25% compared to robust linear and fine tree methods. The ensemble boosted tree method does not improve the accuracy compared to neural network methods.

Table 2. Experimental results

Methods	RMSE	R2 value
Robust Linear	10615	0.64
Fine Tree	10504	0.65
Linear SVM	14402	0.34
Ensemble-Bagged Tree	14128	0.36
Neural Networks - Medium NN	7130	0.84

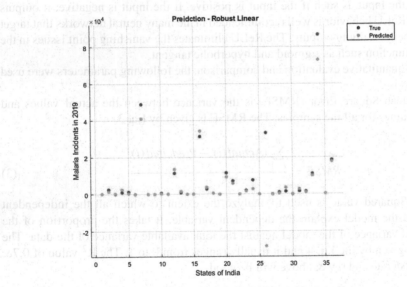

Fig. 2. Prediction of malaria incidents – robust linear

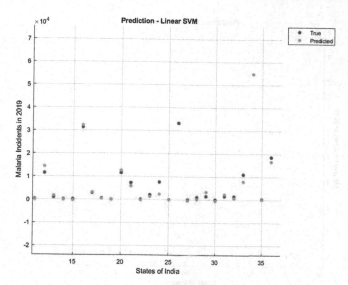

Fig. 3. Prediction of malaria incidents – linear SVM

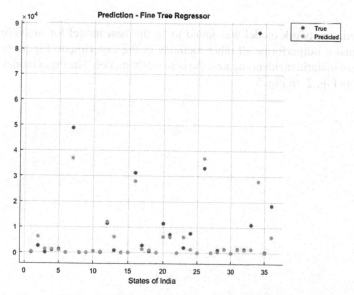

Fig. 4. Prediction of malaria incidents – fine tree

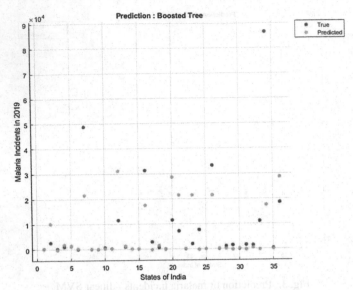

Fig.5. Prediction of malaria incidents – boosted tree

The neural network model was found to be the best model for predicting malaria incidents and it outperforms all other methods in the experiment. Figure 6 shows the prediction of malaria incidents against the actual NN model. The other model results are illustrated in Fig. 2. to Fig. 5.

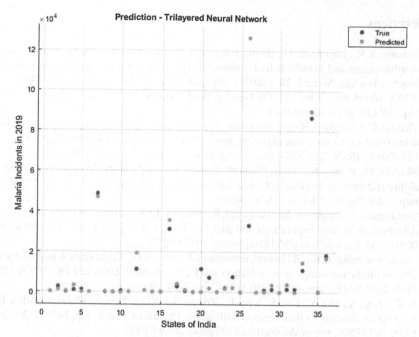

Fig. 6. The prediction of malaria incidents – NN Model

The experimental results show that simple linear relationship models are a good t fit for only around 60% of the total population. The remaining 40% of the population is not into linearity and it fits into non-linearly. To train the non-linearly the boosted trees are good enough to cater to the needs. However, it could not train the linearly related samples. It demands a further level of optimization and the neural network becomes the choice. The NN trains the samples in layered optimizations and it could achieve the best fit as shown in Fig. 6.

5 Conclusions

We presented the analysis of malaria incident data prediction for India. It used the case incident data set from the government India open data portal. The five different classes of regression models are chosen to understand the relationship of data. The experimental results showed the linear and decision tree models can produce 65% of accuracy whereas the SVM and ensemble models could produce 35% of accuracy. The neural networks are shown the better performance and produce 85% of accuracy. It is concluded that the malaria incident data contains samples that are linearly and non-linearly connected. The neural network model was found to be the best model to fit all kinds of samples and outperform the other methods in the experiment.

References

1. Sharma, R.K., Rajvanshi, H., Bharti, P.K., et al.: Socio-economic determinants of malaria in tribal-dominated Mandla district enrolled in Malaria Elimination demonstration project in Madhya Pradesh. Malar J. **20**, 7 (2021). https://doi.org/10.1186/s12936-020-03540-x
2. WHO. World malaria report 2020. Geneva: World Health Organization (2020)
3. https://ihip.nhp.gov.in/malaria/#!/
4. Thakur, S., Dharavath, R.: Artificial neural network-based prediction of malaria abundances using big data: a knowledge capturing approach. Clin. Epidemiol. Global Health **7**(1), 121–126 (2019). ISSN 2213-3984. Doi: https://doi.org/10.1016/j.cegh.2018.03.001.
5. Nkiruka, O., Prasad, R.: Onime Clement, Prediction of malaria incidence using climate variability and machine learning. Inf. Med. Unlocked, **22**, 100508 (2021). ISSN 2352-9148. Doi: https://doi.org/10.1016/j.imu.2020.100508.
6. Domingues, R., Filippone, M., Michiardi, P., Zouaoui, J.: A comparative evaluation of outlier detection algorithms: experiments and analyses. Pattern Recogn. **74**, 406–421 (2018). ISSN 0031-3203. https://doi.org/10.1016/j.patcog.2017.09.037.
7. Gárate-Escamila, A.K., El Hassani, Emmanuel Andrès, A.H.: Classification models for heart disease prediction using feature selection and PCA. Inf. Med. Unlocked **19**, 100330 (2020). ISSN 2352-9148. https://doi.org/10.1016/j.imu.2020.100330.
8. Ji, C., Zou, X., Hu, Y., Liu, S., Lyu, L., Zheng, X.: XG-SF: an XGBoost classifier based on shapelet features for time series classification. Procedia Comput. Sci. **147**, 24–28 (2019). ISSN 1877-0509. https://doi.org/10.1016/j.procs.2019.01.179
9. https://data.gov.in/resources/stateut-wise-malaria-incidence-2016-2019-ministry-health-and-family-welfare
10. Lee, Y.W., Choi, J.W., Shin, E.-H.: Machine learning model for predicting malaria using clinical information. Comput. Biol. Med. **129**, 104151 (2021). ISSN 0010-4825. https://doi.org/10.1016/j.compbiomed.2020.104151
11. Tompkins, A.M., Thomson, M.C.: Uncertainty in malaria simulations in the highlands of Kenya: relative contributions of model parameter setting, driving climate and initial condition errors. PloS One **13**(9), 1–27 (2018). https://doi.org/10.1371/journal.pone.0200638
12. Balding David, J.: Linear models and time-series analysis: regression. ANOVA, ARMA, and GARCH (2019)
13. Adeola, A.M., et al.: Predicting malaria cases using remotely sensed environmental variables in Nkomazi, South Africa. Geospatial. Health **14**(1) (2019). https://doi.org/10.4081/gh.2019.676
14. Mopuri, R., Kakarla, S.G., Mutheneni, S.R., Kadiri, M.R., Kumaraswamy, S.: Climate-based malaria forecasting system for Andhra Pradesh. J. Parasit. Dis., India (2020)
15. Anwar, M.Y., Lewnard, J.A., Parikh, S., Pitzer, V.E.: Time series analysis of malaria in Afghanistan: using ARIMA models to predict future trends in incidence. Malar J. **15**(1) (2016). https://doi.org/10.1186/s12936-016-1602-1
16. Makinde, O.S., Abiodun, G.J., Ojo, O.T.: Modelling of malaria incidence in Akure, Nigeria: negative binomial approach. GeoJournal **86**(3), 1327–1336 (2020). https://doi.org/10.1007/s10708-019-10134-x
17. V Le, P.V., et al.: Malaria epidemics in India: role of climatic condition and control measures. PloS One **14**(2), 1–15 (2020). https://doi.org/10.1016/j.cities.2019.01.009
18. Tompkins, A.M., Colon-González, F.J., Di Giuseppe, F., Namanya, D.B.: Dynamical malaria forecasts are skillful at regional and local scales in Uganda up to 4 months ahead. GeoHealth **3**(3), 58–66 (2019). https://doi.org/10.1029/2018GH000157
19. V Le, P.V., Id, P.K., Ruiz, M.O., Mbogo, C., Muturi, J.: Predicting the direct and indirect impacts of climate change on malaria in coastal Kenya, pp. 1–18 (2019)

20. Kim, Y., et al.: Malaria predictions based on seasonal climate forecasts in South Africa: a time series distributed lag nonlinear model. Sci. Rep. **9**(1), 1 (2019). https://doi.org/10.1038/s41598-019-53838-3

21. Wiemken, T.L., Kelley, R.R.: Machine learning in epidemiology and health outcomes research. Ann. Rev. Public Health **41**(1), 21–36 (2020). pmid:31577910

22. Alghamdi, M., Al-Mallah, M., Keteyian, S., Brawner, C., Ehrman, J., S S. Predicting diabetes mellitus using SMOTE and ensemble machine learning approach: The Henry Ford Exercise Testing (FIT) project. PLoS ONE **12**(7) (2017)

23. Delahanty, R.J., Alvarez, J., Flynn, L.M., Sherwin, R.L., Jones, S.S.: Development and evaluation of a machine learning model for the early identification of patients at risk for sepsis. Ann. Emerg. Med. **73**(4), 334–344 (2019). pmid:30661855

24. Reddy, G.T., Reddy, M.P.K., Lakshmanna, K., Rajput, D.S., Kaluri, R., Srivastava, G.: Hybrid genetic algorithm and a fuzzy logic classifier for heart disease diagnosis. Evol. Intel. **13**(2), 185–196 (2019). https://doi.org/10.1007/s12065-019-00327-1

25. Gadekallu, T., Khare, N., Bhattacharya, S., Singh, S., Maddikunta, P., Ra, I., et al.: Early detection of diabetic retinopathy using pca-firefly based deep learning model. Electronics **9**(2), 1–16 (2020)

26. Rehman, Z.U., Zia, M.S., Bojja, G.R., Yaqub, M., Jinchao, F., Arshid, K.: Texture based localization of a brain tumor from MR-images by using a machine learning approach. Med. Hypotheses. **141**, 109705 (2020). pmid:32289646

27. Hammoudeh, A., Al-Naymat, G., Ghannam, I., Obied, N.: Predicting hospital readmission among diabetics using deep learning. Procedia Comput. Sci. **141**, 484–489 (2018)

28. Reddy, B.G., Ofori, M., Jun, L., Ambati, L.S.: Early public outlook on the coronavirus disease (COVID-19): a social media study. In: AMCIS 2020 Proceedings (2020)

29. Kalipe, G., Gautham, V., Behera, R.K.: Predicting malarial outbreak using machine learning and deep learning approach: a review and analysis. In: 2018 International Conference on Information Technology (ICIT), pp. 33–38 (2018)

30. Fang, G., Annis, I.E., Elston-Lafata, J., Cykert, S.: Applying machine learning to predict real-world individual treatment effects: insights from a virtual patient cohort. J. Am. Med. Inf. Assoc. **26**(10), 977–988 (2019). pmid:31220274

31. Flaxman, A.D., Vos, T.: Machine learning in population health: opportunities and threats. PLOS Med. **15**(11), 1–3 (2018). pmid:30481173

32. Boray, F.T., Andrew, G.D., Izzet, K.: Computer vision for microscopy diagnosis of malaria. Malaria J., 1–14 (2009)

33. Zhaohui, L., et al.: Cnn-based image analysis for malaria diagnosis. In: 2016 IEEE International Conference on Bioinformatics and Biomedicine (BIBM). IEEE (2016)

34. Yuhang, D., Zhuocheng, J., Hongda, S., David, P.W.: Classification accuracies of malaria-infected cells using deep convolutional neural networks based on decompressed images. In: SoutheastCon 2017. IEEE (2017)

35. Jane, H., et al.: Applying faster r-CNN for object detection on malaria images. In: 2017 IEEE Conference on Computer Vision and Pattern Recognition Workshops (CVPRW). IEEE (2017)

Meta-learning for Few-Shot Insect Pest Detection in Rice Crop

Shivam Pandey$^{(\boxtimes)}$, Shivank Singh, and Vipin Tyagi

Department of Computer Science and Engineering,
Jaypee University of Engineering and Technology, Guna, Madhya Pradesh, India
pandeyshivam1423@gmail.com

Abstract. Recent advancements in the field of Deep learning have helped in predicting and locating pests in agricultural field images accurately. A drawback of this approach is that it requires a large training dataset for each sample, which is not feasible. Since there is a wide variety of pests, collecting thousands of training images for each sample is impractical. To deal with this issue, a pest detection meta-learning technique based on Few-shot is proposed in this paper. In this work, pests from rice crops are considered for experiments. Two pest-image datasets: IP102 as a supported dataset to perform meta-learning and an image library for insects and pests known as the Indian Council of Agricultural Research-National Bureau of Agricultural Insect Resources (ICAR-NBAIR) are taken to perform Few-shot learning. In meta-learning phase, the proposed model is trained on a variety of pests, and hence the proposed system is capable of learning new categories of pests with very few training images.

Keywords: Few-shot learning · Meta-learning · CNN · Object detection

1 Introduction

The Indian economy relies heavily on agriculture since it employs over 60% of the people and provides enormous amounts of revenue. India's total Gross Domestic Product (GDP) includes 23% contributions from the agriculture sector alone [21]. The world's plants are harmed by over thousands of insect species, various weed species, millions of illnesses (caused by bacteria, viruses, fungus, and other microorganisms), and thousand nematode species [22]. Rice is a staple food for about half of the global population [3]. Rice production is primarily concentrated in Asia, where it is the main source of food. The Asia-Pacific region produces and consumes over 90% of the world's rice, and it is frequently a source of economic importance. Being the largest rice producer in the world after China, India produced whopping 173 million metric tons production in 2021. However, due to the diverse climate of India, chances of threats from pests in crop production grow exponentially. In terms of monetary value, insect pests currently cause an annual loss of approximately 863 billion rupees in Indian agriculture [22].

Every year, farmers observes a loss of approximately 37% of their rice crop due to pest attacks and diseases [4]. Apart from good crop management, timely and accurate early diagnosis of pest insects can help to reduce losses significantly. Moreover, the effect of climate change is growing at an alerting rate because the damage caused by pests to the crops is critical and unforeseeable [20]. Since most of the Indian farmers use conventional methods and anticipate pest attacks based on past experiences which might result in a wrong diagnosis and result in a huge loss than expected. Therefore, it becomes necessary to timely identify pests in crops.

In this paper, a technique that requires very few training samples to train a predicting model to identify pest insects in crops is proposed, that would be practically feasible to be implemented in real. Also the proposed model can be extended to detect various other types of pest insects based on the requirement of the farmers.

2   Related Work

In recent years, there has been a significant advancement in the field of computer vision to build algorithms that can detect pests effectively and accurately. Moreover, advanced and accessible drones, robots have made these algorithms practically viable to use. Image processing, machine learning, deep learning, and convolutional neural network are some of the technologies that have contributed majorly to solving problems such as plant disease detection, weed detection, and pest detection [23].

For instance, Fuentes et al. [5] proposed robust tomato plant diseases and pest detection using meta-learning of Faster R-CNN, R-CNN, and SSD with the help of two base model VGG-Net and ResNet. Lu et al. [6] used a deep convolutional neural network (DCNN) for developing a rice disease identification model. Lin et al. [7] introduced a region based convolutional neural network which was anchor-free with a Fast R-CNN model used for classification of 24 pest categories. Zhong et al. [8] developed a vision-based pest's recognition system deployed on Raspberry Pi with YOLO as base network to detect number of pests, and the SVM model was trained for the classification of pests.

The vast majority of existing work in this area is focused on classification, which cannot be directly applied to other applications. In machine learning and computer vision, few-shot learning is a fundamental and lesser used area. For example, Yang et al. [17] proposed classification technique for few-shot cotton pest recognition. Yang li et al. [18] developed a plant disease classification technique using Semi-supervised few-shot learning approach. Nuthalapati et al. [19] introduced a Multi-Domain Few-shot Learning technique for pests and diseases recognition, however this approach was also restricted to classification as well.

Even though, many novel research has already been carried out using object detection techniques for pest recognition. However, to achieve higher accuracy and efficiency mostly researchers have used the traditional approach by using a dataset with a large number of samples. These approaches are practically not suitable for new species of insect pests as manually annotation of large dataset is a tedious task to perform.

Few-shot based object detection has gained wide attention from researchers in recent times that uses small dataset for training. In this paper, a meta-learning few-shot based pest detection technique has been proposed to achieve high accuracy without need of a large image training dataset.

3 Proposed Work

In this work, detection of insect pests found in rice crops is performed. The work is divided into two parts. Initially, a meta-learning model is trained on a large pest dataset (IP102). Secondly, with the help of this small training dataset detection of insect pests is performed.

Meta-Learning incorporates two stages, 1) Meta-training and 2) Meta-testing. As mentioned in Fig. 1, the model is trained using the entire dataset in the first place to generate a base pre-trained weight to be used in further steps.

To achieve desired results with few training images, meta-training was executed. Meta-training is a phase where the model is trained in a few-shot environment extracted from a large annotated image dataset (IP102 in this case). In this paper, a model has been proposed that performs meta-training in two stimulation of 14 ways-5 shots and 14 ways-3shots, where 14ways represents the number of classes and N (3 or 5) shots denotes the number of sample images used for each class. Therefore, for 14 ways-5 shots and 14ways-3 shots, the total numbers of training images are 70 and 42 respectively. These N (3 or 5) images are taken randomly from the IP102 dataset. To validate, a query set is created with an equal number of images for each category. These images from the query set are fed into the proposed model to extract features and predict the result with a bounding box around the detected pest.

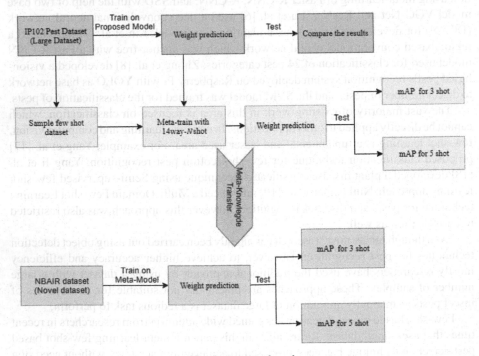

Fig. 1. Proposed work

Thereafter completing meta-training, the pre-trained weight is generated which is capable of mapping features related to pests and the meta-model trained with a fabricated few-shot environment. These are the meta-knowledge that is transferred to perform meta-testing. Further, few-shot object detection is performed on the dataset (ICAR-NBAIR) with similar steps. Since this dataset has a whole new class of pests, training of the meta-model is performed again and results based on query set images fed to the train meta-model.

In this work, the process is divided into two phases as shown in Fig. 1. In the first phase, a large volume dataset is used as the input to train the proposed model and output is compared with known results. Dataset is divided into N way-3shot and N way-5shot. Further, meta-training and meta-testing is performed on both the datasets and the weights are predicted to be used as meta-knowledge on ICAR-NBAIR (few shot dataset). Finally, in the second phase, the model and weights are used to carry forward meta-training and meta-testing on the few shot dataset.

Proposed model combines SSD (VGG-16 as backbone network) [11] and YoloV3 (DarkNet-53 as backbone network) [13].

To train the proposed model, initially, the input image needs to be resized according to the networks, 512×512 pixels size for SSD and 416×416 pixels size for Yolo-V3. Further, these resized training images are fed to both the networks, and resultant features are merged with the fully connected layer. Finally, the non-max suppression algorithm are used to prevent overlapping and multiple detected bounding boxes in the final results.

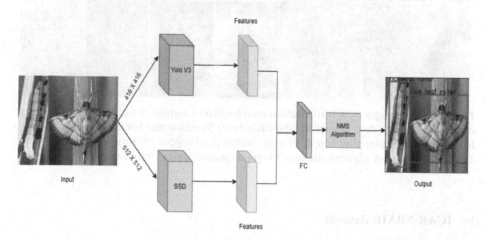

Fig. 2. Proposed model

4 Experiments and Discussion

4.1 Datasets Used

a) IP102 insect pest image dataset:

IP102 is a large scale benchmark dataset for insect pest recognition with more than 75,000 images for classification and 18,983 annotated images in PASCAL VOC format [14] for detection, containing 102 species of insects. [1] These images are annotated with respective species names and bounding boxes for object detection.

For experiments, this dataset was filtered out with 14 different rice pest species, samples are shown in Fig [3].

To test the meta-learning model, a rough split of data for each class was done in a ratio of 6:2:2 for training, testing and validation phases.

Fig. 3. Sample images from IP102 Dataset used for Meta-Learning 1) rice leaf roller 2) rice leaf caterpillar 3) paddy stem maggot 4) asiatic rice borer 5) yellow rice borer 6) rice gall midge 7) Rice Stemfly 8) brown plant hopper 9) white backed plant hopper 10) small brown plant hopper 11) rice water weevil 12) riceleafhopper 13) grain spreader thrips 14) rice shell pest

b) ICAR-NBAIR dataset:

To perform few-shot detection, a dataset is required with a lesser number of images for each class. Furthermore, no class should be the same as the class in the dataset used to train the meta-learning model (IP102).

In this work, rice pest images taken from the Indian Council of Agricultural Research-National Bureau of Agricultural Insect Resources (ICAR-NBAIR) data, which has a collection of common pests images found in the Indian agroecosystems [16] has been used. This collection contains images of 50 species of rice pests. In experiments for training and testing at least ten image samples of each species are required. Out of those 50 pest species images, only for 14 species more than ten different samples are available. For that reason, these 14 species images are filtered from this dataset and used for experiments. Sample images are shown in Fig. 4 [16].

Fig. 4. Sample Images of Insect Pests from ICAR-NBAIR Dataset for Few-Shot Learning [16] 1) Chilo polychrysa 2) Cnaphalocrocis medinalis 3) Cnaphalocrocis trapezalis 4) Creatonotos gangis 5) Leptocorisa oratorius 6) Melanitis leda 7) Mocis frugalis 8) Mythimna loreyi 9) Nezara viridula 10) Parapoynx stagnalis 11) Psalispennatula 12) Pyrilla perpusilla 13) Scirpophaga incertulas 14) Sesamia inferens

ICAR-NBAIR is essentially a knowledge source, not an official dataset for pest detection. So, annotation of images is not available. To use annotation in this work, annotation of images for training and testing of proposed method has been done manually in Pascal VOC format (XML files) [14].

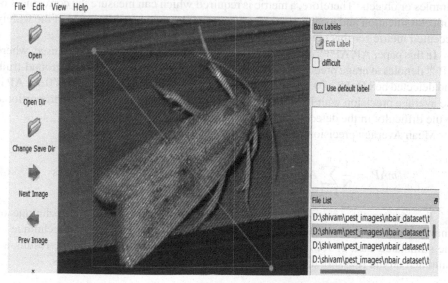

Fig. 5. Annotation example of a pest image taken from ICAR-NBAIR Dataset

For annotation, a graphical image annotation tool LabelImg is used as shown in Fig. 5.

4.2 Evaluation Metrics Used

To validate the performance of the proposed model, different metrics are used in various phases of the work viz. precision, recall, F1 score, Average Precision (AP), $AP^{.75}$ and $AP^{.50}$ and mean average precision (mAP).

The precision provides incite about the model, not labeling false negative samples and can be calculated by:

$$Precision = \frac{TruePositives}{(TruePositives + FalsePositives)} \tag{1}$$

The recall anticipates locating all the positive samples by the model and is calculated as follows:

$$Recall = \frac{TruePositives}{(TruePositives + FalseNegatives)} \tag{2}$$

The F1 score helps in calculating a trade-off between precision and recall. Higher the F1 score more robust the model is. F1 score can be calculated as:

$$F1score = \frac{(2 * Precision * Recall)}{(Precision + Recall)} \tag{3}$$

As object detection has two prime activities: classification and localization of the samples or objects. Therefore, a metric is required which can measure the correctness of both classification and localization. Mean Average Precision (mAP) is one of the widely accepted metric for this purpose.

In this paper AP, $AP^{.75}$, and $AP^{.50}$ and mean average precision (mAP) are used, where $AP^{.50}$ denotes average precision for an intersection over the union between ground-truth and detected box [IoU] of 0.5. $AP^{.75}$ indicates average precision for an IoU of 0.75. AP is the average precision with IoU = [.50: .95]. The amount of IoU is directly proportional to the difficulty in the detection.

Mean Average precision can be calculated as follows:

$$mAP = \frac{1}{N} \sum_{k=1}^{n} \text{Average precision}_k, \text{ where n } = \text{Number of classes} \tag{4}$$

To evaluate the proposed technique, initially results on traditional technique on images taken from IP102 dataset are obtained. Then results of proposed technique on the stimulated data of few-shot on IP102 dataset are obtained. Lastly, the results are validated on the few-shot dataset on the basis of multiple evaluation metrics.

4.3   Experimental Results on IP102 Dataset

To evaluate the proposed model trained on the entire IP102 dataset and compare the result with some of the state-of-the-art object detection methods was based on average precision with different IoU shown in Table 1. Average Precision with IoU $= .50$ of the model is 57.56%, which is 4% higher than FPN model [10] with ResNet-50 as backbone. The difference in the result for the proposed model and FPN becomes even higher for average precision with IoU of 0.75 and [0.5:0.95], which are 23.43% and 16.15% higher.

Table 1. Comparison of performance of Pest detection methods based on Mean Average Precision under different IoU thresholds.

Method	Backbone	$AP^{.50}$	$AP^{.75}$	AP
FRCNN [9]	VGG-16	47.87	15.23	21.05
FPN [10]	ResNet-50	54.93	23.30	28.10
SSD300 [11]	VGG-16	47.21	16.57	21.49
RefineDet [12]	VGG-16	49.01	16.82	22.84
YOLOv3 [13]	DarkNet-53	50.64	21.79	25.67
Proposed model	VGG-16 + DarkNet-53	**57.56**	**28.76**	**32.64**

Table 2. Performance on IP102 large dataset for meta-learning model based on Precision, Recall and F1 score.

Model	Precision %	Recall %	F1 score %
Baseline model	31.5	30.0	30.4
Proposed Model	43.33	44.63	43.97

Further, to evaluate the robustness and performance of the proposed model, the precision, recall, F1 score and mAP are calculated, to compare it with the baseline model of IP102 dataset. Table 2 shows the results, in which the proposed model outperforms the baseline model by 43.33%, 44.63%, 43.97% in the precision, Recall and F1 score respectively. The mAP and F1 score with respect to epochs of the complete test set is shown in Fig. 6.

4.4   Experimental Results on IP102 Stimulated Data

In this section, the proposed model is tested using the stimulated few-shot data from IP102 dataset to perform meta-learning. As discussed earlier, there are two stimulations of data 14 ways-5 shots and 14 ways-3 shots, which are evaluated based on average precision with IoU of 0.5, 0.75 and [0.5:0.95] Table 3. Despite training the model with just 3 and 5 sample images for each class respectively, the model's performance was depreciated by 9.82% and 2.87% in average precision with IoU of [0.5:0.95] respectively.

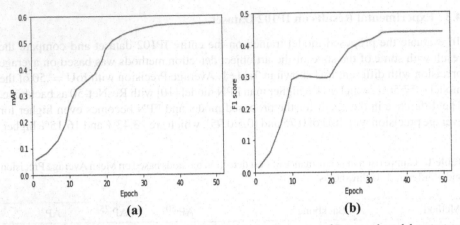

Fig. 6. (a) Mean Average Precision and (b) F1 score performance of Proposed model on test-set with respect to epochs

Table 3. Performance of meta-learning model based on Mean Average Precision under different IoU thresholds on IP102 stimulated data.

	AP.50	AP.75	AP
14 ways-3 shots	37.23	18.34	22.82
14 ways-5 shots	44.51	25.03	29.77

4.5   Experimental Results on Few-Shot Data Taken from ICAR-NBAIR Dataset

In this section, meta-testing is done to measure the performance for the meta-knowledge. Similar to the stimulated data, there were 14 classes of pest, so 14 ways-5-shots and 14 ways-3shots testing is performed. The proposed model achieves 26.68% of mAP in 3-shots and 33.43% of mAP in 5-shots. Table 4 shows the performance of the proposed model for this few-shot data. Due to unavailability of similar work, results could not be compared with other techniques.

Table 4. Performance of meta-learning model based on Mean Average Precision under different IoU thresholds on few-shot data taken from ICAR-NBAIR dataset.

	AP.50	AP.75	AP
14 ways-3 shots	26.68	13.56	16.07
14 ways-5 shots	33.43	17.21	21.32

5 Conclusion

In this work, authors acknowledged the current problems due to pest and difficulties in performing pest insect detection, such as annotation of large dataset and new species of pest. To solve this issue, this approach of few-shot pest detection using meta-learning. The proposed model was trained using pest images from a large dataset (IP102) and a few-shot data stimulation of this large dataset (IP102). The meta-knowledge was used to train the novel few-shot dataset that was annotated manually. The proposed model is evaluated using experimental results for 3-shots and 5shots methods. Experimental results show that the proposed method achieved promising results.

References

1. Wu, X., Zhan, C., Lai, Y.-K., Cheng, M.-M., Yang, J.: Ip102: a large-scale benchmark dataset for insect pest recognition. In: IEEE CVPR, pp. 8787–8796 (2019)
2. Dhaliwal, G.S., Jindal, V., Dhawan, A.K.: Insect pest problems and crop losses: changing trends. Indian J. Ecol. **37**(1), 1–7 (2010)
3. Sharma, S., Kooner, R., Arora, R.: Insect pests and crop losses. In: Breeding Insect Resistant Crops for Sustainable Agriculture, pp. 45–66. Springer, Singapore (2017)
4. Sparks, A., Nelson, A., Castilla, N.: Where rice pests and diseases do the most damage. Rice Today **11**(4), 26–27 (2012)
5. Fuentes, A., Yoon, S., Kim, S.C., Park, D.S.: A robust deep-learning-based detector for real time tomato plant diseases and pests recognition. Sensors **2017**, 17 (2022)
6. Lu, Y., Yi, S., Zeng, N.: Identification of rice diseases using deep convolutional neural networks. Neurocomputing **267**, 378–384 (2017)
7. Jiao, L., et al.: AF-RCNN: an anchor-free convolutional neural network for multi-categories agricultural pest detection. Comput. Electron. Agric. **174** (2020)
8. Zhong, Y., et al.: A vision-based counting and recognition system for flying insects in intelligent agriculture. Sensors 18(5), 1489 (2018)
9. Ren, S., He, K., Girshick, R., Sun, J.: Faster R-CNN: Towards real-time object detection with region proposal networks. In: NIPS (2015)
10. Lin, T.-Y., Doll'ar, P., Girshick, R., He, K., Hariharan, B., Belongie, S.: Feature pyramid networks for object detection. In: CVPR (2017)
11. Liu, W., et al.: SSD: single shot MultiBox detector. In: Leibe, B., Matas, J., Sebe, N., Welling, M. (eds.) ECCV 2016. LNCS, vol. 9905, pp. 21–37. Springer, Cham (2016). https://doi.org/10.1007/978-3-319-46448-0_2
12. Zhang, S., Wen, L., Bian, X., Lei, Z., Li, S.Z.: Single-shot refinement neural network for object detection. In: CVPR (2018)
13. Redmon, J., Farhadi, A.: YOLOv3: an incremental improvement. arXiv preprint arXiv:1804.02767 (2018)
14. Everingham, M., Van Gool, L., Williams, C.K.I., Winn, J., Zisserman, A.: The pascal visual object classes (VOC) challenge. Int. J. Comput. Vision **88**(2), 303–338 (2010)
15. Tzutalin, D.: https://github.com/tzutalin/labelImg (2015)
16. Indian Council of Agricultural Research-National Bureau of Agricultural Insect Resources. https://databases.nbair.res.in/insectpests/pestsearch.php?cropname=Rice
17. Li, Y., Yang, J.: Few-shot cotton pest recognition and terminal realization. Comput. Electron. Agric. **169**, 105240 (2020)
18. Li, Y., Chao, X.: Semi-supervised few-shot learning approach for plant diseases recognition. Plant Methods **17**(1), 1–10 (2021)

19. Nuthalapati, S.V., Tunga, A.: Multi-domain few-shot learning and dataset for agricultural applications. In: Proceedings of the IEEE/CVF International Conference on Computer Vision, pp. 1399–1408 (2021)
20. Skendžić, S., Zovko, M., Živković, I.P., Lešić, V. and Lemić, D.: The impact of climate change on agricultural insect pests. Insects **12**(5), 440 (2021)
21. https://www.fao.org/india/fao-in-india/india-at-a-glance/en/
22. Dhaliwal, G.S., Jindal, V., Dhawan, A.K.: Insect pest problems and crop losses: changing trends. Indian J. Ecol. **37**(1), 1–7 (2010)
23. Tyagi, V.: Understanding Digital Image Processing, 1st edn. CRC Press (2018). https://doi.org/10.1201/9781315123905

Analysis and Applications of Biogeography Based Optimization Techniques for Problem Solving

Gauri Thakur(✉) and Ashok Pal

Department of Mathematics, Chandigarh University, Mohali, India
thakurgauri98@gmail.com

Abstract. Computational intelligence helps in detecting erroneous decisions and fastens the whole process of decision making by applying various techniques. In this study, we will discuss the Biogeography Based Optimization (BBO) and further it will be divided into three parts, firstly, we will describe natural biogeography of BBO with the help of model and algorithm, secondly, we will compare BBO with other optimization methods. Thirdly, we will detect the best location for decomposing solid waste and we will also analyze the benchmark functions with BBO and some recently developed algorithms. At the end of the paper, we will also examine some applications and future possibility of Biogeography Based Optimization.

Keywords: Computational intelligence · Optimization · Migration · Algorithm

1 Introduction

Computational intelligence plays a very pivotal role in the field of decision making by its various techniques [1]. Computational intelligence is the ability of computer to combine smarter techniques, computational methodologies and nature inspired optimization techniques to analyse complex real world problems [2]. Computational methodologies characterize the series of methods that begin from a whim of taking a decision and continue forward by selecting particular trajectory until the decision is made, the method recognized and examined [2].

Decision making problem based Computational intelligence techniques are applied in various fields like in Healthcare system, Autonomous vehicles, Bp plc- Energy, Financial and Defence services and many more [3]. Unquestionably computer intelligence is the future of decision making for the future inventions or events by selecting what we need according to status and computational intelligence selects what is best for us by using optimum methodology [4].

The historical backdrop of computerized reasoning follows right back to antique with logicians thinking about how conceivable it is that counterfeit creatures, mechanical men, and various robots had existed or could exist in some plan. Scholars considered how human thinking could be misleadingly mechanized and constrained by clever non-human machines [5]. Alan Turing proposed a test that thinks a machines ability to copy

M. Singh et al. (Eds.): ICACDS 2022, CCIS 1614, pp. 415–429, 2022.
https://doi.org/10.1007/978-3-031-12641-3_34

human exercises partially that was dubious from human lead. Sometime from that point, the field of artificial intelligence research was set up all through a pre-summer gathering at Dartmouth School during the 1950s, where John McCarthy instituted the articulation in artificial intelligence [4, 5]. The group of researchers in 1980 made the Neural Networks Council who were concerned in the expansion of ANN (artificial neutral network) and biology (natural selection). Bezdek and Marks in 1993 defined the Computational intelligence and they differentiated it with artificial intelligence [5]. IEEE Neural Networks Council turn into the IEEE Computational Intelligence in 2001 by adding novel fields such as evolutionary algorithm, fuzzy set, and optimization techniques based on natural selection [6]. In this paper we will discuss the technique Biogeography based optimization for solving decision making problems. In 2008, Dan Simon was recognized for constructing natural optimization process named as Biogeography based Optimization. It is dependent upon the hypothesis of island biogeography as a computational technique using nature inspired algorithm in solving out decision making problems [4]. BBO has been shown to be an incredible chasing procedure, since it incorporates both investigation and exploitation systems which dependents on relocation or migration [5].

2 Literature Review

Puligilla Prashanth Kumar (2018), [6] the author presents the optimal solution for decisions making using diverse variations of CGP, LP, PWGP models. He concluded that PWGP model gives best optimal solution as compared to another models like it gives feasible solution to optimize small scale enterprises. Dr. Jennifer S. Raj (2019), [7] the author highlights the computational techniques for getting optimal results [7]. The author proposed a variety of computational techniques to locate finest solution of detecting fraudulent access. Emelia Opoku Aboagye and Rajesh Kumar (2019), [8] the author presented the deep neutral modelling with multimodal datasets for combined proposal scheme. The framework was projected with MTF and MLP with stochastic algorithm and tensor factorization for modelling collaborative decisions in a novel way. Touqeer Ahmed Jumani, Mohd Wazir Mustafa, Nawaf N. Hamadneh, Samer H. Atawneh, Madihah Md. Rasid, Nayyar Hussain Mirjat, Muhammad Akram Bhayo and Ilyas Khan (2020) [9] all these authors presented the study on computational intelligence based optimization techniques specifically ANN, GA, FL, ANFIS with their methodology, demerits and merits and explored techniques to deal with power quality and ac micro grids and analysed the limitations of computational intelligence based techniques. Juan M. Sanchez, Juan P. Rodríguez, and Helbert E. Espitia (2020) [10] the authors presented the study that the artificial intelligence has helped lot in decision making by reducing uncertainty, which can be seen in various decision making problems. Agent Based Model, cellular automata and genetic algorithm are the techniques of artificial intelligence which are used for decision making in agricultural public policies.

Haiping Ma, Dan Simon (2017) [11], analyzed and abridged the literature of BBO for past 10 years in their paper. They reviewed the BBO algorithms, its modification and hybridizations, applications in various fields with its future scope. They added the graphical representation for the number of published papers of BBO in IEEE, Springer, Elsevier, Taylor & Francis, SAGE, Hindawi and number of BBO publication by year, and compared BBO with other algorithms. Due to the limitations of BBO there was desire to modify the BBO [11]. Recently, Zhishuang Xue and Xiaofang Liu (2021) [12] proposed an off-line and robust trajectory planning algorithm for computing fixed-wing UAV. The author highlighted an enhanced trajectory planning algorithm which is based on BBO, it is projected to form a finest trajectory linking the begin node and the target node for obstruction accident prevention. Dr. S.Sangeetha, Dr.P.Shanthakumar and S.Abirami, (2020) [13] the author proposed that the hybridization of Parallel Fuzzy Systems, Multi-point decision building algorithm and anticipated Enhanced BBO is used to choose a preeminent RAT with a advanced user's contentment rate [13]. From the results computed based on the simulations accepted out for the measured number of mobile users, it is experiential that the anticipated hybrid PFS – EBBO system execute better with the other methods engaged for RAT choice. Meiji Cui, Li Li, and Miaojing Shi [14] studied the execution improvement to calculate the contribution. Author, proposed threshold plan, as an extra restraint, is taken to compute to check the subgroup whether is inert or not. They analyzed the resource allocation scheme total role which improves the EAs' execution when entrenched into the DC framework.

3   Characteristics of BBO

BBO is one of the fastest growing meta-heuristic algorithm which comes under evolutionary algorithms and motivated by the species migration between different habitats [14]. Dan Simon also correlated BBO to different benchmark functions and to sensor selection problem and by getting the good performance and concluded that BBO can be efficiently used to solve the real life troubles [15]. It can be implemented for solving continuous and discontinuous functions, optimizing multi-dimensional functions, travelling sales man problem, power system problems, it has become vital device for parameter estimation and control, scheduling problems, antenna and network problems, image processing [16]. Being population based optimization technique, BBO has prevailed momentous achievement in many areas whether it related to engineering problems, healthcare system or in any public sector [15, 16].

BBO was modified by using various migration and mutation model. Ma et al. [17] introduced three types of BBO which were named as total immigration, total emigration and partial immigration based BBO models. These modals work efficiently in unimodal and multimodal problems. Gong et al. [18] incorporated three types if mutation operator and showed the more effectiveness of their modifications. Chaos theory and mutation approaches [19] from DE played the vital role in BBO. Many more modifications were done to improve BBO which had made it more useful by overcoming its drawbacks [20]. Hybridization strategy is growing in vide range to develop the standard algorithms by integrating with other algorithms [21]. In earlier researches BBO is hybridized with local search algorithms [22] and population based algorithms like PSO [23], FA [26], HS [28], BFA [29] and many more meta-heuristic optimization algorithms. Hybridization can be done with other algorithm also like K. Jayaraman and G. Ravi integrated BBO with artificial neutral network for forecasting sector oriented energy, likewise there are many researches which shows that hybridization is making the algorithm work more easily and efficiently [30].

Table 1. Parallelism between BBO and other some algorithms

Algorithm	Year	Parameters included	Crucial properties
BBO [21]	2008	Population based, habitat, SIV, HSI, Migration and mutation operator	Inspired by biogeography science, meta-heuristic algorithm, convergence speed is fast, generally used to optimize multidimensional real valued problems, it has been extended to discontinuous, constrained, noisy, combinatorial functions
GA [22]	1966	Population based, chromosome, gene, fitness, crossover operator mutation operator	Search based algorithm, meta-heuristic algorithm, exploiting random search provided with historical data, robust in nature, convergence speed is slow, optimizes constrained and unconstrained
PSO [23]	1995	Population size, weighting factor, cognitive learning coefficient	Stimulate social behavior, meta-heuristic algorithm, depends upon the intelligence and movement of swarms, convergence speed is slow, used to optimize multidimensional real valued problems
BA [24]	2010	Population size, loudness, pulse rate, frequency	Inspired by micro-bats for their echolocation behavior, swarm based meta-heuristic algorithm, used for optimizing discrete and continuous functions
CSA [25]	2009	Netsize, discovery rate	Inspired from cuckoo for their obligate brood parasitism behavior, meta-heuristic algorithm, used for optimizing continuous non linear optimization
FA [26]	2007	Population size, gamma	Inspired from fireflies for their flashing behavior, meta-heuristic algorithm, convergence speed is fast, parallel implementation, used for multi-objective, continuous and discrete, dynamic and noisy optimization problems

(continued)

Table 1. (*continued*)

Algorithm	Year	Parameters included	Crucial properties
DE [27]	1995	Population size, weighting factor, crossover rate	Evolutionary computing algorithm, meta-heuristic algorithm, convergence speed is fast, used for multi-objective optimization and constrained optimization, nonlinear differentiable continuous space function
HS [28]	2001	Memorize, consolidation rate, pitch adjust rate	Inspired by the improvisation of the harmony by musicians, meta-heuristic algorithm, used optimize fitness function of the problems
BFA [29]	2002	Cellsnum, Ned, Nre, Nc, Ns, stepsize, Ped	Novel swarm intelligence optimization algorithm, inspired from E coli. Bacteria for its foraging behavior, parallel distributed processing, initial value insensitivity, used in constrained numerical optimization

4 Methodology and Structure

Biogeography-based optimization is a sub part of evolutionary algorithm which increases the capacity by solving iteratively and randomly and optimizes the problem [30].

Emigration: The emigration is defined as the leaving of a place or habitat with the intention of living in another place or habitat. Emigration differs from immigration although both terms are related to migration only [31].

HSI and SIV: High suitability index (HSI) is defined as when the Islands are favorable to the lives. Characteristics which are connected through HSI add downfall, geographic assortment, vegetative assortment, temperature, surface part. Suitableness index variables (SIVs) occur as the options which are confirmed. Habitability is described by the variables which are known as SIV. HSI are often measured as dependent variables and SIV are the independent variables [31, 32].

Migration: With the help of rates λ and μ of the island SIV is migrated by the migration operator. SIV starts emigrating islands into immigrating islands which is arbitrarily elected. BBO migration is probabilistic in nature and shares stochastically the solution of migration rate [32].

Mutation: SIV is redesigned arbitrarily by mutation machinist stated by islands fixed position probability p of group count. Mutation purpose is to extend diversity among the species or population [32].

Geometric models of biogeography based optimization illustrate movement about from one island to a different and coming up of novel class of species and how they become vanished. The island word is used as explanatory in-spite of precisely (Fig. 1).

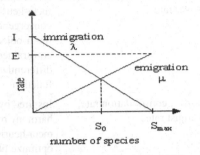

Fig. 1. Species habitat model [30]

Consider λ and μ be emigration probabilities. Stability species count is denoted by S_O, and greatest number of species which island can contain is denoted by S_{max}. Greatest emigration and immigration rate is denoted by E and I [33]. Emigration would increase if there is high HSI (high habitat suitability index) rate. HSI habitat will have high opportunity if the numbers of species are amplified to migrate to neighboring island. Since the crowd between the populations decreases the habitats which have little HSI will elevated immigration rate [34]. The highest number of species that the habitat can sustain is S_{max}, where the immigration rate is zero. If there is no species in the island then there would be zero emigration rate [34, 35]. Islands becomes congested when number of species are greater than before and due to which more species can depart from the island as a effect there is boost in emigration rate. Emigration rate will be in its peak value denoted as E when there would be utmost number of species S_{max}. As habitat suitability index (HSI) increases, number of species increases, smaller amount of species come into the habitat which implies that Immigration decreases. More species exit the habitat which implies that Emigration increases [35].

$$\text{Probability for Emigration rate} = \frac{U_i}{\sum_{i=1}^{N} U_i} \text{ for i} = 1, 2, \ldots N \tag{1}$$

where N is denoted as the number of solutions within inhabitants. Probability for S species in the particular habitat denoted as P_S and is changes from time to time the mathematical equation as following: [35, 36]

$$P_s(t + \Delta t) = P_s(t)(1 - \lambda_s \Delta t - \mu_s(t) + P_{S-1}\lambda_{S-1}\Delta t + P_{S+1}\mu_{S+1}\Delta t \tag{2}$$

Emigration rate denoted as λ_s, Immigration rate denoted as μ_s. Equation (1) holds for time $t + \Delta t$ in the following ways:

- When there is S species then there would be no occurrence amid t and $t + \Delta t$ of immigration and emigration rate.

- If a single species is immigrated then there would be S-1 species on time t.
- If a single species is emigrated then there would be S + 1 species on time t.
- There can be ignorance of immigration and emigration probabilities of more than one when we consider Δt small [37].

Emigration and immigration rate is computed with the help of following equations:

$$\lambda_i = I\left(1 - \frac{K}{S_{max}}\right) \tag{3}$$

$$\mu_i = E\left(\frac{K}{S_{max}}\right) \tag{4}$$

Here, uppermost immigration rate and emigration rate is denoted as I and E, For ith habitat number of groups and largest number of groups are denoted by K and S_{max}, (3) and (4) given emigration and immigration rate [36]. The optimal solution will have more HSI and they are parallel to a habitat with many species which has high emigration and low immigration [37]. The vital attribute of migration operator is to share to data between diverse solutions, whereas fine solution divide their feature with reduced answers and as a result habitat's SIV is updated by migration operator by acquiring good features from good habitats as follows:

$$H_i(SIV) \leftarrow H_k(SIV) \tag{5}$$

Here H_i is denoted as immigration habitat and H_k is the emigration habitat which is obtained from roulette wheel selection [36, 37]. Species probability and mutation rate is inversely proportional to each other. The mutation rate is computed as follows:

$$m_i = m_{max}\left(1 - \frac{p_i}{p_{max}}\right) \tag{6}$$

Here p_i is the BBOs species probability, maximum mutation rate denoted by m_{max} which is user defined parameter [37]. The mutation process makes low HSI solutions likely to mutate which results in some improvement in them. Elitism approach is used to accumulate the finest solution so that in case if mutation ruins HSI, due to which we have our previous results and we can go back again [36]. Due to migration and mutation BBO has well-built local and global search ability. But there also seems some drawbacks of BBO some of them are like Incapable in utilizing the solutions in profitable way. Best candidate can't be selected from each generation. Infeasible results are produced, as during immigrating the characteristics consequential fitness not thinks about the habitat. In nutshell, these drawbacks are improved by modified or hybrid BBO [37].

4.1 Classical BBO Algorithm

1) Create the migration and mutation probability.
2) Initialize the contender's population including BBO parameters.
3) For each contender in the population compute the immigration and emigration rate.

4) Decide the modification of island which would be built on the immigration rate.
5) For the emigration rate choose the island which is to be emigrated for SIV with the help of roulette wheel selection.
6) From the decided island which is to be emigrated choose SIV arbitrarily.
7) For every island with the help of mutation probability carry out the mutation.
8) For each island compute fitness value.
9) If step 9 condition is not fulfilled follow step3 again. Repeat the process until optimal solution is not obtained.

5 Detecting Best Location for Decomposing Solid Waste

Solid waste management is crucial service for sustaining solid waste materials comprises range of unwanted garbage material which is generated from residential, industrial or commercial activities [38]. The solid wastes are disposal area can be located in land-fills. For locating the area some of the factors are required which includes land slope, groundwater depth, and stability of soil, vicinity of surface water and susceptibility of flooding. These all factors will come under SIV. The best disposing site selection for solid waste will be depending upon the factors and the cost (transfer station, planning the route for vehicles which will carry waste, appropriate designing of landfills) [39]. In this paper the optimum disposal location will be carried out by BBO to get the results we have taken help from the (Yarpiz) https://yarpiz.com/ (Tables 1, 2)

Table 2. Finding out optimum location decomposing solid waste.

Maximum iteration	Number of habitat	Best location	
		cost	Position
50	50	0.0013	0.0107, 0.0083, −0.0089, 0.0061, 0.0035, 0.0064, −0.0023, 0.0118, 0.0211, 0.0162, 0.0117
100	50	2.0854e−04	−0.0042, −1.1769e−04, 0.0059, −0.0023, 0.0078, 0.0017, 0.0029, −0.0033, −0.0082, 1.7210e−04, 0.0012
1000	100	1.1422e−08	−3.1793e−05, −4.6202e−06, −2.4997e−07, 1.0701e−05, 9.8193e−06, −7.5602e−06, −1.9311e−05, 3.3441e−06, −8.1430e−07, −1.3838e−05, −9.7701e−05

Above results (Table 2) gives us the optimum location which includes best cost and optimum position (which is in direction of coordinates). The variables for BBO SIV (Md Mainul Sk, Sk Ajmin Ali, Ateeque Ahmad, 2020) which are landfill suitability including

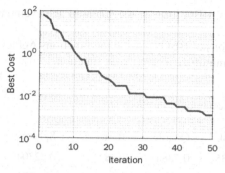

Fig. 2. Best location for 50 iteration

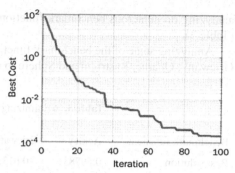

Fig. 3. Best location for 100 iterations

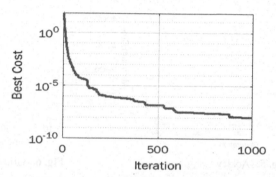

Fig. 4. Best location for 1000 iterations

land elevation, slope, soil, distance to surface water and road, land cost value are used as decision making variables in BBO algorithm [39]. The results compiled MATLAB compiler on Windows 8 operating system with 4 GB RAM, INTEL R core processor to find out best location for locating landfills (Figs. 2, 3 and 4).

6 Analysis of Benchmark Results

Dan Simon analyzed the efficiency of Biogeography Based Optimization in place of various different optimization methods such as probability based incremental learning, evolutionary strategy, genetic algorithm, [22] Stud genetic algorithm (SGA), differential evolution and particle swarm optimization [23] applying fourteen benchmark functions (like Ackley, Rosenbrock, Sphere, Quartic, Rastrigin etc.) [40]. Among the fourteen benchmark SGA and BBO stood out the easiest between the seven benchmarks. Amid the eight algorithms BBO takes the fifth quickest position. The benchmark outcomes specify that Biogeography Based Optimization is very competing with alternative optimization algorithms. By analyzing the data [41] artificial bee colony technique, glowworm Swarm optimization, particle Swarm optimization and genetic algorithm convergence speed is time consuming whereas in BBO the convergence speed is rapid. By analyzing on numerous benchmark functions, BBO leads to the splendid achievement (Table 3).

Analyzing some of the benchmark functions of BBO by using the functions Ackley, Griewank, Quartic, Rastrigin and Sphere to minimize a continuous function.

Table 3. Minimizing continuous function

Benchmark functions	Ackley	Griewank	Quartic	Rastrigin	Sphere
Best solution	0.50783	−0.61335	0.73881	0.6871	0.22304

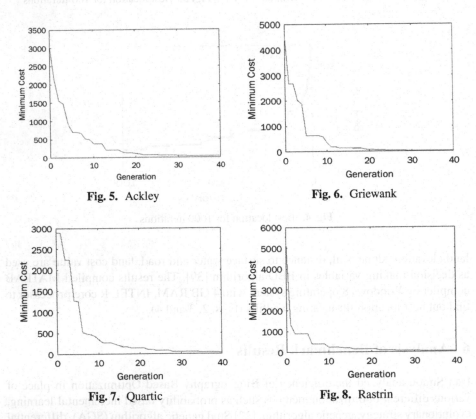

Fig. 5. Ackley

Fig. 6. Griewank

Fig. 7. Quartic

Fig. 8. Rastrin

Analyzing the above table (Table 3), we can conclude that when Ackley, Griewank, Quartic, Rastrigin and Sphere are used to minimize a continuous function using BBO gives very valuable solution which is less than one depicts that less the value more the optimum result. Whereas in many meta-heuristic optimization algorithms gives the values for the same benchmark functions more than one. In this we have taken generation limit 40, population size 60, dimension 10, mutation probability 0.07, total number of elites were 3. We have taken help from the (Yarpiz) https://yarpiz.com/ for getting results (Figs. 5, 6, 7, 8 and 9).

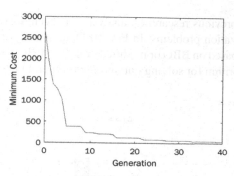

Fig. 9. Sphere

Table 4. Analyzing some of the benchmark function with BBO and recently developed other algorithms.

Algorithm	Sphere	Ackley	Rosenbrock
BBO	0.0084731	0.00057949	0.0095646
FA	0.04501	1.3021	13.7431
CGO	1.0479e−36	7.2532e−37	5.3556e−38
SFLA	5489.7989	6200.1692	6219.9353
CA	767.8401	663.1003	509.1459
MA	0.012619	0.11237	20.7443

By analyzing the above table (Table 4) we can conclude that BBO gives one of the best optimal solutions for optimization problems. We had compared BBO with recently developed algorithm in past 10 years i.e. FA, CGO, SFLA, CA, and MA. CGO gives better results than BBO but if we compare BBO with other algorithm, it gives much better results. The results are gathered using MATLAB software we gave used the same variables and functions for each algorithm and then we had combined the results (Table 4).

7 Applications of BBO

BBO has been appreciably explored to resolve several issues like producing system programming, provide chain style optimization and hub competitive location [41]. Due to the potency, it has been widely used in several engineering and technical decision making tasks. It is used in many decision making problems like allocating timetable, adaptive neighborhood, resource allocation for large scale optimization [42]. Biogeography Based Optimization is a population based algorithm which plays an important role in decision problems such as it is used in flexible job scheduling problem, [41] overlapping communities within social network with automatic detection, patient admittance scheduling problem and many more [42], resource reallocation for universal optimization [43].

With the help of previous researches, analysis has been done to show the broadly used BBO for optimization problems. In Fig. 10 Data is analyzed from 2014 to 2021 researches which are based on BBO or modified or hybrid BBO which depicts that BBO is very beneficial algorithm for solving out optimization problems related to real life.

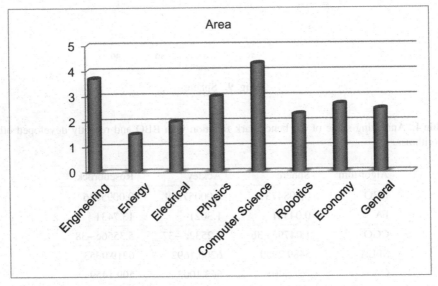

Fig. 10. BBO applications in optimization problem based on various areas.

8 Conclusion

Computational intelligence helps in detecting erroneous decisions and fastens the whole process of decision making by applying various techniques. In this research, we have highlighted the Biogeography Based Optimization, which plays a significant role in decision making problems. BBO integrates an uncomplicated process to search for the finest solution for the uneven or uncertain decision making problems by the procedure of migration and mutation. In this paper we have applied BBO in detecting location for decomposing solid waste. Because of its top notch overall performance at the benchmarks offers several evidence that BBO can be profitably carried out to realistic problems and thus, discloses a massive variety of potentialities for future analysis.

8.1 Future Scope

1) BBO can be used to resolve several sensible issues of engineering.
2) Biogeography based Optimization incorporates nice scope in Dynamic/noisy fitness operate.
3) Study may be created in future on however species age affects extinction and expatriation.

4) Study may be done on migration varies with species quality.
5) Compatible value for stagnation evaluation with complete accuracy can be searched through BBO.

References

1. Secinaro, S., Calandra, D., Secinaro, A., Muthurangu, V., Biancone, P.: The role of artificial intelligence in healthcare: a structured literature review. 021-01488-9 (2021)
2. Buchanan, B.G.: Brief History of Artificial Intelligence. AI Magazine, vol. 26 Number 4 (2006)
3. Mijwel, M.M.: History of Artificial Intelligence. Computer science, college of science, University of Baghdad, Iraq (2015)
4. Wallace, A.: The Geographical Distribution of Animals. Boston, MA Adamant Media Corporation, Two Volumes (2005)
5. Simon, D.: Biogeography-based optimization. IEEE Trans. Evol. Comput. (2008)
6. Kumar, P.P.: An Optimization Techniques on the Managerial Decision Making. Int. J. Mech. Prod. Eng. Res. Dev. (IJMPERD), ISSN(P): 2249-6890, 8(6), 507–516 (2018)
7. Raj, J.S.: A comprehensive survey on the computational intelligence techniques and its applications. J. ISMAC 01(03), 147–159 (2019)
8. Aboagye, E.O., Kumar, R.: Simple and Efficient Computational Intelligence Strategies for Effective Collaborative Decisions. Computer Science Department, UESTC (2019)
9. Jumani, T.A., et al.: Computational intelligence-based optimization methods for power quality and dynamic response enhancement of ac microgrids. Energies 13(16), 1–22 (2020)
10. Sánchez, J.M., Rodríguez, J.P., Espitia, H.E.: Review of Artificial Intelligence Applied in Decision-Making Processes in Agricultural Public Policy 8(11), 1374 (2020). https://doi.org/10.3390/pr8111374
11. Ma, H., Simon, D., Siarry, P., Yang, Z., Fei, M.: Biogeography-based optimization: a 10-year review. IEEE Trans. Emerging Topics Comput. Intell. 1(5), 391–407 (2017)
12. Xue, Z., Liu, X.: Trajectory planning of unmanned aerial vehicle based on the improved biogeography-based optimization Algorithm. In: Advances in Mechanical Engineering 2021, vol. 13(3), pp. 1–15 (2021)
13. Sangeetha, S., Shanthakumar, P., Abirami, S.: RAT Selection in Heterogeneous Wireless Networks Using a Hybrid Fuzzy-Enhanced Biogeography Based Optimization. Department of CSE. Karpagam Academy of Higher Education Tamilnadu, India (2020)
14. Cui, M., Li, L., Shi, M.: A selective biogeography-based optimizer considering resource allocation for large-scale global optimization. College of Electronics and Information Engineering, Tongji University, Shanghai 201804. China (2019)
15. MacArthur, R., Wilson, E..: The Theory of Biogeography. Princeton Univ. Press, Princeton 1967 (2014)
16. Ma, H.: An analysis of the equilibrium of migration models for biogeography-based optimization. Inf. Sci. 180(18), 3444–3464 (2010)
17. Ma, H., Simon, D., Fei, M., Xie, Z.: Variations of biogeography-based optimization and Markov analysis. Inf. Sci. 220(1), 492– 506 (2013)
18. Gong, W., Cai, Z., Ling, C., Li, H.: A real-coded biogeography based optimization with mutation. Appl. Math. Comput. 216(9), 2749–2758 (2010)
19. Niu, Q., Zhang, L., Li, K.: A biogeography-based optimization algorithm with mutation strategies for model parameter estimation of solar and fuel cells. Energy Convers. Manage. 86, 1173–1185 (2014)

20. Simon, D., Omran, M., Clerc, M.: Linearized biogeography-based optimization with re-initialization and local search. Inf. Sci. **267**, 140–157 (2014)
21. Ma, H., Simon, D.: Evolutionary Computation with Biogeography-Based Optimization. Hoboken, NJ, USA (2017)
22. Katoch, S., Chauhan, S.S., Kumar, V.: A review on genetic algorithm: past, present, and future. Multimed. Tools Appl. **80**, 8091–8126 (2021)
23. Pal, A.: Decision making in crisp and fuzzy environments using particle swarm optimization. Ph.D. thesis, Department of Mathematics, Punjabi University, Patiala-India (2015)
24. Liu, J., Ji, H., Liu, Q., Li, Y.: A bat optimization algorithm with moderate orientation and perturbation of trend. J. Algorithms Comput. Technol. **15**, 1–11 (2021)
25. Rajabioun, R.: Cuckoo optimization algorithm. Appl. Soft Comput. **11**(8), 5508–5518 (2011)
26. Zhang, P., Wei, P., Yu, H.: Biogeography-based optimization search algorithm for block matching motion estimation. IET Image Process. 6(7), 1014–1023 (2012)
27. Qin, A.K., Huang, V.L., Suganthan, P.N.: Differential evolution algorithm with strategy adaptation for global numerical optimization. IEEE Trans. Evol. Comput. **13**(2) (2009)
28. Lin, J.: Parameter estimation for time-delay chaotic systems by hybrid biogeography-based optimization. Nonlinear Dyn. **77**(3), 983–992 (2014). https://doi.org/10.1007/s11071-014-1356-7
29. Lohokare, M., Pattnaik, S., Devi, S., Panigrahi, B., Das, S., Bakwad, K.: Intelligent biogeography-based optimization for discrete variables. World Congr. Natural Biol. Inspired Comput, pp. 1087–1092 (2009)
30. Jayaraman, K., Ravi, G.: Long-term sector-wise electrical energy forecasting using artificial neural network and biogeography-based optimization. Electr. Power Compon. Syst. **43**, 1225–1235 (2015)
31. Hanski, I., Gilpin, M.: Metapopulation Biology. Academic,. New York (1997)
32. Wesche, T., Goertler, G., Hubert, W.: Modified habitat suitability index model for brown trout in southeastern Wyoming. North Amer. J. Fisheries Manage. **7**, 232–237 (1987)
33. Ammu, P.K., Sivakumar, K.C., Rejimoan, R.: Biogeography-based optimization - a survey. Int. J. Electron. Comput. Sci. Eng. (2013). ISSN- 2277-1956
34. Gong, W., Cai, Z., Ling, C.: DE/BBO: a hybrid differential evolution with biogeography-based optimization for global numerical optimization. Soft Comput. **15**(4), 645–665 (2010)
35. Du, D., Simon, D., Ergezer, M.: Biogeography-based optimization combined with evolutionary Strategy and immigration Refusal. In: IEEE International Conference on Systems, Man, and Cybernetics, pp. 997–1002 (2009)
36. Lohokare, M.R., Pattnaik, S.S., Devi, S., Bakwad, K.M., Jadhav, D.G.: Biogeography based optimization technique for block based motion estimation in video coding. National Conference on Computational Instrumentation, CSIO Chandigarh, INDIA, pp. 19–20 (2010)
37. Zhang, X., et al.: Improved Laplacian Biogeography-Based Optimization Algorithm and Its Application to QAP. Article ID 782478 (2020)
38. Storn, R., Price, K.: Differential evolution - a simple and efficient heuristic for global optimization over continuous spaces. J. Global Optim. **11**(4), 341–359 (1997)
39. Md Mainul, S.k., Sk Ajim, A., Ahmad, A.: Optimal sanitary landfill site selection for solid waste disposal in Durgapur city using geographic information system and multi-criteria evaluation technique. J. Cartography Geograph. Inf. **70**, 163–180 (2020)
40. Ma, H., Simon, D., Fei, M.: On the convergence of biogeographybased optimization for binary problems. Math. Probl. Eng., **2014**, Art. no. 147457 (2014)

41. Bhattacharya, A., Chattopadhyay, P.K.: Biogeography-based optimization for different economic load dispatch problems. IEEE Trans. Power Syst. **25**(2), 1064–1077 (2010)
42. Hammouri, A.I.: A modified biogeography based optimization algorithm with guided bed selection mechanism for patient admission scheduling problems. Department of Computer Information Systems, Al-Balqa Applied University, Al-Salt, Jordan (2020)
43. Cui, M., Li, L., Shi, M.: A selective biogeography-based optimizer considering resource allocation for large-scale global optimization. College of Electronics and Information Engineering, Tongji University, Shanghai, 201804, China (2019)

A Proposed Model for Precision Agriculture

Jyotiraditya Singh Tomar, Parishkrat Mishra, Akash Gupta, Kunj Bihari Meena(✉),
and Vipin Tyagi

Department of Computer Science and Engineering, Jaypee University of Engineering and
Technology, Raghogarh, Guna, MP, India
kunjmeena@gmail.com

Abstract. The concept of precision agriculture mainly focuses on collecting,
processing, and analyzing data of inter- and intra-field variables in a farm so that
in minimum resources optimum returns can be obtained from agriculture. This
paper proposes a model for precision agriculture that automates data collection of
intra-field variables such as soil moisture, sunlight exposure, and relative humidity
across farms through a network of sensors. The proposed model provides actionable information to the farmers for their crops through which they can optimize
irrigation and nutrient management, hence saving costs on agricultural inputs.
This also ensures an environmentally sustainable approach to agriculture by being
resourceful. The hardware-level implementation of the proposed model is also
suggested so that the model can be implemented and deployed easily. At last,
the paper also explores various challenges in the adoption of the proposed model
along with improvements that can be made in the future as the model scales.

Keywords: Precision agriculture · Precision map · Innovation in agriculture ·
Wireless Sensor Network · IoT

1 Introduction

With the world's population expected to reach 9.7 billion by 2050 [10], agriculture faces
the challenge of closing a gap of 56% between the amount of food available today and
that required by 2050 [7]. Sustainable agricultural practices need to be adopted to ensure
maximum output from limited resources. This is where precision agriculture promises
to be a viable solution. Precision agriculture treats a farm as a heterogeneous field with
intra-field variability in crops instead of a uniform homogenous field as in the case of
traditional agricultural methods. A farm is certain to have variability in soil conditions
such as soil moisture, pH, nutrient content, etc. because of variations in soil content and
topography [4] which means that different areas of the farm will have different irrigation
and nutrient requirements. Traditional irrigation and nutrient management may result in
over-application or under-application of resources in specific locations within the farm.
This may lead to a negative impact on crops and the environment, at the same time
resulting in higher input costs or lower outputs. All of these issues can be prevented
by precision agriculture. Precision agriculture methods address the crops' needs on a
requirement basis [11]. A farmer only needs to irrigate an area of the farm which is below

© The Author(s), under exclusive license to Springer Nature Switzerland AG 2022
M. Singh et al. (Eds.): ICACDS 2022, CCIS 1614, pp. 430–441, 2022.
https://doi.org/10.1007/978-3-031-12641-3_35

the desired range of soil moisture or stop applying fertilizer where there is an excess. Similarly, various other variables (soil pH, relative humidity, etc.) can be accounted for accordingly.

Precision agriculture can be easily understood with the help of a succulent plant. Succulents can survive in arid climates and soil conditions. They are favored as house-plants for their attractiveness and ease of care, in theory. For many succulent owners however, over-watering and associated infections regularly result in death of succulents. All plants have an ideal soil moisture range that they thrive in. Over-watering results in root rot and then infections whereas under-watering results in water stress, eventually drying the plant. If the soil moisture levels can be monitored using sensors, plants can be irrigated according to their specific requirements, eliminating any guess-work. This is what precision agriculture deals with. The soil and environmental conditions are moni-tored and agricultural inputs (irrigation and nutrition) are adjusted according to the ideal growing conditions of the crop. This ensures the best yield while preventing excess of any input.

To enable precision agriculture, various technological solutions are being devel-oped and refined such as wireless sensor networks, agricultural robots, drones, satellite imagery, etc. If precision agriculture is to be adopted on a global scale, automating the data collection process becomes a necessity for it to be viable and efficient. The proposed model for precision agriculture automates data collection using a network of sensors and IoT-related solutions. This data is then presented to the farmer through a dashboard through which they can make informed decisions. The data collected over a period of time can then be used to determine ideal irrigation and nutrient requirements for specific crops across different soil conditions and geographies. This can reduce costs and increase agricultural productivity [3]. The proposed model also aims to address the impact on crops due to changes in weather conditions by climate change. This is done by accounting for rainfall and adjusting irrigation accordingly as well as deciding on the best time for the application of fertilizers so that a run-off of nutrients by sudden rain can be prevented.

The remaining sections of the paper are organized as follows: related existing systems for precision agriculture are discussed in Sect. 2 along with their pros and cons. Section 3 provides a detailed discussion of the proposed model. Section 4 describes various areas where the proposed model can be applied successfully. Section 5 discusses the possible challenges while adopting the proposed model in the real world situations. At last, Sect. 6 provides a conclusion of the paper with possible future research directions in this area.

2 Existing Work

There is a lot of work and research going on in the field of precision agriculture. Microsoft Azure FarmBeats [8] is a great example of IoT applications for precision agriculture. This system uses sparsely placed sensors across a farm and aerial imagery to gather data and generate precision maps by extrapolating data obtained from the sensors to the aerial image of the farm using machine learning pipelines [18]. However, this project has only shown precision maps for when the farm was void of crops, which raises the question of how the system will generate precision maps when there is significant vegetation cover on the field. One way to address this would be to analyze the vegetation color, plants that lack water would appear dry. This means taking a reactive approach, so while the Microsoft Azure FarmBeats system of using fewer sensors and applying machine learning to extrapolate data does seem to work well on the surface it remains to be seen as to how effective it will be once the crops cover the field. The model proposed in this paper does have a reliance on a high number of sensors because as of now, only that seems to be a way that is proactive in terms of addressing the crops' needs with the required frequency of data collection to enable precision agriculture. One way by which the number of sensors can be reduced is briefly discussed in Sect. 5. Then there is AgriFusion [20] which talks about the architecture for IoT based on a Precision Agriculture Survey. It tells about the Network architecture, which is a necessary element for the proposed model as it may be using a wireless network between microcontrollers and base station(s).

Another company that is benefiting from data collection and automation is Bowery Farming [12], a New York-based vertical farming and digital agriculture company. This company has vertical farms across the US that grow leafy greens using some of the most advanced agriculture technologies and methods including the use of sensors for precision monitoring of nutrients. They go a step further by combining precision agriculture with vertical farming, saving on arable land as well as being in complete control of environmental variables, for the most part. This allows Bowery Farming to have produce all year round and provide off-season crops in the market. Bowery Farming's technology relies on automation, sensors, robotics, AI, and a proprietary operating system, BoweryOS, that takes photos of crops and analyzes data in real-time. The BoweryOS benefits from historical data gathered from thousands of growth cycles for each crop grown at a Bowery farm and continuously tweaks growing conditions to further hone in on the most effective growing conditions which they term "recipes". The "recipes" with the highest rates of success (i.e., the healthiest plants and crop yields) are further refined and then scaled automatically throughout their farms.

Other solutions that enable precision agriculture include selective use of herbicides such as Blue River's See & Spray technology [13] which uses computer vision technology to selectively apply herbicides on a field, reducing the amount of herbicide required. Some solutions are proactively looking out for crop disease outbreaks such as Plantix [14], a mobile crop advisory app that can diagnose pest damages, plant diseases, and nutrient deficiencies affecting crops and offers corresponding treatment measures. There are many more examples of solutions being worked upon that enable precision agriculture, trying to place it as a better, more cost-effective, and environmentally sustainable alternative. This paper proposes one such solution and briefly discusses the challenges that it may face in its implementation.

3 Proposed Model

This section provides a detailed discussion of the proposed model.

3.1 Overview

Precision agriculture is the result of the third wave of the agricultural revolution. Since its inception, various tools and technologies have been developed that enable farmers to collect data from their farms such as soil pH, and nutrient levels using sensors and other measuring techniques [11]. However, this data collection has mostly been manual and not automated at a large scale. For precision agriculture to be effective at a large scale, automating data collection becomes necessary.

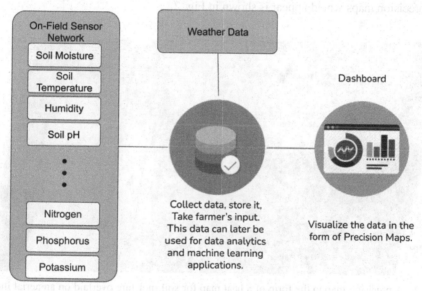

Fig. 1. Overview of the proposed model.

This paper proposes a model (Fig. 1) for precision agriculture that employs multiple sensors connected to microcontrollers which then form a network along with a base station for each farm as shown in Fig. 5. The sensors collect data of various intra-field variables including but not limited to soil moisture, soil pH, nutrients(nitrogen, phosphorus, and potassium) [4] along with inter-field variables such as wind speed, relative humidity, incident sunlight, etc. The data collected is then forwarded to a base station. The base station then processes the data and uploads it to a cloud database using broadband or cellular connectivity or satellite internet, whichever is viable. A dashboard (Fig. 3) is then presented on the farmer's device(s), providing the farmer with actionable information for their farm in the form of precision maps. The dashboard also displays the weather forecast for the farmer to anticipate rain and make decisions accordingly for irrigation and nutrition management. Over time, data collected can be analyzed for different crops over varying soil conditions and topographies. This enables the model to better understand the requirements of individual crops and provide an ideal, economical and environmentally sustainable approach with each new crop cycle.

3.2 Precision Maps

As data can be better interpreted with the help of visualization, hence, the data collected from sensors is visualized in the form of precision maps, akin to heat maps, for all the variables that are accounted for such as soil moisture, pH, etc. across the farm. The sensors will be geo-tagged during deployment across the farm providing spatial data, making it possible to generate precision maps for the farm. These precision maps present the farmer with visualized data that they can use to make informed decisions such as which area of the farm needs to be irrigated and by how much [15]. An example of how the precision maps would appear is shown in Fig. 2.

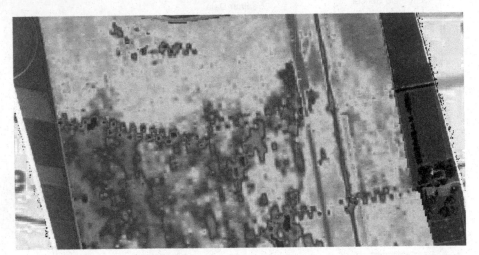

Fig. 2. A precision map in the form of a heat map for soil moisture overlaid on an aerial image of a farm [15]. Greener shades indicate regions with ideal soil moisture whereas regions marked with red indicate low soil moisture and blue indicate excess soil moisture.

Precision maps enable the farmer to be precise with application of resources such as water for irrigation and fertilizers for nutrition. Thereby, reducing costs by minimizing chances of over-application of any resource while ensuring that the crops have irrigation and nutritional requirements sufficiently fulfilled for healthy growth.

3.3 Precision Irrigation

Soil moisture across a farm varies due to varying soil conditions across a single farm itself, such as when an area of the farm has soil with higher moisture-holding capacity and others with lower or when the farm is on a slope. Traditional irrigation methods are not efficient in such cases. Soil moisture sensors are geo-tagged and deployed across the farm after an intensive analysis of the soil conditions and topography of the farm. An ideal density of sensor deployment is determined and sensors are placed to measure soil moisture at different regions of the farm at varying depths. With the help of the spatial data of soil moisture from sensors, a precision map can be prepared for irrigation. Farmers can then irrigate the field with the least water consumption, this irrigation system is commonly known as precision irrigation [1]. This enhances the implementation of modern methods of irrigation such as micro-irrigation and fertigation. They may also invest in automatic irrigation systems that respond to the data gathered and all that the farmer would need to do is monitor the system. With time, these systems can be refined to anticipate rainfall by weather forecasts and schedule watering to be more resource-efficient while trying to maintain the ideal soil moisture for healthy crops. Irrigation systems also need to account for the application of fertilizers to prevent a run-off of the nutrients which in case of over-watering may end up too deep inside the soil to be of any benefit to the crop or simply run-off if the farm is on a slope.

3.4 Precision Nutrition

Uniform application of fertilizers across the farm is not resourceful as it would result in under-application or over-application of nutrients in different regions of the farm. However, using data collected from macronutrient sensors [4] and creating precision maps (primarily for nitrogen, potassium, and phosphorus), gives the farmer the data they need to be resourceful in their application of fertilizers [6]. This reduces the amount of fertilizer needed. This would save input costs as well as be more sustainable while also reducing the problem of surface run-offs by accounting for irrigation and rainfall.

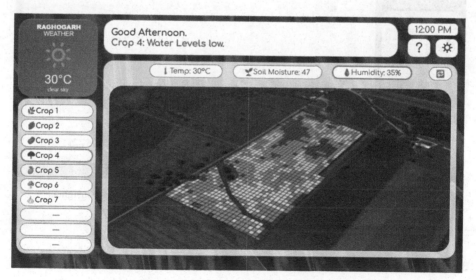

Fig. 3. The proposed dashboard design.

3.5 Dashboard

The dashboard will provide the farmer with a graphical interface that would include the precision maps along with other important information including but not limited to the local weather conditions (current and forecasted), such as rainfall, relative humidity, cloud cover, etc. The dashboard presents visualized data of the soil conditions across the farm so that the farmer can use it to make informed decisions.

An example of what a dashboard may look like is shown in Fig. 3. In the proposed dashboard, the farmer is provided with an alert window and a precision map in the form of a heat map overlaid on an aerial image, where green cells indicate ideal soil moisture and other colors are used to mark cells that need farmer's attention [16]. The dashboard can also provide current and forecasted weather information to the farmers. Weather data is aggregated from various weather stations that provide data to weather services [22] such OpenWeather and The Weather channel. The model will also monitor and record input variables such as water consumption over a crop cycle and visualize the recorded data as shown in Fig. 4. Additionally, the proposed dashboard comprises of different element to visual data which going to be important for doing of precision agriculture such as- Weather Widget, Alert box, Graph Selection menu Timestamp, setting button, and learn more button.

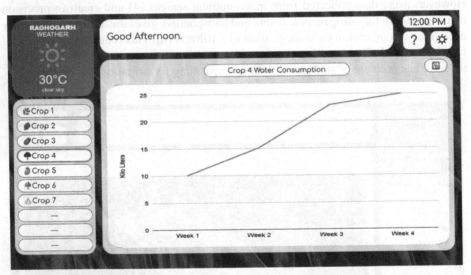

Fig. 4. Dashboard showing visualized data for agricultural inputs for insights.

3.6 Suggested Hardware Implementation of the Proposed Model

This section suggests a hardware implementation of the proposed model (Fig. 5). Several wired and wireless sensors deployed across the farm at varying depths will be used to monitor the soil and environmental conditions. For measuring soil moisture, there primarily exist two types of sensors: resistive soil moisture sensors and capacitive soil moisture sensors [21]. Capacitive soil moisture sensors are more suitable for the model proposed as they are more robust and do not corrode whereas resistive soil moisture sensors are prone to corrosion, which renders them useless for our proposed model. The ideal number of sensors and their deployment across the farm would vary based on soil type and topography. These sensors will be connected to microcontrollers that will form a network, collecting data from sensors that they will forward to a base station. The base station then processes the data and uploads it to a cloud database using broadband or cellular connectivity or satellite internet, whichever is viable. This allows us to automate the data collection process, a necessary requirement to be fulfilled in order to scale the system. The farmer is then presented with a dashboard on their device of choice such as a standard computer or tablet. There exist many protocols and technologies for such sensor networks, such as low-power wide-area network based DASH7 Alliance Protocol (D7A) [2] or MIoTy [17]. The sensors are rapidly advancing in capabilities such as precise monitoring of soil macronutrients [4]. Key factors to be considered for such a sensor network are energy efficiency and zero downtime. To provide power to the sensor modules, solar-powered batteries can be used along with small form factor Vertical Axis Wind Turbines (VAWT) [9], which should be enough for powering the sensors.

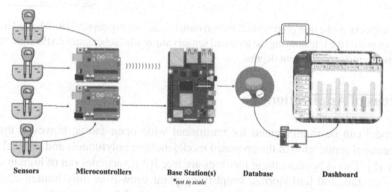

Sensors   Microcontrollers   Base Station(s)   Database   Dashboard
not to scale

Fig. 5. A simplified overview of implementation of the proposed model.

Measurements of soil moisture from varying depths are required to better analyze the moisture content of soil, accounting for evapotranspiration. Key factor to be considered for irrigation is ensuring ideal soil moisture for the roots, which vary with crops and grow to different depths. Surface mounted soil moisture sensors alone are not effective for determining if the soil moisture content is ideal, making it necessary to measure soil moisture at varying depths using special probes such as the one shown in Fig. 6.

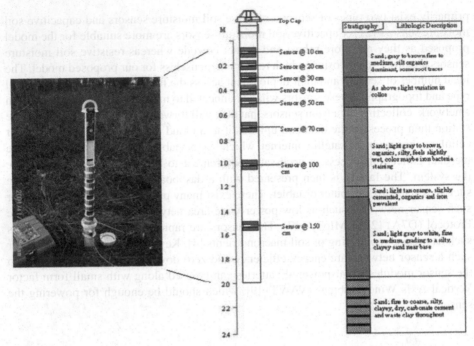

Fig. 6. A special probe designed to measure soil moisture at varying depths [19]. The soil moisture probe is shown on the left showing the mounted sensors along with schematics on the right showing sample stratigraphy at different depths.

4   Areas of Application

This model can be implemented for traditional wide-open farms, however, the best-suited areas of application of the proposed model include polytunnels and vertical farms, initially [5]. This is because these facilities are free from variables out of human control such as rainfall, and hail storms, simplifying plant growth as only human-controlled inputs have to be taken care of. Polytunnels and vertical farms have the equipment to control temperature, humidity, and ventilation. They allow crops to be grown off-season and are used for season extension. Such practices prevent price spikes for off-season crops. Some of these facilities even have space-heating systems as well as soil-heating systems to purify the soil of unwanted viruses, bacteria, and other organisms. These facilities would greatly benefit by employing precision farming techniques and would help in refining the model. The solutions proposed by the model would eventually spill over into traditional farms as the technology becomes more affordable and scalable, thereby ensuring that it can function on a global scale.

5 Challenges

Following are some of the challenges for the proposed model:

Slow Adoption: Due to high start-up costs the adoption of technological solutions for precision agriculture is very slow. This can be changed by increasing awareness of the many benefits that precision agriculture provides to farmers. Such as by saving costs on irrigation and nutrition management. Farmers should carefully assess the return on investment that precision agricultural practices offer over traditional agricultural practices.

Implementation: Technological solutions for precision agriculture are complicated to implement due to various factors, for example, the ideal density of sensor deployment on a farm has to be determined for particular soil types and geographies. Only a limited number of sensors can be deployed on a farm, cost-effectively. To address this issue, agricultural robots equipped with a suite of sensors can be used to scout the fields and take measurements, thereby reducing the number of sensors required. However, these robots will have to be autonomous and capable enough to navigate across unfavorable terrain. Initially such robots will be expensive but with time they will improve and also become more affordable and cost-effective. The expenses required from farmer would include the cost of hardware and setup as well as a recurring cost for using cloud services along with the cost of an active internet connection, these costs can be recovered in the form of saving from reduced agriculture inputs.

Data Handling: Gathering, storing, and processing vast amounts of data if such systems were to be implemented on a large scale, pose another challenge that could be addressed using cloud computing and cloud storage. This eliminates the need for having high performance computers on-site, provided that internet connectivity can be ensured which in itself poses another challenge.

Internet Connectivity: If cloud solutions are to be used, internet access from even remote farms becomes mandatory, a requirement that can be fulfilled by satellite internet access if other means of connecting to the internet are not accessible. Satellite internet services are still new and not accessible to all for an affordable price. These systems would need to expand for more availability as well as reduce costs for them to be of economic value to the farmer.

Hardware Malfunctions: It has to be ensured that all sensors/systems are working properly which requires routine checks, most of which can be automated. However, several types of sensors are prone to malfunction by the very nature they have been designed. For example, resistive soil moisture sensors corrode over a short period of time and are rendered useless. Capacitive soil moisture sensors prove to be a better alternative in this case. Similarly, all other sensors that are to be used, will have to be tested in different conditions to ensure that hardware malfunctions do not stop the model from working properly.

As with the adoption of any new technological solution, there will be more challenges as the system is implemented and scaled. Rapid advancements in sensor technology and robotics will greatly impact the scalability of the proposed model. Affordability of systems will also play a crucial role in wide scale adoption of the proposed model.

6   Conclusions and Future Research Directions

Precision agriculture is a growing area for research; several researchers are paying attention to this area. This paper proposes a model for precision agriculture by the means of generating precision maps for intra-farm variables such as soil moisture, nutrient content. The proposed model can eliminate some of the limitations of existing systems. The paper also suggests hardware implementation of the proposed model so that it can be implemented. The paper also discusses some of the possible challenges in implementation. The proposed model can collect data related to farm that can be analyzed to suggest suitable crops according to soil and related parameters.

However, even with current technological advances, a farmer's intuition combined with the proposed model is necessary for the model to function optimally. While Computer Vision, Artificial Intelligence, and Machine learning are rapidly advancing, there is still room for improvement in applications such as detecting crop diseases to prevent an outbreak. A farmer can designate an area of the farm which needs special attention such as if the crops show signs of disease. Farmer's actions for the affected crops could then be monitored and evaluated in the future to better combat the disease. They can also provide better indicators of plant health and contribute to the growth of a system that eventually achieves near perfect automation. Further experimentation and study is required to establish an ideal number of sensors for a given unit area.

In future, with data collected over time, artificial intelligence could be applied to suggest irrigation and nutrition guidelines for specific crops across different geographies and soil conditions. Sensor technology can be improved and should be made widely available. Better robots and probe designs could be tested. The possibilities are endless when it comes to redefining agricultural practices for the ever so quickly developing world and its growing needs.

References

1. Adeyemi, O., Grove, I., Peets, S., Norton, T.: Advanced monitoring and management systems for improving sustainability in precision irrigation. Sustainability 9(3), 1–29 (2017)
2. Ergeerts, G., et al.: DASH7 alliance protocol in monitoring applications. In: Proceedings - 2015 10th International Conference on P2P, Parallel, Grid, Cloud and Internet Computing, 3PGCIC 2015, pp. 623–628 (2015)
3. Godfray, H.C.J., et al.: Food security: the challenge of feeding 9 billion people. Science 327(5967), 812–818 (2010)
4. Kim, H.J., Sudduth, K.A., Hummel, J.W.: Soil macronutrient sensing for precision agriculture. J. Environ. Monit. 11(10), 1810–1824 (2009)
5. Le, T.D., Ponnambalam, V.R., Gjevestad, J.G.O., From, P.J.: A low-cost and efficient autonomous row-following robot for food production in polytunnels. J. Field Robot. 37(2), 309–321 (2020)

6. Robert, P.C.: Precision agriculture: a challenge for crop nutrition management. Plant Soil **247**(1), 143–149 (2002)
7. Searchinger, T., et al.: Creating a sustainable food future. A menu of solutions to sustainably feed more than 9 billion people by 2050 (2014)
8. Vasisht, D., et al.: Farmbeats: an IoT platform for data-driven agriculture. In: Proceedings of the 14th USENIX Symposium on Networked Systems Design and Implementation, NSDI 2017, pp. 515–529 (2017)
9. Zemamou, M., Aggour, M., Toumi, A.: Review of savonius wind Turbine design and performance. Energy Procedia **141**, 383–388 (2017)
10. World population prospects 2019 - Highlights (2019)
11. Precision Agriculture An International Journal on Advances in Precision Agriculture (2022). https://www.springer.com/journal/11119
12. Agriculture Technology: How It's Changing The Future of Farming (2019). https://bowery farming.com/agriculture-technology. Accessed 15 Mar 2022
13. Blue River Technology (2022). https://bluerivertechnology.com/our-products/. Accessed 06 Apr 2022
14. Reap higher yields with the help of Plantix App: Your Crop Doctor (2022). https://plantix. net/en/. Accessed 02 Apr 2022
15. GIS and Precision Agriculture (2013). https://blogs.lincoln.ac.nz/gis/2013/03/19/gis-and-pre cision-agriculture/. Accessed 01 Apr 2022
16. Mapping as a path to success in precision agriculture (2019). https://www.farmmanagement. pro/mapping-as-a-path-to-success-in-precision-agriculture/. Accessed 04 Apr 2022
17. STACKFORCE: Smart Farming with mioty (2022). https://mioty-alliance.com/2022/02/18/ stackforce-smart-farming-with-mioty/. Accessed 25 Mar 2022
18. Tyagi, V.: Understanding Digital Image Processing. CRC Press (2018)
19. Shah, N., Ross, M., Trou, K.: Using soil moisture data to estimate evapotranspiration and development of a physically based root water uptake model. Evapotranspiration-Remote Sensing and Modeling, edited by: Irmak, A., IntechOpen (2012)
20. Singh, R.K., Berkvens, R., Weyn, M.: AgriFusion: an architecture for iot and emerging technologies based on a precision agriculture survey. IEEE Access **9**, 136253–136283 (2021)
21. Kumar, R., Paiva, S. (eds.): Applications in ubiquitous computing. EICC, Springer, Cham (2021). https://doi.org/10.1007/978-3-030-35280-6
22. Building visual agro service based on weather and satellite data I Part 1: Agro Dashboard (2019). https://openweather.co.uk/blog/post/agro-dashboard-agricultural-monitoring-i-part-i-about-agro-dashboard1. Accessed 22 Mar 2022

Author Index